Global Issues in Water Policy

Volume 20

More information about this series at http://www.springer.com/series/8877

Hilda R. Guerrero García Rojas
Editor

Water Policy in Mexico

Economic, Institutional and Environmental Considerations

Springer

Editor
Hilda R. Guerrero García Rojas
Facultad de Economía
Universidad Michoacana de San Nicolas
de Hidalgo (UMSNH)
Morelia, Michoacan, Mexico

ISSN 2211-0631 ISSN 2211-0658 (electronic)
Global Issues in Water Policy
ISBN 978-3-319-76113-8 ISBN 978-3-319-76115-2 (eBook)
https://doi.org/10.1007/978-3-319-76115-2

Library of Congress Control Number: 2018940503

Printed on acid-free paper

This Springer imprint is published by the registered company Springer International Publishing AG part of Springer Nature.
The registered company address is: Gewerbestrasse 11, 6330 Cham, Switzerland

To my father, who left us during the process of preparing this book. His absence forced pauses for reflection but not for abdication.

To my mother, who in her silent style, kept me focused on finishing the book.

To my sisters, Ruth, Gloria, Vero, and Nora, and brothers, Migue, Cesar, Sergio, Alex, Jorge, and Marco, for always keeping us in this positive mood and for motivating us to go on with our projects.

To my dear friend Ricardo, you left us before the conclusion of the book. Chapter 9 has been maybe one of your last writing. This goes for your Alma and your Valentina.

To all of them, thanks!

Acknowledgments

I deeply appreciate all contributors to this book for their solidarity and trust in this project, for their support and understanding in difficult moments of my life, and for patiently waiting for the book to come to a close. Here it is, to all of you. Thanks.

I recognize my friend Ariel Dinar for pushing this project and for remaining patient while waiting for the product. Here it is, to all those interested in the study of water policy in Mexico.

Thanks to Gloria Gutierrez for reading the texts over and over and for her support in the translation when needed, as well as acting as a language reviewer. All mistakes are mine.

Abbreviations and Acronyms

ANEAS	National Association of Water and Sanitation Utilities
ASF	Superior Audit of the Federation
BANOBRAS	National Bank of Works and Services
BANXICO	Bank of Mexico
BOD5	Five-Day Biochemical Oxygen Demand
CBC/CGC	Communal Goods Commissary
CDI	Nationa l Commission for the Development of Indigenous Peoples
CEAS	State Water and Sanitation Commission
CENAPRED	National Disaster Prevention Center
COD	Chemical Oxygen Demand
CONABIO	National Commission for Knowledge and Use of Biodiversity
CONAFOR	National Forestry Commission
CONAGUA	National Water Commission
CONANP	National Commission of Protected Natural Areas
COTAS	Technical Groundwater Commission
DF	Federal District (Mexico City)
DNM	Mean Natural Water Availability
DOF	Official Government Gazette
DQO	Chemical Oxygen Demand
EPA	Environmental Protection Agency
ES	Environmental Services
FAO	Food and Agriculture Organization
FG	Federal Government
FINFRA	National Infrastructure Investment Fund
FMCN	Mexican Fund for the Preservation of Nature
FONADIN	National Fund for Infrastructure
FONDEN	National Natural Disaster Fund
GDP	Gross Domestic Product
GEF	Global Environmental Facility

GIS	Geographic Information System
HARs/RHA	Hydrological-Administrative Regions
IBRD	International Bank for Reconstruction and Development
IBWC/CILA	International Boundary and Water Commission
ID	Irrigation District
IMTA	Mexican Institute of Water Technology
INALI	National Institute of Indigenous Languages
INECC	National Institute of Ecology and Climate Change
INEGI	National Institute of Statistics and Geography
IPCC	Intergovernmental Panel on Climate Change
IWA	International Water Association
IWRM	Integrated Water Resources Management
LCMOPFIH	Law of Contribution of Improvements for Public Federal Works of Hydraulic Infrastructure
LFD	Government Fees Federal Law
LGEEPA	Law of Ecological Balance and Environmental Protection
LIF	Income Law of the Federal Government
MA	Metropolitan Area
MASL	Meters Above Sea Level
MDGs	Millennium Development Goals
NADBANK	North American Development Bank
NDP/PND	National Development Plan
NDVI	Normalized Difference Vegetation Index
NMX	Mexican Standard
NOM	Official Mexican Standard
NPV	Net Present Value
NWL/LAN	National Water Law
NWP/PNH	National Water Program
OET	Ecological Zoning
PAMIC	Action Plans for Basin Management
PES	Payment for Ecosystem Services
PINE	Ecologic Net Domestic Product
PMPMS	Programs for Preventive Measures and Drought Mitigation
PNRA	National Program of Water Reserves
PRODDER	Program for Reimbursing Duties
PROFEPA	Attorney General's Office for Environmental Protection
PROMAGUA	Water Utility Modernization Project
PRONACOSE	National Program Against Drought
PSA-CABSA	Environmental Services Program for the Capture of Carbon, Biodiversity and Agroforestry Systems
REPDA	Public Registry of Water Duties
SAH	Hydrological Environmental Services
SCFI	Ministry of Trade and Industrial Development
SDGs	Sustainable Development Goals

SEMARNAT	Ministry of the Environment and Natural Resources
SFA	Water's Financial System
SICC	Information System on Climate Change
SINA	National Water Information System
SPI	Standard Precipitation Index
SRRAR	Wastewater Regeneration and Reuse Systems (in Spanish)
TD	Tropical Depression
TSS	Total Suspended Solids
UMSNH	Universidad Michoacana de San Nicolás de Hidalgo
UNDP	United Nations Development Program
UNESCO	United Nations Education, Science and Culture Organization
URDERALES	Irrigation Units
VAEAS	Annual Volume of Superficial Water Extraction
VHI	Vegetation Health Index
WB	World Bank
WHO	World Health Organization
WMO	World Meteorological Organization
WRRS	Wastewater Regeneration and Reuse Systems
WWF	World Water Forum
WWTP	Wastewater Treatment Plant
ZMVM	Metropolitan Zone of the Valley of Mexico

Contents

Contributors

Patricia C. Aguilar-Mendez Independent Consultant, Mexico City, Mexico

Luis Alberto Seguí Amórtegui EAE Business School, Barcelona, Spain

Álvaro A. Aldama Independent Consultant, Jiutepec, Mexico

Hector Manuel Arias-Rojo Consultant on Watershed Management and Environmental Planning, Hermosillo, Sonora, Mexico

Andrés de Andrés Mosquera EAE Business School, Barcelona, España

Diego Garcia-Vega Facultad de Economía, Universidad Michoacana de San Nicolas de Hidalgo (UMSNH), Morelia, Michoacan, Mexico

Faustino Gomez-Santiz Independent Consultant, Morelia, Mexico

Ignacio Daniel González Mora Mexican Network of Hydrographic Basins (Remexcu), World Wildlife Fund (WWF México), Ciudad de México, México

Daniel Iura González Terrazas National Institute of Ecology and Climate Change (INECC), Ciudad de México, México

Hilda R. Guerrero García Rojas Facultad de Economía, Universidad Michoacana de San Nicolas de Hidalgo (UMSNH), Morelia, Michoacan, Mexico

Cipriana Hernández-Arce Instituto Mexicano de Tecnología del Agua, Jiutepec, Mexico

Ricardo Hernández-Murillo World Bank (in memory), Mexico, DF, México

Hugo Amador Herrera-Torres Facultad de Economía, Universidad Michoacana de San Nicolas de Hidalgo (UMSNH), Morelia, Michoacan, Mexico

Jorge Arturo Hidalgo-Toledo Instituto Mexicano de Tecnología del Agua, Jiutepec, Mexico

Edith F. Kauffer Centro de Investigaciones y Estudios Superiores en Antropología Social-Sureste, Chiapas, México

Erandi Maldonado-Villalpando Facultad de Economia, Universidad Michoacana de San Nicolas de Hidalgo (UMSNH), Morelia, Michoacan, Mexico

Gabriela Moeller-Chávez Universidad Politécnica del Estado de Morelos, Progreso, Mexico

Sazcha Marcelo Olivera-Villarroel Departamento de Teoría y Procesos del Diseño, Universidad Autónoma Metropolitana – Cuajimalpa, Ciudad de México, Mexico

Gustavo A. Ortiz Rendón Independent Consultant, Jiutepec, Morelos, Mexico

Luis Enrique Ramos Bustillos Independent Consultant, Jiutepec, Morelos, Mexico

Eduardo Ríos Patrón Mexican Network of Hydrographic Basins (Remexcu), Ministry of Environment and Natural Resources (SEMARNAT), Morelia, Michoacán, Mexico

Roberto Fernando Salmón-Castelo Commissioner of Mexico, Comisión Internacional de Límites y Aguas (CILA), Ciudad Juárez, Chihuahua, Mexico

Vicente Sánchez Munguía El Colegio de la Frontera Norte, Tijuana, Mexico

Ricardo Sandoval-Minero Sextante Consultores SC., Mexico City, Mexico

Sergio Vargas-Velázquez Universidad Autónoma del Estado de Morelos, Cuernavaca, Mexico

Antonio Yunez-Naude El Colegio de Mexico, A.C, Mexico City, Mexico

Part I
Mexican Water Policy As Introduction

Chapter 1
Introduction and Lessons Learned

Hilda R. Guerrero García Rojas

Abstract This chapter provides the overall context and setting for the volume. In this chapter the scope of the book is established along with the way each chapter addresses its objectives and reach. It also provides an overview of the chapters underlining the links and continuity among them. A lesson learned section is provided, summarizing the lessons from the different chapters in the book in a way that readers can learn about the concluding notes presented by the authors in each section.

Keywords Mexico · Water policy · Environmental policy · Institutional policy · Water economics

1.1 Introduction

Environment, institutions, economics, and water, each of these words means a lot by itself. Here, in this book we combine everything around the great country, Mexico, which has a unique ecosystem for its variety; strong institutions by its laws; an economy among the 15th best in the world; and enough water for its people, that is, water availability at a greater level of the international standard requirements; but mainly it has researchers, executives, legislators, officials, and workers, people who deal with the study of water, its functions, its uses, and its limitations; yes, as a human being because it is so, water is life.

Water has always been present in the cosmovision[1] worldview of peoples, from ancient times until now, from the mythical Tlalocan, Tlaloc's Paradise, to the legends about supernatural beings that produce water in current existing towns. In pre-Hispanic Mexico, water was an element to live with; therefore

[1]Neologism that describes the general idea of the origin, existence, reality, and place within the cosmos that a person, society, or culture has in a certain time

H. R. Guerrero García Rojas (✉)
Facultad de Economía, Universidad Michoacana de San Nicolas de Hidalgo (UMSNH), Morelia, Michoacan, Mexico

H. R. Guerrero García Rojas (ed.), *Water Policy in Mexico*, Global Issues in Water Policy 20, https://doi.org/10.1007/978-3-319-76115-2_1

we can say that there was a communication between man and nature: Gods were included in this relationship.

In *Dresden Codex*,[2] it is mentioned that water is the origin of life; the Bacabs emerged from water to support the heaven, in ancient Mayan worldview; after the universal flood, human beings appeared, according to several Mesoamerican cultures. Tlaloc and Chalchiuhtlicue ruled the celestial and underground water, while ancient Mayans talked about Cauac, the monster of the heaven, and the ancient Red Goddess, who, in the presence of a black Bacab, poured the waters over the earth (IMTA 2006).

The creation of the world in ancient peoples was always linked to water; in fact, the world was over water, the Cemanahuac, for ancient and current Nahuas, but not just in the Nahua culture but in other peoples as well.

Water is a vital resource for mankind, since it is present in all social, economic, and environmental activities, in such a way that it is an inherent requirement for life in our planet. So water fulfills several functions as a strategic factor for cooperation, development, and well-being, but it can also act as a key trigger in the creation and development of different kinds of conflicts.

Water resources management in a sustainable and equitable way is basic for supporting life, as well as people's health and dignity; and it is one of the essential bases of our social and economic development. It involves many sectors and interest groups and expands from local to global levels.

Since 2007, the Board of Directors of the United Nations Environment Program (UNEP) approved its policy and strategy on water resources. The general objective of the strategic policy on water resources is to substantially contribute to environmental sustainability in the management of all water resources through an integrated ecosystem approach of water and socioeconomic development agreed upon international level (PNUMA 2012).

As shown in Fig. 1.1, the analysis of the natural water resource implies superior efforts than the simple description of its intrinsic properties and characteristics as an element of nature. Water is a factor that under any social context may acquire endless functions and roles, since its appropriation will be in function of economic, institutional, and environmental aspects. Nevertheless, the interaction among these three aspects tends to trigger problems, confrontations, and severe complexities, to a degree of opposing and affecting the dynamics of water resources.

Economic Considerations

In contrast to the prediction arguments of the economic policy regarding the efficiency of resources under a scarcity context, the use of water resources might seem efficient and effective; but it is not; water is constantly being used in an inefficient way, even while resource keeps showing signs of scarcity and availability.

[2]*Dresden Codex*, also known as *Codex Dresdensis*, is a book of the Mayans, dating from the eleventh or twelfth century. The *Dresden Codex* contains astronomical tables of extraordinary precision.

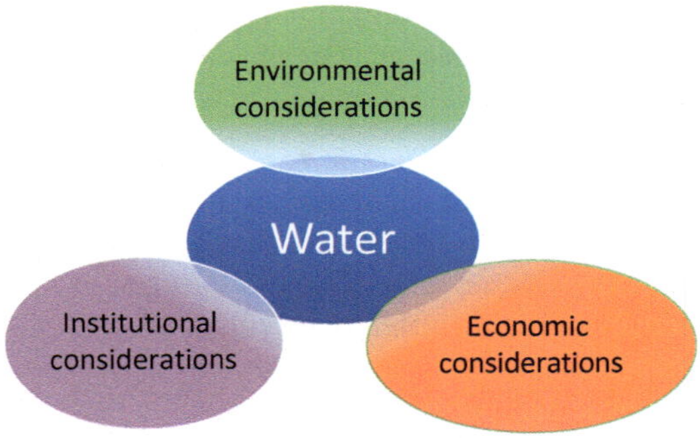

Fig. 1.1 Economic, institutional, and environmental interrelations around water. (Source: own elaboration)

Why? Perhaps, we might find the answer in water sector policies, in their approach, and within the framework in which they are formulated and implemented.

Available evidence from several analysis shows that water demand is sensitive to the evolution of income and population growth, its price and some other prices, demographic and socioeconomic characteristics of households, and also climate, including temperature and rainfall. It is worth mentioning the inelasticity of water demand as a function of income and price, particularly price. This suggests that water consumption will increase though in a lesser proportion than the GDP growth. Galindo et al. (2014) mention that the estimation of demand functions, in water resources, allows the identification of a consumption path in the future. It is worth noting that an increase in temperature and a modification in rainfall patterns will have an impact on water consumption especially as an increase in water demand, which will intensify stress over this resource (CEPAL 2015). The current scenario is critical, since the impact of climate change on water resources is framed in the context of a growing water demand, both for economic activities and population, which will intensify the pressure over that resource. This is exemplified with the accelerated evolution of water demand intended for human consumption.

In search for an interaction or conduct with the environment under lesser conflict schemes, the use of economic instruments has been turned to as licenses, grants, users' rights, use taxes, permits, fees, tariffs, and prices, among others, toward an integrated management of resources, specifically water. Use of economic instruments is important to control the consumption, but it has its limitations and must include other social considerations.

It is important to mention that water problems are just the result of an imbalance between water needs, arising from the economic and demographic growth and the decreasing capabilities of an increase in the supply, determined by nature, public and private policies of appropriation and overexploitation and the inefficient conditions in the distribution network of the water resource.

We can say that from the moment water shifts to the economic sphere, this vital resource turns merely into an input for production or raw material, adopting commercial and transactional characteristics, and sometimes subject to market forces via supply and demand.

Therefore, it is really important to articulate an increasingly efficient and flexible model for the management of water resources, in keeping with its surroundings, without limiting the economic and social development of the country.

Institutional Considerations
In 2010, the General Assembly of the United Nations explicitly acknowledged the human rights to water and sanitation, reassuring that clean drinking water and sanitation are essential for the fulfillment of all human rights. It establishes that the human rights to water is essential for a dignified human life, as well as having sufficient, healthy, acceptable, physically accessible, and affordable water for domestic and personal use. Actually, is that so?

In Mexico, until 2012 the constitutional Article 4 was amended with section six, which recognizes every person's right to water access, disposition, and sanitation for domestic personal consumption in a sufficient, healthy, acceptable, and affordable manner.

The entity, which coordinates and plans everything related to the water sector, is the National Water Commission (CONAGUA). According to its own definition, the function of CONAGUA is to manage and preserve the national waters and inherent goods, with the purpose of attaining a sustainable use of this resource. This entity is a decentralized agency of the Ministry of Environment and Natural Resources (SEMARNAT), which is the highest authority of the federation regarding the management of natural resources. Also, CONAGUA is the entity in charge of water policy, where the governing document is the National Water Program (PNH) formulated every 6 years. At the same time, Mexico has a system of concessions and grants for superficial and underground waters formalized in the National Water Law (NWL).

The political intentions, or the water management objectives, must be translated into laws and regulations, with responsibilities assigned to diverse stakeholders. The results of the policies largely depend on the way in which those responsibilities are applied, in all levels, considering costs (UN-Water 2014). Therefore, the water sector requires a process of reforms complementing the predominantly technical hydraulic approach, for making more efficient the process of integrated water management through the incorporation of proposals developed under a social and environmental perspective, within the framework of sustainable development.

Environmental Considerations
Water is an essential element for life; most of human activities are conditioned with the use of this good, as well as the ecological systems, which need it to be in balance; and cultures have created different and complex relations with it. Therefore, water management must be conducted to fulfill human and natural needs and to ensure its sustainability over time; so we are talking about water security. Its management must be analyzed in an integrated manner in terms of exchange relations among social groups with the natural environment from a historic perspective in order to design sustainable strategies for the future.

Climate change scenarios foresee a worsening in spatial and temporal variations in the dynamics of the water cycle, so the gap between supply and demand will be increasingly sharpened. De Miguel and Tavares (2015) state that climate change threat is added to human and economic costs of other forms of environmental degradation, which are frequently absorbed by those mostly vulnerable. The use of water resources must be globally addressed in a wide and integrated manner, so with humankind efforts toward attaining a sustainable development and economic growth, let's not destroy the resource that makes all that possible, water.

"The complexity of water resources problems requires solutions that are beyond the scope of a single discipline (*e.g.* hydraulic engineering or hydrology). A holistic approach including the concurrent participation of many disciplines is often indispensable" (Aldama Rodriguez 2018, Chap. 15). This book looks out to attend this vision, taking into consideration the economic, institutional, and environmental aspects of the water policy in Mexico.

1.1.1 Mexican Water Policy As Introduction (This Chapter and Chap. 2)

Each chapter can be read in an independent manner, and all of them are related. However, each one was written providing enough information to have a framework of the Mexican water policy, so that people interested only in one topic do not need to read all chapters to get the needed or wanted knowledge. Of course, it is recommended to read the entire book to have a good glance of the water policy in Mexico.

The book is organized into five parts. *The first one is an introductory section* of the Mexican water policy, which contains two chapters: in this chapter the reader has an introduction including a section with lessons learned from each chapter of the book.

The second part is dedicated to sectorial issues of water, so it touches topics for agriculture, wastewater, and industrial water use and water utilities.

The third part addresses institutional issues, that is, it deals with the policy of water prices, basin organizations, payment for environmental services, water law, and indigenous right to water.

Mexico has a broad border in the north and a smaller border in the south, both divided by bodies of water and rivers, which must be considered in politics, as in the economic activity of the territories involved. The *fourth part deals with international water issues*, in the south the Mexico-Guatemala and Mexico-Belize borders and in the north the United States-Mexico border.

Part five attends challenges in the present and in the future with the topics of climate change and the role of technology and science for integrated water resources.

Arias and Salmon (2018) begin Chap. 2 pointing out the fact that Mexico is one of the countries with the most biological diversity in the world. This diversity is due to its geographical location and other geographical features. Mexico is divided in the middle by the Tropic of Cancer: in the northern part, there are two of the largest and driest deserts of the world due to high-pressure centers, the southern part is in the humid tropic and the center is on highlands with temperate climate. With this framework, they offer a brief review of the history of Mexican water sector, going from the pre-Hispanic period to date. In this chapter we can learn the evolution of technological changes and how the population made institutional arrangements in the organization and utilization of water resources in Mexico. They present the current situation of water resources, surface water and groundwater, the stress over them, availability by users, legislation and administrative organization evolution, and so on to point out the creation of CONAGUA, as a decentralized administrative institution in charge of the water sector in Mexico.

1.2 Sectorial Issues (Chaps. 3, 4, 5, and 6)

Water conservation and its efficient use in agricultural production is a challenge that has to be addressed by the Mexican government. In Chap. 3, Yunez and Aguilar (2018) present the results of a research related to water availability and water pricing. They use a general equilibrium model (CGEM) applied to Mexico, focused on agricultural and food production. First, they present the main features of the CGEM and the social accounting matrix (SAM) for Mexico. They offer a description of the Mexican economy and its agricultural sector according to the SAM, to assess the probable effects of two water-related shocks: a decrease in water availability for irrigation and a rise in the price charged by the Mexican state to farmers with access to superficial water for irrigation.

In Chap. 4 Seguí et al. (2018) deal with the challenges and opportunities that wastewater treatment and reuse mean. They discuss about the need to reinforce the reuse of regenerated water regarding direct or indirect potable reuse. A model of recycled water management is proposed to assure a continuous water supply from the technical, health, economy, legal, and environmental perspective, in order to preserve quality and increase stock in the region. Knowing that regeneration and reuse activities are of paramount importance, many countries follow this practice with successful results.

Guerrero et al. (2018b) in Chap. 5 analyze the efficiency of industrial water use in Mexico throughout water price elasticity. In this chapter, water demand of the industrial sector per hydrological regions is studied. An examination of changes in the demand toward variations in tariff structure is made. Analyzed data take into account a 14-year period where irregular variations can be seen, both in tariff and in demand for water in hydrological regions.

Mexico guarantees access to water and sanitation infrastructure, but the quality of service provision is still unsatisfactory. Besides facing the growing problems due to the scarcity and bad quality of water sources, the gap between demand and supply also widens because of rapid and messy urban growth in metropolitan areas, a lack of investment in infrastructure rehabilitation and extension, and the limited capacities for managing and operating the water and sanitation systems. These topics are analyzed in Chap. 6 where Sandoval (2018) establishes the question, in the Mexican case, if, for water utilities, their financial efficiency is achievable in a sustained way.

1.3 Institutional Issues (Chaps. 7, 8, 9, 10, and 11)

Guerrero et al. (2018a) in Chap. 7 performed an analysis on the reform of the federal law of rights that establishes four availability zones in Mexico for water use, catchment or exploitation, and its role in the tariff policy as an economic instrument for water resources management. Before the Reform until 2013, Mexico was classified in nine availability zones, where Zone 1 presented lesser availability levels of water and hence a higher rate or price and vice versa in Zone 9, besides the fact that this zone classification complies with a municipal geographic division criteria with no distinction between hydrological basins and aquifers. They deal with the topic of water pricing as a policy and economic instrument for the management of water resources, addressing the current structure of water price related to availability water zones.

In Mexico, the integrated water resources management, in practice, is far from being crystallized. However, the process that started with the amendments of the law in 2004, in terms of a decentralization process, is not considered a proper route, but as the beginning of a process that needs to be consolidated and settled. In Chap. 8 Hidalgo et al. (2018) review the experience of Mexico regarding integrated water resources management (IWRM) and look to answer if the river basin organisms could be the best path to achieve it.

Environmental services from the perspective of the state have as public policy objective the conservation of natural environments as well as the coverage of vegetation that provides environmental services to society. From this perspective, Hernandez and Olivera (2018) in Chap. 9 analyze the current institutional framework that regulates environmental services, such as the recognition of the ecological flow and system of payment for environmental services giving a contrastive emphasis between forest resources management and institutional building.

Ramos and Ortiz (2018) in Chap. 10 address the water management instruments in the National Water Law. They debate on the need to rethink regulation with respect to what the current National Water Law defines as "gestión del agua." The concept "gestión" can be translated as "management," but, as it is explained in the chapter, in current Mexican Law, it has yet other meanings that need to be acknowledged. They point out that clarifying this definition might foster a better regulation of different instruments that may improve the public policy of water, since many of these instruments are currently scarcely and dispersedly regulated in the law.

The number of Indian communities in Mexico is imprecise, but the importance of the way they manage water as a common pool resource is critical to understand many other rural sites, where water is also a common resource under a regulatory framework of legal pluralism. Vargas (2018) in Chap. 11 deals with the topic of indigenous people and their rights to water. It is known that the social situation of the indigenous peoples has historically had the greatest backwardness and marginalization, as well as a lack of recognition of their cultural specificity. Currently, in the indigenous regions, the scarcity of jobs, low wages, and asymmetrical relations between indigenous and non-indigenous population prevail, a situation that exerts an enormous pressure toward its dissolution as organized peoples; the loss of control of a resource such as water can be very adverse to their political and economic organization ways.

1.4 International Water Issues (Chaps. 12 and 13)

Kauffer (2018) confronts us with the paradox of the borders of Mexico with Guatemala and Belize, through which most of the water resources of Mexico and Guatemala flow, against the lack of studies about it. She places the question about the way to articulate local and international dimensions into the political water analysis in the boundary regions that Belize, Guatemala, and Mexico share. Starting from the description of transboundary dimension of water in the studied region, Kauffer (2018) in Chap. 12 proposes a new concept of "hydropolitics" as a multiple to analyze diverse international and local dynamics regarding water in this region of multiple borders.

Sanchez (2018) in Chap. 13 points out the fact that water has been a central issue on the bilateral relationship between Mexico and the United States since the border between both countries was established; in the conflicting character of their relationships, the countries have reached agreements to secure access to water in their shared watersheds. In recent decades, the governments of both countries have promoted the construction of a new institutional framework aimed at cooperation; and this has been the result of pressure from border communities and environmental problems arising from changes in the border economy and trade integration proposals, which facilitated the mobility of goods across the border. In a broader sense, Sanchez (2018) shows that despite the asymmetries, the governments of the two countries have managed to give sense and meaning to the institutional framework, which, through a bargaining process, have overcome conflicts and strengthen cooperation.

1.5 Present and Future Challenge (Chaps. 14 and 15)

Rios et al. (2018) in Chap. 14 address the issue of climate change by analyzing the vulnerability that water resources have under its effects; and they place the challenge that implies the correct basin management approach to attend impacts of climate change. This challenge implies adapting and generating governance conditions to changes in water availability, as well as against an increase in extreme meteorological events and its consequences from the acknowledgment of basin functionality and the relation of hydrological environmental services (SAH) with the local and regional, economic, social, and environmental dynamics.

In Chap. 15 science and technology for integrated water resources management in Mexico, Aldama (2018) shows the importance that the use of correct instrument of policy has to achieve the goals of integrated water resources management (IWRM), among which science and technology play a crucial role, particularly in river basins where water conflicts are commonplace. He speaks about the case of two of the most strategically important river basins in Mexico, the Lerma-Chapala basin and the Río Bravo/Río Grande basins, regarding a conflict between upstream and downstream water users due to the resource scarcity: for the first case, through simulation and optimization models to build consensus between water users and to define rules for the integrated operation of the system, according to the availability of water, and, for the second case, through innovations that have been adopted by water users in the Rio Bravo Basin and have been used to define water policy in the region.

1.6 Lessons Learned As Concluding Notes

At end of each chapter, authors bring in reflections, in some as concluding notes and proposals in others, looking forward and placing challenges. In this way, the reader can have information regarding possible scenarios on the path to follow in each topic addressed in the chapters. Therefore, this section seeks to point out the main lessons learned in each chapter.

Arias and Salmon (2018) devoted Chap. 2 to give a brief historical travel from pre-Hispanic times until now, presenting a review of water sector history and offering the evolution of agendas to current situation telling about the transition that legislation, institutions, governments, and water situation have had through the time. After this review, it is possible to locate the announcement of some commitments to analyze the future of the water sector in Mexico, establishing that in general, it looks promising, but there are still several issues to be considered, looking to reduce water conflicts and inequities. In this scheme, fortunately, Mexico has signed several international agreements where water is the strategic resource. Authors specify that the Millennium Development Goals and the Global Change Strategy are two of them, and the commitments made were considered in the CONAGUA 2030 Water Agenda. But the principal lesson from them is when they point out that the best response is in the establishment of new water governance. Certainly, this topic is considered in legislation, but it is necessary to do more to reduce water conflicts and improve water efficiency.

In Chap. 3, Yunez and Aguilar (2018) give a final reflection, regarding water policies in Mexico for the agricultural sector from the results of the general equilibrium model they applied to this sector. In their analysis they let us know that the current federal administration does not have an explicit purpose to reduce irrigation water supply, although the use of irrigation technology is promoted for a more efficient use of water in agriculture, accompanied of greater public investment in infrastructure for irrigation. Another scenario is to keep in mind the fact that before the possibility of having less water for irrigation, either due to climate change or changes in water policy, the production of food and crop in the country would be affected, forcing larger imports to ensure food security. They are expectant that the results discussed in this chapter will contribute to a better understanding of the policy options that could lead to a more efficient use of water for agricultural production, with no major negative effects on food security in Mexico.

The main learning from Chap. 4 is to know the Segui et al. (2018) proposal of recycled water management, which can assure a continuous water supply. They present a model from the technical, health, economic, legal, and environmental perspective to preserve quality and increase stock in the region. They report that although Mexico is an international benchmark in the field of wastewater regeneration and reuse, the practice of recharging aquifers with regenerated water is incipient. Their teaching focuses fundamentally on a paradigm shift, in which more importance should be given to water quality and not to the origin of the water. In this way, challenges will be accomplished, risks will be minimized, and opportunities will be improved.

In Chap. 5 Guerrero et al. (2018b) made an analysis of the efficiencies of water using water price elasticity for the Mexican industry. They study if tariff by zones, as allocation mechanisms, could make a more efficient water use in the industry, but they do not find significant variations in the demand when the price or tariff changes. This can be explained, in part, because the expenses for water consumption do not represent a significant variable within the cost function. Even though the price of water is defined by an economic principle of scarcity according to the water availability zones, in fact this has no impact on consumption modifications, since the price elasticity of the water demand estimated by hydrological region oscillates in very wide ranges, which is not reflected in low water consumption, limiting the efficient use of the water resource, not by the economic instrument (elasticity) but as a result of the management instrument (tariff).

Sandoval (2018) in Chap. 6 addresses the issue of water utilities, considering that the goal is to have an equitable access to water and sanitation. He establishes challenges in four aspects, which oscillate from the need to set up an effective resource regulation system; a deep institutional reform that ensures an appropriate relationship between political authorities, utility operators, and citizens; an efficient financial system to water utilities that could be part of the political effort to achieve a more effective land use planning and development. In fact, he proposes that integrated urban water management needs to become the model for urban water management in Mexico.

In Chap. 7 Guerrero et al. (2018a), through an analysis on the reform of the federal law of rights, admit that while it is true that until 2013 the classification of nine ZDs did not respond to quantitative criteria or estimations that would classify water availability per hydrological basin and aquifer, even in the same territory, it does not necessarily mean that since 2014, with the new calculation methodology (through the proposed algorithm), a more efficient water management is obtained. They found it important to mention the inelasticity of water demand as a function of income and price, especially price. This suggests that water consumption will continue increasing. So, the use of economic instruments is important to control consumption, but it also has its limitations and should include other social and environmental considerations.

Hidalgo et al. (2018) in Chap. 8 present a thoughtful analysis of River Basin Organization, looking to know if that could be the best path toward integrated water resources management. They finish with a specific section of lessons learned; there they manifest that IWRM implies the redistribution of the decision-making, financial resources, and knowledge related to water systems functioning, considering no matter how it was done. But the question is still in the air, regarding knowing if all actors involved are really helping to the design of River Basin Organization along with its centralized way of working. That question should include if the institutional agreement is, in fact, competent for regulating the dispute over water, since it moves different interest groups with dissimilar levels of power and action capabilities.

Hernandez and Olivera (2018) in Chap. 9 analyze the payment system for environmental services, emphasizing the contrast with the management of forest resources by forest owners and managers, considering the case study of Capulalpam de Mendez Oaxaca. Although economic development regularly is accompanied by deforestation as well as other harmful activities, in the community of Capulalpam, the CONAFOR'S program for the payment of environmental services (PES) is not a determining factor in the owner community's decision to preserve the forest and to continue providing environmental services. The existence of the program and the participation of the community in the announcement did not modify the social or productive conducts since the community strategy of making traditionalist exploitation, improving the forest, and keeping it for future generations was a decision made before the PES program. Therefore the acceptance of the payment for environmental services and the decision-making process of the community owner can only be understood by distinguishing the importance of the community institution both formal and informal.

From the debate that Ramos and Ortiz (2018) describe in Chap. 10 regarding the need to rethink regulation with respect to what the current National Water Law defines as "gestión del agua," they establish, as concluding notes, that there is a need to integrate current disperse regulations regarding "gestión del agua" recognized as a separate process within the water management field in order to provide a better framework for regulation of relevant topics regarding water that go beyond exclusive government participation in water management. They bring out, on table 10, points to be considered as some aspects that would need to be reformed in the law to improve government fees' enactment and collection. To mention some, they propose to support an incentive technology transfer for more efficient, clean, and water-saving technologies, via deduction of other taxes, e.g., income tax; they also suggest allocating at least 1% of the total tax or duty collection in water for research and development of the water sector.

The topic regarding the right to water for indigenous people is addressed by Vargas (2018) in Chap. 11. In his final notes, he establishes that it is very complicated to summarize in just a few pages a process as intricate as the permanence/

dissolution of indigenous societies in Mexico, even more in its relationship with water, their struggles, and collective action to have their right to water and water justice recognized. Besides, the problem of accessing to safe water and quality water is an important issue for indigenous peoples, as they are indicators of poverty and lack of access to public services. He determines five main theses regarding this situation where the many conflicts between peoples for water access and distribution are present. Vargas lays down the fact that what remains under discussion are the local water management models that should be developed in Mexico in the coming years.

Kauffer (2018) in Chap. 12 suggests a new concept of "hydropolitics" as a multiple to analyze diverse international and local dynamics regarding water in the borders of Mexico with Guatemala and Belize region. The proposal is centered in extending the study of hydropolitics in the region by articulating two major scales: international and local. It is important to note her statement that it is necessary to go from one hydropolitics to multiple hydropolitics, as a result of the diversity and complexity of the analysis of water-related political problems in the region, along with diverse visions and perspectives resulted from different disciplinary approaches as political sciences, anthropology, sociology, economy, and history. She concludes that the essence of politics, regarding the relationship between conflict and regulation, converges in the heart of multiple hydropolitics, from a combination of bordering and transboundary perspectives and as a continuum from local to international water issues.

Chapter 13 deals with the relationship between Mexico and the United States where water has been a central issue; here, Sanchez (2018) highlights the fact, in his conclusion notes, that some of the problems related to water in the border and of the relationship between the two countries are related with the inadequacies and omissions since the signing of the international treaty. Even so, he mentions that it is worth highlighting the persistent ability to reach agreements between the two countries on issues of water management and conflict resolution at different times, derived from the asymmetric condition among them and the availability of water and from omissions or failures of institutional problems due to the lack of more integrated water policies in the border area. Above all, Mexico needs to improve its capacities in planning, management, and collection of water, but also in the application of the law.

The study of the vulnerability that water resources have under the effect of climate change is addressed by Rios et al. (2018) in Chap. 14, where they review the role that basin management could have to counteract impacts. In their conclusions explain that integrating in an effective manner, in governance spaces, the participation and coordination of water policy and climate change policy, the water resources supply, and demand approach in the context of climate change is a full challenge for the articulated operation of the Mexican environmental normativity. Solutions imply a paradigm shift in the design and implementation of public policies integrating a common vision of the territory for all sectors.

Aldama (2018) in Chap. 15 shows the importance that science and technology have to achieve the goals of integrated water resources management (IWRM) in Mexico. At the end of the chapter, he presents a concluding remark, highlighting that it has been shown that science and technology may be true catalysts of IWRM and good water governance. By way of example, the case studies of the Lerma-Chapala basin and the Río Bravo (Grande) basin have demonstrated that sound scientific results and effective technological tools may be used to end stalemates in negotiations between conflicting water users, particularly when stakeholders base their positions on opinions rather than on facts. Aldama says that the key idea in using science and technology is to shift the discussion to technical issues, which can be objectively settled. This serves as a solid base to build trust among stakeholders and in water authorities, thus facilitating the consensus construction process, an essential step in achieving IWRM.

1.7 Final Remarks

This book looks to provide knowledge on water policy in Mexico taking into consideration, if not all, almost all elements involved with it. The book treats first a sectorial aspect since users of water in agriculture and industry and people in cities are all part of the principal branch that should be considered for water demand-side analysis and the way users use the resource, to have information for establishing any constraint, requirement, legislation, or economic instrument, and for mentioning some on the supply side, to achieve an integrated water management; these are topics addressed in second place in the book. Water makes frontiers but at the same time has no frontier; in this way water creates relationships, conflicts, and agreements. Hence, it is necessary to attend the international aspect of water to analyze how they are for considering when taking a decision is needed. Third section focused on it. One of the factors that produces impacts on water, hopefully more positive than negative, is climate change; and science and technology could help to guarantee the positive effects and counteract the negative effects. Knowing the significant role that this factor has on water resources, it is incorporated in the section of present and future challenges.

Any publication should begin with an introduction; to comply with this, we turned to a review of the extensive experience that Mexico has had since pre-Hispanic times in the use of water, making an exercise in the history of water in Mexico, its institutions, legislation, infrastructure, and uses. Also, in the introduction, in this chapter, a section with lessons learned is considered.

Here you have in your hands this book made with the intention to offer comprehensive information regarding water policy in Mexico from the economic, institutional, and environmental considerations, with the conviction that it will meet expectations of all those interested in the study of water.

References

Aldama Rodriguez, A. (2018). Science and technology for integrated water resources management in Mexico (Chapter 15). In H. R. Guerrero García Rojas (Ed.), *Water policy in Mexico* (Global Issues in Water Policy 20, pp. 309–329). Cham: Springer.

Arias Rojo, H, & Salmon Castelo, R. F. (2018). Mexican water sector: A brief review of its history (Chapter 2). In H. R. Guerrero García Rojas (Ed.), *Water Policy in Mexico* (Global Issues in Water Policy 20, pp. 19–51). Cham: Springer.

CEPAL. (2015). La economía del cambio climático en América Latina y el Caribe. Paradojas y desafíos del desarrollo sostenible. *Comisión Económica para América Latina y el Caribe (CEPAL). Euroclima. Comisión Europea. Cooperación Alemana. Cooperación Española. Cooperación Regional Francesa.*

De Miguel, C., & Tavares, M. (2015). El desafío de la sostenibilidad ambiental en América Latina y el Caribe. *Páginas selectas de la Comisión Económica para America Latina y el Caribe (CEPAL).*

Galindo, L. M., Samaniego, J. L., Alatorre, J. E., & Carbonell, J. F. (2014). Reflexiones metodológicas del análisis del cambio climático. Una visión desde América Latina. *Estudios del cambio climático en América Latina. Comisión Económica para América Latina y el Caribe (CEPAL). Euroclima. Comisión Europea.*

Guerrero García Rojas, H., García Vega, D., & Herrera Torres, H. (2018a). Water price policy and its institutional role as an economic instrument for water management (Chapter 7). In H. R. Guerrero García Rojas (Ed.), *Water policy in Mexico* (Global Issues in Water Policy 20, pp. 137–152). Cham: Springer.

Guerrero García Rojas, H., Gomez-Santiz, F., & Maldonado-Villalpando, E. (2018b). Industrial water use in Mexico: Analysis of efficiencies using water price elasticity (Chapter 5). In H. Guerrero García Rojas (Ed.), *Water policy in Mexico* (Global Issues in Water Policy 20, pp. 89–113) Cham: Springer.

Hernández Murillo, R., & Olivera-Villarroel, S. M. (2018). Payment for environmental services: Between forest resources management and institutional building (Chapter 9). In H. Guerrero García Rojas (Ed.), *Water policy in Mexico* (Global Issues in Water Policy 20, pp. 171–185). Cham: Springer.

Hidalgo Toledo, J.A., Hernández Arce, C., & Vargas Velázquez, S. (2018). River basin organization, the best path towards integrated water resources management? (Chapter 8). In H. Guerrero García Rojas (Ed.), *Water policy in Mexico* (Global Issues in Water Policy 20, pp. 153–169). Cham: Springer.

IMTA. (2006). Leyendas del Agua en México. *Coordinación de Desarrollo Profesional e Institucional. Instituto Mexicano de Tecnología del Agua (IMTA).*

Kauffer, E.F. (2018). From the abundance of waters to the scarcity of studies: Contemplating hydropolitics in Mexico-Guatemala and Mexico-Belize borders (Chapter 12). In H. R. Guerrero García Rojas (Ed.), *Water policy in Mexico* (Global Issues in Water Policy 20, pp. 237–262). Cham: Springer.

PNUMA. (2012). Aguas saludables para el desarrollo sostenible. Estrategia operativa del PNUMA para el agua dulce (2012–2016). *Programa de las Naciones Unidas para el Medio Ambiente.*

Ramos Bustillos, L. E., & Ortiz Rendón, G. A. (2018). Water management instruments in the National Waters Law (Chapter 10). In H. R. Guerrero García Rojas (Ed.), *Water Policy in Mexico* (Global Issues in Water Policy 20, pp. 187–211). Cham: Springer.

Ríos Patrón, E., González Terrazas, D. I., & González Mora, I. D. (2018). Climate change and vulnerability of water resources in Mexico: Challenges from basin management (Chapter 14). In H. R. Guerrero García Rojas (Ed.), *Water policy in Mexico* (Global Issues in Water Policy 20, pp. 283–307). Cham: Springer.

Sánchez Munguía, V. (2018). The US-Mexico border: Institutional weaknesses in meeting the growing demand for water (Chapter 13). In H. R. Guerrero García Rojas (Ed.), *Water policy in Mexico* (Global Issues in Water Policy 20, pp. 263–279). Cham: Springer.

Sandoval-Minero, R. (2018). Water utilities: Is their sustained financial efficiency achievable? – The Mexican case (Chapter 6). In H. R. Guerrero García Rojas (Ed.), *Water policy in Mexico* (Global Issues in Water Policy 20, pp. 115–134). Cham: Springer.
Seguí Amórtegui, L.A., Moeller Chávez, G., & De Andrés Mosquera, A. (2018) Mexico, the water stress: Challenges and opportunities in wastewater treatment and reuse (Chapter 4). In H. R. Guerrero García Rojas (Ed.), *Water policy in Mexico* (Global Issues in Water Policy 20, pp. 75–87). Cham: Springer.
UN-Water. (2014). Un Objetivo Global para el Agua Post 2015. Asegurando agua sostenible para todos. *Síntesis de las principales conclusiones y recomendaciones de ONU-Agua.*
Vargas Velázquez, S. (2018). Indigenous people and the right to water (Chapter 11). In H. R. Guerrero García Rojas (Ed.), *Water policy in Mexico* (Global Issues in Water Policy 20, pp. 213–234). Cham: Springer.
Yunez Naude, A., & Aguilar Mendez, P. (2018). Effects of water availability and policy changes for irrigated agriculture (Chapter 3). In H. R. Guerrero García Rojas (Ed.), *Water policy in Mexico* (Global Issues in Water Policy 20, pp. 55–74). Cham: Springer.

Hilda R. Guerrero García Rojas is a Professor of Economics and Environment in the Facultad de Economia, Universidad Michoacana de San Nicolas de Hidalgo (UMSNH), Morelia, Michoacán, Mexico. She holds a PhD in Economics from the Toulouse School of Economics. She is among the leading water policy experts in Mexico and contributed to policy work in Mexico. Her PhD thesis focused on "economic assessment of policy interventions in the water sector; industrial water demand in Mexico: econometric analysis and implications for water management policy." She worked for Instituto Mexicano de Tecnología del Agua (IMTA), and as a consultant, she has been collaborating with the World Bank, Colegio de México (COLMEX), OECD, and with the Inter-American Development Bank (IDB). Publication topics involving Guerrero include "Water pricing reforms in Mexico: The case of manufacturing sector"; "Applying water demand: Case for industrial use on Mexico"; "Water in Mexico: Impact of policy interventions in the sector"; and "Water pricing reforms in Mexico: The case of the manufacturing sector", among others. Guerrero's areas of interest are natural resources and environmental economics, integrated water management (IWM), economic instruments applied to natural resources, water price and its reforms, applied economics, environmental economic valuation, micro-econometrics, climate change in water sector, and economics of climate change.

Chapter 2
Mexican Water Sector: A Brief Review of Its History

Hector Manuel Arias-Rojo and Roberto Fernando Salmón-Castelo

Abstract In Mexico, there are wide conditions in terms of the sources and utilization of water resources. A brief description is presented, on the first part, to show technology level, water institutions, and water legislation through different periods of Mexico, showing the evolution of the water sector in the country. The current situation of the water sector is described based on an administrative zoning. Since agriculture is the main user, an emphasis is made on the description of its hydraulic infrastructure available. Water governance is accentuated due to its importance in terms of the continuous decrease of water availability and continuous increase in water demand. The future of the water sector is analyzed in terms of the commitments of Mexico in international treaties (Millennium Development Goals and Climate Change) and water governance; this one is considered for a proposal about what water sector needs to focus on in the future, searching poverty alleviation, efficient water use, and resolution of water conflicts.

Keywords CONAGUA · Mexico's water statistics · Water legislation in Mexico · Evolution of water institutions in Mexico · Hydraulic infrastructure in Mexico · Water governance

2.1 Introduction

All living organisms need water. Since water is not homogeneously distributed on earth, living organisms develop adaptation mechanisms to local conditions. The same is true for humans. When resources are scarce, conflicts develop, and to reduce

H. M. Arias-Rojo (✉)
Consultant on Watershed Management and Environmental Planning, San Luis Obispo 11, Capistrano Residencial, Hermosillo, Sonora 83249, Mexico
e-mail: hector.arias@prodigy.net.mx

R. F. Salmón-Castelo
Commissioner of Mexico, Comisión Internacional de Límites y Aguas (CILA), Ciudad Juárez, Chihuahua, Mexico

H. R. Guerrero García Rojas (ed.), *Water Policy in Mexico*, Global Issues in Water Policy 20, https://doi.org/10.1007/978-3-319-76115-2_2

conflicts, cultures have developed rules to minimize problems on the utilization of resources.

The right for water is a key element in human rights and is the main goal of global initiatives. This chapter describes the physical conditions of Mexico and how its population has responded to the environmental conditions. In the first part, we want to acknowledge that our country is one of the most diverse in terms of biological resources and our ancestors dealt with these extreme conditions with technology and ways of organization to endure the wide range of environmental conditions they found.

Later, we describe the institutional arrangements made to face new challenges and new technologies. Although the environmental conditions are as extreme as in the past, new technological developments have influenced on the way society organizes in creating new rules and/or new approaches.

Water resources are key elements in civilization and its survival. At the end of the chapter, the story is not finished, because new technology will bring new challenges, but we want to address what the situation is, how can we cope with the future, and what are the best possible approaches.

2.2 Context

The geographical location of Mexico has influence on the country diversity, not only biological but cultural and economic as well. Geographically, the southern part of Mexico, from 14°32′27″ N latitude up to the Tropic of Cancer, is in the tropical region, while the northern part of the country, from the Tropic of Cancer up to 32° 43′ 06″ latitude, is in the temperate region.

2.2.1 Climate

Although temperate and tropical climates are expected from the geographical location of Mexico, the northern part is in the high-pressure zone where the driest deserts of the world are located that include the deserts of Sonora and Chihuahua, which are shared with the United States (CONAGUA 2016, p. 13). The climate in the temperate region, in more than half the land, is dry, and only a small portion, the highlands, is temperate. Most of Southern Mexico has a tropical climate.

Climate is defined as the average annual atmospheric conditions in a given region over 30 years or more. In Mexico, we use the Köppen classification (INEGI 2000). The classification is based on the native vegetation and uses mean annual temperature and mean annual rainfall. According to those two indicators, the main climate units are tropical, dry, and temperate.

Figure 2.1 shows climate units according to Köppen classification adjusted to Mexican conditions (INEGI 2000). On the average, a good climate will have an

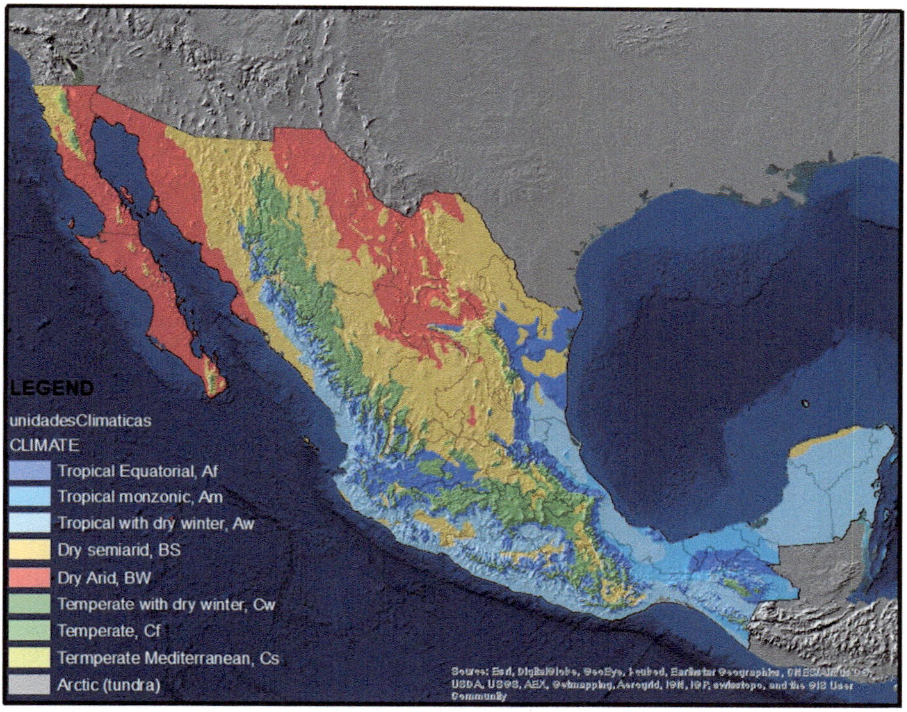

Fig. 2.1 Climate units of Mexico (modified from INEGI 2000)

average annual rainfall of 800 mm, and a wet climate is above 2000 mm; Northern Mexico is dry, Southern Mexico is wet, and Central Mexico is temperate. The average annual rainfall for the country is 760 mm, close to the ideal precipitation; however, annual rainfall distribution shows a bimodal distribution: 65% of rainfall in the summer (May–October) and 35% of rainfall in the winter months (November–February).

Climate in part is related to extreme events, which are a main concern at the national level. Droughts and floods are almost the rule in Mexico. Droughts are less frequent than floods, but their effects are devastating due to the lack of water in a significant portion of the nation. Floods are associated with atmospheric disturbances, tropical cyclones, in the Pacific and Atlantic oceans.

Damage assessment is conducted with the Saffir-Simpson scale, which has been used to measure tropical cyclones, as described in Table 2.1.

It is possible to add tropical depressions and tropical storms to moderate (H1 and H2) and intense (H3–H5) cyclones, since some of the damages have also been related to the first two categories. Table 2.2 shows that in 45 years, Mexico has had 224 major storms, most of them in the Pacific side, as well as the most destructive ones.

Table 2.1 Saffir-Simpson scale to assess tropical cyclones

Category	Maximum winds (km/h)	Storm tide (m)	Description	Further information
H1	119–153	1.2–1.8	Very dangerous winds will produce some damage	Well-constructed framed homes could have damage to roof, shingles, vinyl siding, and gutters. Large branches of trees will snap and shallowly rooted trees may be toppled. Extensive damage to power lines and poles likely will result
H2	154–177	1.8–2.5	Extremely dangerous winds will cause extensive damage	Well-constructed framed homes could sustain major roof and siding damage. Many shallowly rooted trees will be snapped or uprooted and block numerous roads. Near-total power loss is expected with outages that could last from several days to weeks
H3	177–208	2.5–4.0	Devastating damage will occur	Well-built framed homes may incur major damage or removal of roof decking and gable ends. Many trees will be snapped or uprooted, blocking numerous roads. Electricity and water will be unavailable for several days to weeks after the storm passes
H4	209–251	4.0–5.5	Catastrophic damage will occur	Well-built framed homes can sustain severe damage with loss of most of the roof structure and/or some exterior walls. Most trees will be snapped or uprooted and power poles downed. Fallen trees and power poles will isolate residential areas. Power outages will last weeks to possible months. Most of the area will be uninhabitable for weeks or months
H5	>252	>5.5	Catastrophic damage will occur	A high percentage of framed homes will be destroyed, with total roof failure and wall collapse. Fallen trees and power poles will isolate residential areas. Power outages will last for weeks to possible months. Most of the area will be uninhabitable for weeks or months

Source: Modified from CONAGUA (2016, p. 37)

Table 2.2 Tropical cyclones that have impacted Mexico in the period 1970–2015

Ocean	Tropical depressions (TD)	Tropical storms (TS)	Moderate cyclones (H1 and H2)	Intense cyclones (H3–H5)	Subtotal
Atlantic	27	31	14	12	84
Pacific	32	49	46	13	140
Total	59	80	60	25	224

Source: CONAGUA (2016)

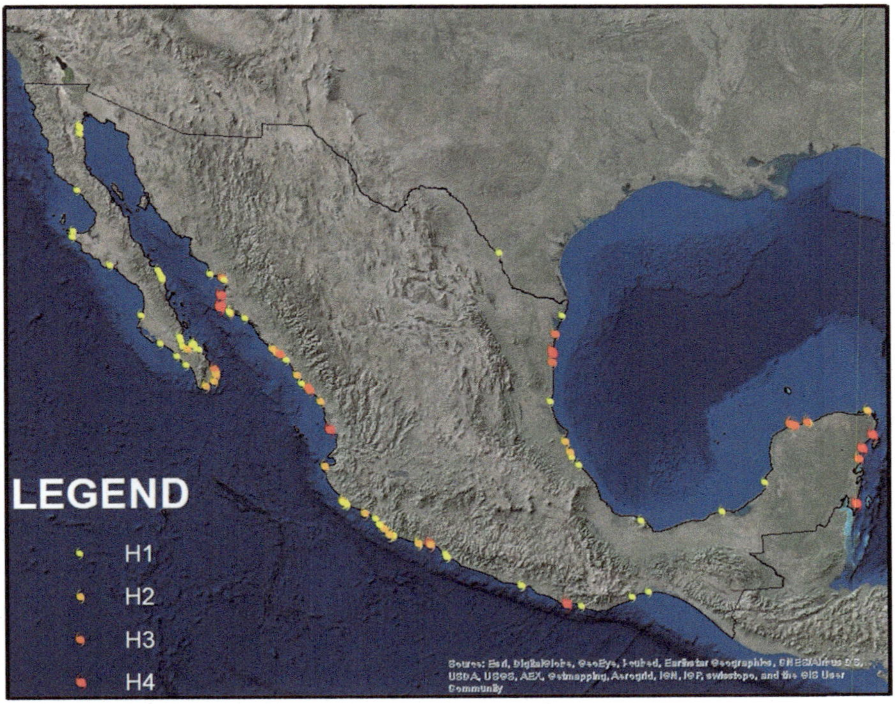

Fig. 2.2 Tropical cyclones impact location in the coasts of Mexico (CONAGUA 2016, p. 38)

Figure 2.2 shows the impact zones of tropical cyclones from 1970 to 2015 in the coasts of Mexico showing that the cyclones in the Pacific Ocean are more frequent and destructive.

A very important portion of the national budget is spent to control and manage those catastrophic events.

2.2.2 Land Resources

Climate and relief are factors that contribute the most to Mexico's biological and cultural diversity, but land and water resources play a significant role in the economic conditions due to the ability of man to transform the land.

2.2.2.1 Terrain

Mexico's terrain is rough due to the presence of large mountain ranges that cross south-north, *Sierra Madre Occidental* and *Sierra Madre Oriental*, and mountain ranges east-west in the south, *Sierra Juárez*, *Sierra del Sur*, and *Sierra del Lacandon*. The rough country is about 40% of the surface, and it is combined with high plains and coastal plains where human activities are concentrated. Fortunately, rough terrain is also responsible for the presence of rivers that provide water for human consumption and economic activities.

2.2.2.2 Soils

Relief and climate influence soils, and as a result, about 40% of the land has good productive soils, especially in the dry zones and the high plains. On the average, high-clay content soils or vertisols predominate in the plains, independently of the climate, and are the basis for agriculture and livestock production. Desert soils are productive, too, but water is a limiting factor.

2.2.3 Water Resources

The presence of mountain ranges has a very strong influence on Mexico's water resources. Rainfall is converted in runoff, surface water, and some portion infiltrates the soil and percolates to aquifers, turning into groundwater. Surface and groundwater resources play a major role in the regional development. Surface water and groundwater are separately described, but they are connected. Also, since water quality is as important as water quantity, a brief description is included, too.

2.2.3.1 Surface Water

Fortunately, even the largest deserts of Mexico have a river system, which in the arid and semiarid regions are ephemeral or intermittent, while in the temperate and tropical zones, the rivers are perennial. In the Sonoran Desert, the Colorado River, shared with the United States, is the main supply; while the Chihuahuan Desert has

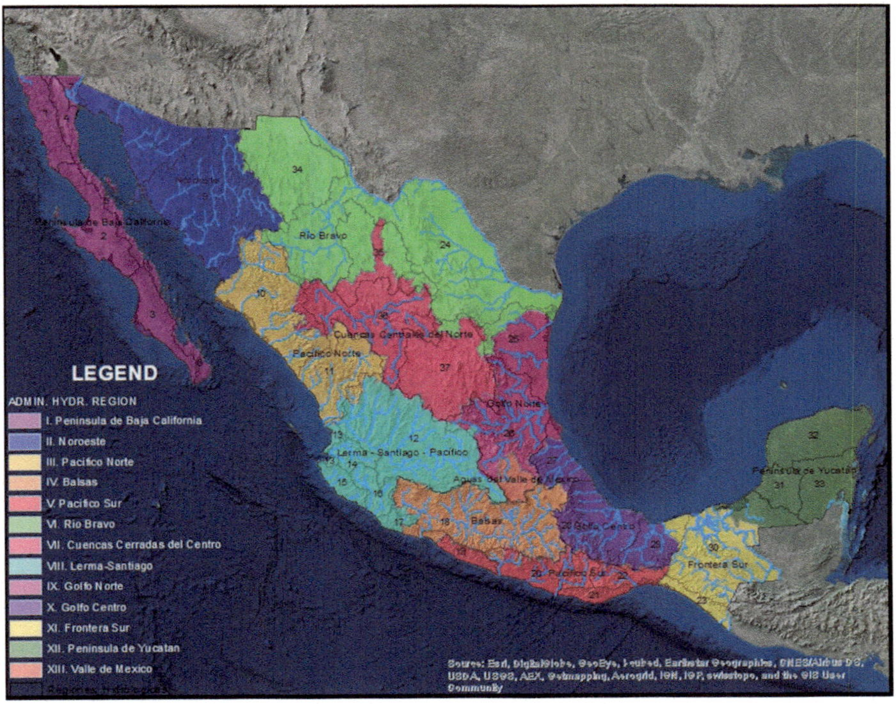

Fig. 2.3 Hydrologic Administrative Regions (RHA in Spanish) and 37 subregions (CONAGUA 2016, p. 29)

the Rio Bravo (Rio Grande) and its tributaries, as the main source of water. Lakes are more common along the Lerma and Balsas basins that drain into the Pacific Ocean, in the temperate areas. Finally, in the tropics, the rivers are not just perennial but have large discharges, enough to house large hydropower plants in their reservoirs. Although there are 1471 river systems, 731 basins are managed for hydrological balances, grouped in 37 hydrological regions (Fig. 2.3). Some river systems are shared with neighboring countries, three with the United States (Tijuana, Colorado, and Bravo) (CILA 1944), four with Guatemala (Grijalva-Usumacinta, Suchiate, Coatan, and Candelaria), and one with Belize and Guatemala (Río Hondo).

Figure 2.3 shows that the Hydrologic Administrative Regions (RHA in Spanish) of the two peninsulas are different than in the continental regions. In Peninsula de Baja California (RHA I), streams are short and abrupt in slope, while Peninsula de Yucatan (RHA XII) has no surface streams due to the geology of the region, karstic. Overall, the annual surface water adds up to 80% of the water availability as will be later described.

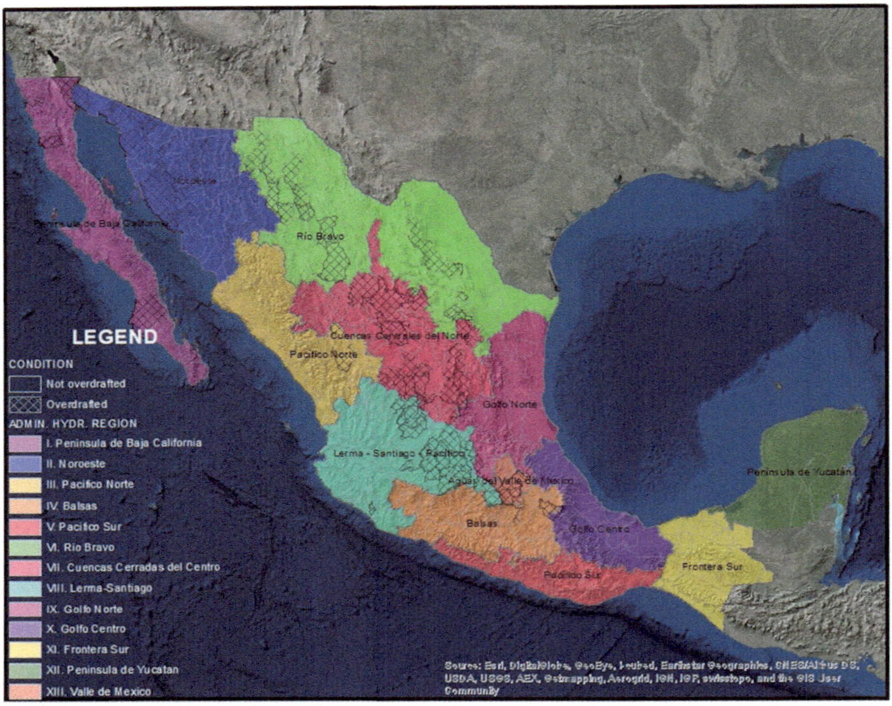

Fig. 2.4 Location of aquifers in Mexico by Hydrologic Administrative Regions (RHAs) showing overdrafted aquifers (modified from CONAGUA 2016, p. 30)

2.2.3.2 Groundwater

There are 653 aquifers (Fig. 2.4) that provide with about 20% of the water supply on an average annual basis. Aquifers started to become a secure water source in the arid regions with the arrival of pumps in the 1940s. Since then and with the support of power lines to agricultural areas, in the 1970s, many areas were heavily pumped. By 1975 there were 32 overdrafted aquifers; by the 1980s there were 80. Currently there are 105 overdrafted aquifers, 17 of which have already saline intrusion and 32 are responsible for soil salinity and/or with low-quality regimes.

Figure 2.4 shows the location of the aquifers, and it can be seen that most of the overdrafted aquifers are in the arid and semiarid regions as well as in the most populated areas.

2.2.3.3 Water Quality

Water quality has been classified based on three indicators, biochemical oxygen demand (BOD5), chemical oxygen demand (COD), and total suspended solids

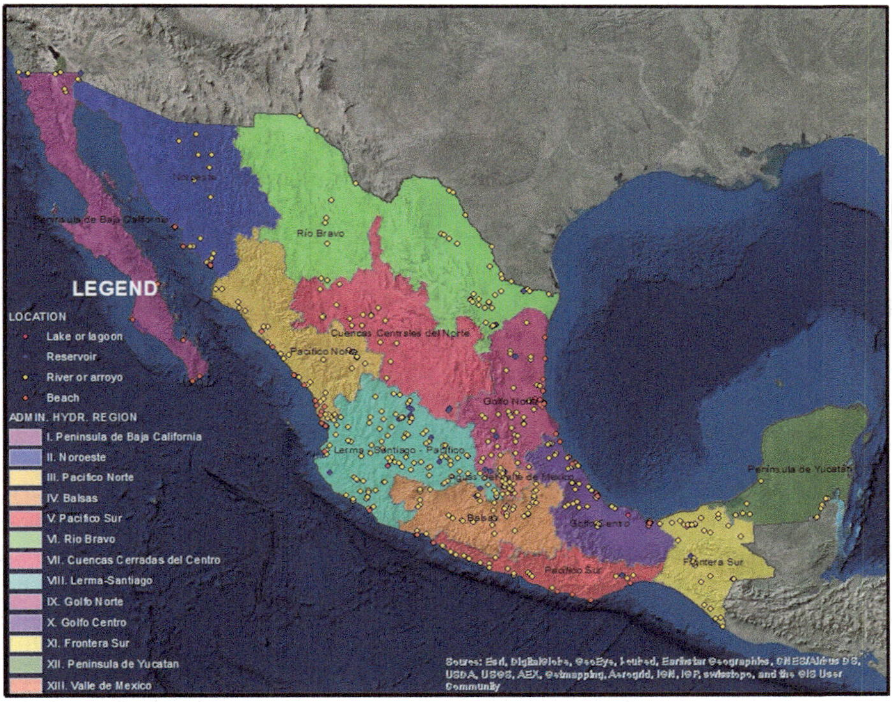

Fig. 2.5 Surface water quality monitoring sites in Mexico (Source: modified from CONAGUA 2016)

(TSS), and 21 basins have segments where one or several indicators do not comply with the limits established on the Official Mexican Norm (SEMARNAT 2003).

Figure 2.5 shows rivers, lakes, and coastal monitoring sites. The Hydrologic Administrative Regions with most problems due to water quality are RHA XIII, RHA IV, and RHA VI, located in Central Mexico, the most densely populated area.

2.3 History of Water Legislation and Institutional Framework

In the history of the social organization to cope with food and water challenges, the Mexican cultures have evolved. This is a description of the arrangements and main techniques used from pre-Hispanic to current times. Water has played a relevant role in the history of Mexico. Dams, aqueducts, canals, wells, and other water works were built before Europeans arrive to the continent. Some works were for flood control, but most of them for domestic use and rainfed or irrigated agriculture. The hydraulic works and the arrangements to its construction, operation, and maintenance have changed since then.

2.3.1 *Pre-hispanic Times: 800 BC–1521*

Climate not only influenced flora and fauna but man, too. Early cultures used technology to provide food for the early settlements. Anthropologists recognize Mesoamerica, the temperate and tropical climate cultures of Mexico, but little is known of Aridoamerica. Braniff Cornejo (2009) called those cultures as "La Gran Chichimeca." It corresponds to Northern Mexico, including Southwestern United States.

Mesoamerican cultures were distributed in the temperate region where Olmec, Aztec, Purepecha, and Zapotecs, as well as Tlaxcallan cultures, were dominant and the tropical region where Mayan cultures dominated. Aridoamerican cultures vary from the Hohokam in Arizona (USA), Pueblo in New Mexico (USA), and Chihuahua (Mexico) to Trincheras culture, Rio Sonora culture (Sonora), and the Yoreme culture (Sonora and Sinaloa) (Braniff Cornejo 2009) that diverted river streams to grow their crops. Mesoamerican cultures have been deeply studied, and there are several references on the hydraulic works used. Rojas Rabiela (2009, pp. 7–20) mentioned hydraulic works that involved surface and groundwater sources for domestic use. The main source was surface water and rainfall. Groundwater sources were more common in the Yucatan Peninsula due to the geology of the area in the Mayan culture. Other than rainfall collection for domestic and/or agricultural use, surface water involved storage, conveyance, and application techniques on gravity irrigation. Subsurface irrigation was mainly in chinampas around Xochimilco, near Mexico City.

In the Mesoamerican or Aridoamerican cultures, (1) water was diverted from streams to irrigate neighboring areas with masonry, rock, or earthen levees, and (2) terraces were formed to collect and distribute rainwater in the land. Variations depended on the level of development of each culture.

In terms of the social organization, Palerm and Wolf (1972) mentioned that most pre-Hispanic cultures have an "Asiatic mode of production," where water technicians played a major role due to the importance of the construction and maintenance of the hydraulic works so dearly needed. Water works were a very intensive work that required coordination and cooperation among the members of those cultures. It is possible to say that the Asiatic mode of production created a specialized group that had an authoritative scale in those cultures in the collection and distribution of water.

2.3.2 Colonial Times: 1521–1824

When the Spaniards came to America, they claimed the lands they visited in the name of the Castile Crown and declared the Catholic faith. Those cultures that represented some gain to the crown's treasure and/or did not accept the Catholic faith were conquered by force. The arid zones were not of special interest, and the harsh conditions were claimed in the name of the King by the Order of Jesus of the Catholic Church. On May 3, 1493, Pope Alexander VI, gave power to the Catholic Kings to the land they claimed (cited by Molina Enríquez 1909, pp. 166–167). The first legislation on land and water resources in Mexico was on the "Compilation of the Laws of the Kingdoms of the Indies" published in 1681, where water was considered a common good. Under this legislation, the crown gave power to their representatives, viceroys, to concede land and water grants ("mercedes") to individuals for a given period of time. Later these grants were changed to ad infinitum on October 15, 1754 (Molina Enríquez 1909, p. 167).

Rarely, the colonial legislation separated water from land. The reported cases belonged to irrigation concessions, industrial use, or water supply for human settlements. Other than those mentioned, land and water belonged to the crown (Molina Enríquez 1909, p. 171).

Colonial hydraulic works were modified with the introduction of wheels, steel, and animal traction in more efficient and powerful tools, but the basic concepts of extraction or diversion, storage, conveyance, and application of water in the domestic services or agricultural activities did not change much but the scale (Rojas Rabiela 2009, pp. 20–21). What changed were the organization and goals of the hydraulic works. The goals changed from collective or communal to private, and the work organization that used to be voluntary and orderly changed into forced labor provided from local indigenous communities for their lords. Indian rights for land and water were provided in the legislation but not observed. Conflicts on water distribution were solved as in Spain with water judges that in some places in Mexico were still in use until very recently. Water judges were elected, and they had the authority to solve water conflicts.

2.3.3 Independent Times: 1824–1880

After the independence from Spain, and to counteract potential foreign invasions, a colonization decree was published on August 18, 1824, aimed at promoting human settlements, especially in the northern arid zones. Early developments diverted streams through canals; they were called *acequias*. However, as more people arrive and in dry years, conflicts arose and became a major problem. In 1843, the decision process was granted to the municipal authorities (ayuntamiento), especially conflicts among different settlements. No major changes than those conducted during the colonial times were arranged. The major progress in those times were:

- Establishment of the metric system on August 2, 1863. Before there were a lot of local measurement systems that complicated calculations. The introduction of the metric system marked a large improvement and eased calculations and engineering works.
- Publication of the "Código Civil" (Civil Code) in 1870 where water was declared property of the nation.

2.3.4 *Porfirian Times, 1880–2011*

The continuous threat of foreign invasions persisted during Porfirian times. Three laws were published that played a major role in the use of land and water resources in Mexico:

1. *Ley sobre Terrenos Baldíos* (*Law of Wastelands*). The law was published on December 15, 1883, and it allowed the presence of foreign settlers on wastelands and the investment in infrastructure to improve the economic conditions of the arid and semiarid unproductive lands in Northern Mexico.
2. *Ley sobre Vías Generales de Comunicación* (*Law of Transportation*). Published in 1888, it not only allowed the terrestrial communication and railroads but the improvement of ports and fluvial and marine transportation. It was modified on June 15, 1888, to include the role of the federal government on the taxation of concessions for the use of marine and fluvial routes and hydropower plants.
3. *Ley General de las Instituciones de Crédito* (*General Law of Credit Institutions*). Issued on March 19, 1897, it allowed foreign capital to invest in productive activities, such as mining, railroads, and large irrigation and hydropower plant projects.

As a result, several colonization projects based on irrigation started and were finished years ahead. Here are five examples:

- In 1865, José de Jesús and Martín and Juan Francisco Salido were granted a loan to build the first channel in Camoa, Sonora, to divert water from the Mayo River. That was the start of the irrigation district DR-038 Río Mayo, Sonora, currently with about 80,000 ha (Kroeber 2009).
- In 1886, a group of American colonizers, led by or Albert Kimsey Owen, arrive to Topolobampo and started the construction of Canal Taxtes, by diverting water from Rio Fuerte, and Benjamin Francis Johnston started the construction of a sugar mill, *Sinaloa Sugar Company* (Infante 2011). Those are the beginnings of DR-074 Río Fuerte, currently with 230,000 irrigation hectares.

- In 1889, Charles R. Rockwood, associated with Guillermo Andrade, through the *Sociedad de Irrigación y Terrenos de la Baja California, S.A.*, was allowed for the construction of Canal El Álamo to divert water from the Colorado River to the Imperial Irrigation District, in California, USA (Hundley Jr. 2000). Irrigation district DR-014 Rio Colorado has currently 180,000 ha with water imported from the Colorado River (CILA 1944).
- In 1890, Carlos Conant Maldonado was granted to irrigate 400,000 ha in rivers Fuerte, Sinaloa, Mayo, and Yaqui in Sonora, through the *Sonora and Sinaloa Irrigation Company*. He could not finish the project and sold his company to the *Richardson Irrigation Company* (Luna Escalante 2007). This was the beginning of irrigation districts DR-018 Colonias Yaquis and DR-041 Río Yaqui, currently covering 245,000 ha.
- On May 7, 1906, a concession was issued to Pablo Ginther and Joaquin Cortazar to build La Boquilla Dam, on the Conchos River, to generate electricity for neighboring mining companies, and the water released was sold for farmers in what is now DR-005 Rio Conchos, Chihuahua (Kroeber 2009).

However, the irrigation and the hydropower plant projects conducted and run by foreign companies increased the conflicts for water, forcing the Porfirio Diaz administration into the creation of an office with trained personnel to make decisions. It also promoted the creation of a hydraulic engineer in the National School of Agriculture, to improve the presence of trained personnel because the water conflicts multiplied (Kroeber 2009).

Diaz started the formation of "Scientific Commissions" with trained personnel who started the field reconnaissance studies and continued on the design and construction of large hydraulic works. Those were the basis for the following stages in the creation of large irrigation and hydropower projects. The "Comisión Geográfica Exploradora" (Geographic Exploration Commission) was responsible for surveys, land adjudication, and zoning in indigenous communities, besides the design and construction of irrigation canals, land clearing, and improvement works.

2.3.5 *Revolution Times: 1911–1926*

In this period, the situation of the country was very unstable, and not much was done until it settled down, by the 1920s. Before that, the basis of the Mexican Constitution was established in 1917. Among some of the regulation changes was article 27 that established that land and water belong to the nation and only an authority invested by the federal government will take decisions about granting concessions to individuals or groups of individuals.

2.3.6 Large Irrigation Projects: 1926–1940

In 1926, the *Ley sobre Irrigación con Aguas Federales* (Law of Irrigation with Federal Waters) was published. This law established the basis for the national agricultural development. Hydraulic works were considered of *public interest*, which allowed the expropriation and purchase of large parcels of land to design and construct large irrigation and hydropower plant projects. To conduct the water policy, the *Comisión Nacional de Irrigación* (National Irrigation Commission) was created. This institution was an administration office that trained personnel, conducted and improved survey studies, as well as conducted the construction and supervision of large irrigation projects.

On August 6, 1929, that law was modified as *Ley de Aguas de Propiedad Nacional* (Law of National Waters), which granted the *Secretaría de Agricultura y Fomento* (Ministry of Agriculture and Economic Development) the authority in water issues but navigation. It started granting water concessions and included "a compensation to the federal government for the use and utilization of national Waters."

The 1929 law is the basis of the current water legal framework. It has only been updated or modified. Among the main issues included there were the rights and obligations of the legally bonded user associations (*sociedades de usuarios*). This is the first time that "national hydraulic reserves" are mentioned to protect and secure water volumes of the irrigation districts and hydropower plants.

On October 19, 1934, it was modified to assign penalties and fines due to unauthorized construction of dams or derivations, especially if there were damages on private property. This modification introduced temporary permits for water use, exploration permits, authorizations for the use of water in humid years on non-granted lands, and the public register of the Water Users Associations. It also introduced water rights suspensions due to pollution and lack of compliance with the volumes assigned.

2.3.7 Ministry of Hydraulic Resources, 1947–1976

The agrarian reform created a group of new farmers with good cropland and irrigation, access to credit, agricultural machinery, and technical assistance. Not only the agricultural land increased from 3.0 Mha in 1930 to 5.2 Mha by 1940 but the land productivity, too. *Ejido* land increased from 800,000 ha to 3.5 Mha.

In 1947, the new presidential administration created the Ministry of Hydraulic Resources (*Secretaría de Recursos Hidráulicos*, SRH), and the government responsibility was concentrated in this institution. Water supply and sanitation were incorporated to the duties of SRH aimed at building a strategy to abate the water services. Multipurpose hydraulic projects were designed, constructed, and operated to generate electricity, flood control, and irrigation. New water management models were instituted based on the recent progress in the United States, the formation of integrated watershed management projects through River Basin Commissions.

Executive commissions were formed to conduct regional development projects in several basins often affected by destructive floods in Southern Mexico. The Papaloapan River Basin Commission was the first to be established by presidential decree; then the Commission for the Tepalcatepec River, later known as the Balsas River Basin Commission; Rio Fuerte River Basin Commission; and Rio Grijalva Basin Commissions in 1952. Those were organizations with administrative autonomy that went beyond hydraulic works into regional economic development. A very special case was the Hydrological Commission of the Mexico Valley Basin in 1951. Within this framework, the Lago de Texcoco Commission was decreed in 1974. Several hydraulic progresses were made, and many personnel trained within this approach conducted several projects once the commissions were repealed.

By 1948, the *Dirección General de Agua Potable y Alcantarillado* (General Direction of Drinking Water and Sanitation) established the regulations for the *Juntas Federales de Agua Potable* (Federal Potable Water Boards), which started the design and construction of wastewater treatment and potable plants throughout the country.

Although groundwater pumping started with the first vapor motors in the early 1900s (Moreno Vázquez 2006, p. 118), the arrival of submersible pumps by the early 1940s, powered with diesel, significantly increased the use of groundwater and, therefore, aquifer depletion. This situation forced regulations to control aquifer pumping (Moreno Vázquez 2006). A law was decreed on December 30, 1947, to control groundwater pumping and to establish aquifer bans. The first aquifer ban was established in 1951 in the Sonoran Desert.

By the 1960s, farmers were organized to introduce power lines to their fields to electrify their pumping systems. This skyrocketed the use of groundwater in Mexico, and consequently, many aquifers have been severely affected and conflicts have surfaced. Over 100 aquifer bans have been established since then.

2.3.8 Federal Waters Law: 1972–1992

On January 11, 1972 the Federal Waters Law was published, where SRH is the water authority. All past water laws were repealed including potable water and sanitation and groundwater. For the first time, surface and groundwater are united in a single general law. Aquaculture was recognized as a water use. The maximum irrigated land tenure was reduced from 50 to 10 ha.

In this water legislation, three hydro-agricultural administrative units were created:

1. *Distritos de Riego* (irrigation districts)
2. *Unidades de Riego para la Producción Rural*, URDERALES (irrigation units)

3. "*Distritos de Drenaje y Protección de Inundaciones*" (drainage and flood protection districts or drainage districts)

The farmers were organized in *Comités Directivos* (Directive Committees) that were created to participate in the hydraulic planning. Aquaculture districts were created, too, but most of them used seawater. Finally, SRH was the sole authority to grant water concessions.

By 1973 SRH serviced 1405 locations through 1120 systems, which weakened the municipal governments.

In 1976, SRH was derogated, and all hydro-agricultural infrastructures were transferred to the new *Secretaria de Agricultura y Recursos Hidráulicos* (Ministry of Agriculture and Hydraulic Resources), and the drinking water treatment and sanitation was transferred to a ministry that dealt with urban services and later was returned to the municipalities by 1983.

2.3.9 Water Decentralization: 1992–2004

In December 1992, the *Ley de Aguas Nacionales* (Law of National Waters) was decreed, and other laws were repealed. The *Comisión Nacional del Agua* CONAGUA (National Water Commission) was created as a decentralized administrative institution part of the SEMARNAT (Secretary of Environment and Natural Resources). The *Registro Público de Derechos de Agua* REPDA (Public Register of Water Rights) was created to have a public register of water concessions as part of CONAGUA. The River Basin Councils were created to promote the participation of water users. For the first time, environment is considered a nonconsumptive water user, and environmental flow is described as the part of river flow that must be maintained to conserve freshwater ecosystems. The integrated river basin management approach was back as one of the strategies in the application of water policies by CONAGUA, the water authority.

Finally, a decentralization process started in 1997 by empowering the 13 Hydrologic Administrative Regions, now called *Organismos de Cuenca* (River Basin Organizations).

2.3.10 Water Democratization: 2004

In less than 12 years, modifications to the National Waters Law were made again and published on April 29, 2004. Most of the modifications were additions and/or clarifications of the old version. For example, the constituency of the River Basin Councils is clearer. The main chapters are as follows: 1st Preliminary Dispositions; 2nd Water Administration; 3rd Water Policy and Planning; 4th User Rights or Utilization of National Waters; 5th Zoning, Bans, or Reserves; 6th Water Uses;

7th Prevention and Control of Water Contamination and Responsibilities on Environmental Damage; 8th Investment on Hydraulic Infrastructure; 8th bis. Water Financial System; 9th National Goods in Charge of CONAGUA; and 10th Infractions, Penalties, and Recourses.

2.4 Current Situation of the Water Sector

Currently, to manage the water resources of the country, administrative divisions have been created, based mostly on hydrologic units, like basins, but in larger areas. For each region, the water availability and its quality are described to analyze the situation of the water sector.

2.4.1 Administrative Division of Water Resources

In 1997 the country was divided in 13 Hydrologic Administrative Regions (RHA in Spanish). Those regions were formed on large basins or the group of several basins, adjusted by state and/or municipal limits to integrate socioeconomic issues (CONAGUA 2007). Table 2.3 lists the 13 RHAs, showing the area, average annual surface and groundwater, population, water availability, and contribution to gross domestic product (GDP). Figure 2.6 shows the location of each RHA.

2.4.2 Water Sources

There are two main sources of water, surface and groundwater. Although they are managed separately, they are connected. Reduction of surface water will reduce the recharge of aquifers, while an excess of surface water will replenish aquifers. The effect may not be instantaneous, but it will be shown depending on the size of the water system.

2.4.2.1 Surface Water

Surface water is by far the most abundant source. Table 2.3 shows that almost 80% of the water consumed in the country comes from rivers or lakes, sometimes stored in reservoirs or diverted from rivers.

Surface water is climate dependent and its effect is almost instantaneous. Rivers in the southeast of the country are perennials, while rivers in Northern and Central Mexico are ephemeral and even intermittent. Therefore, surface water is sufficient in

Table 2.3 Administrative division of water resources and the sources of water, their distribution, and contribution to global domestic product (GDP)

Hydrologic Administrative Regions (RHA)	Continental area (km^2)	Average annual surface water (hm^3/year)	Average annual aquifer recharge (hm^3/year)	Average annual renewable water (hm^3/year)	Population (millions)	Renewable water per capita (m^3/cap/year)	Contribution to GDP (%)
I	154,279	3300	1658	4958	4.45	1114	3.61
II	196,326	5066	3207	8273	2.84	2913	2.86
III	152,007	22,519	3076	25,595	4.51	5675	2.88
IV	116,439	16,805	4873	21,678	11.81	1836	6.14
V	82,775	28,629	1936	30,565	5.06	6041	2.29
VI	390,440	6416	5935	12,351	12.30	1004	14.29
VII	187,621	5529	2376	7905	4.56	1734	4.19
VIII	192,722	25,423	9656	35,079	24.17	1451	19.08
IX	127,064	24,016	4108	28,124	5.28	5327	2.24
X	102,354	90,424	4599	95,023	10.57	8990	5.62
XI	99,094	121,742	22,718	144,460	7.66	18,859	4.93
XII	139,897	4008	25,316	29,324	4.60	6375	7.38
XIII	18,229	1112	2330	3442	23.19	148	24.49
	1,959,247	354,989	91,788	446,777	121.00	3692	100.00

Source: Modified from CONAGUA (2016, p. 20)

Fig. 2.6 Hydrologic Administrative Regions, lately River Basin Organizations, and the renewable water per capita (Source: modified from CONAGUA 2016, p. 21)

the humid tropics but not in the arid and semiarid regions, where groundwater must be pumped to supplement the population needs.

2.4.2.2 Groundwater

Table 2.3 shows that about 20% of the total water available is pumped from aquifers. Three regions are highly dependent on groundwater: RHA XII *Peninsula de Yucatan*, with arid climate and karstic geology, RHA XIII *Valle de Mexico*, and RHA VI *Region Lagunera*, closed basins in semiarid climate.

On the average, groundwater quality is hard (excess of carbonates), but some aquifers may carry potentially harmful substances like fluoride and arsenic, so human consumption must be regulated. The problem with groundwater is that it has been irreversibly depleted causing a series of problems that varies from high pumping cost, high treatment cost, and land subsidence, among others.

2.4.3 Water Availability and Water Users

Table 2.3 shows the water availability per hydrologic region. The area with least available water is RHA XIII, *Valle de Mexico*, where the most populated city is located, Mexico City, as well as several industries and services.

The influence of large cities on the gross domestic product contribution can also be detected: Mexico City is in RHA XIII, with 24.49% GDP; Guadalajara is in RHA VIII with 19.08% GDP; and Monterrey is in RHA VI, with 14.29% GDP. Those regions are more industrialized than other RHAs, and the size of the cities is correlated with the contribution to the GDP.

2.4.3.1 Water Availability

The annual surface water availability is 354,989 hm^3/year, while groundwater availability is estimated at 91,788 hm^3/year. However, their distribution is not homogeneous: it decreases as population increases. The annual water availability per capita has decreased over time; for example, the average availability in 1950 was 18,035 m^3/cap/year, and it dropped to 4422 m^3/cap/year in 2010. According to international standards, water availability per capita is low.

2.4.3.2 Water Users

Although the total water available is 446,777 hm^3/year, not all the water is used. In Mexico, water belongs to the nation. The nation, through the authority invested in *Comisión Nacional del Agua* (CONAGUA), grants water concessions, through *Registro Público de Derechos de Agua* (REPDA or the Public Register of Water Rights). The titles allow an applicant for the concession of a given volume of water, if available, for a period. Water concessions add 266,560 hm^3/year.

It is very important to recognize that there are two types of water users: consumptive, those where water is consumed during the process, and nonconsumptive, those where water is consumed at zero to nonsignificant amounts. Table 2.4 describes the classification of water users in Mexico and grouped water users, either consumptive or nonconsumptive use, as well as the annual figures.

Table 2.4 shows that the concessions for consumptive uses add to 85,665 hm^3, while concessions for nonconsumptive users total to 180,895 hm^3, for a total of 266,560 hm^3/year. It can also be identified that agriculture (including livestock, aquaculture, and other users) represents 76.3% and domestic sums 14.6%, adding 91% of water conceded for consumptive users.

Table 2.4 Grouped water uses and water concession volume per water use

Key	Repda classification	Volume of water concessions (Hm^3)	Grouped uses	Definition	Volume of water concessions (Hm^3)
A	Agriculture	58,450	Agriculture	A+D+G+I+L	65,360
B	Agroindustrial	4	Domestic	C+H	12,480
C	Domestic	39	Industrial	B+E+F1+K	3676
D	Aquaculture	1136	Power plants	F2	4149
E	Services	1474	Subtotal consumptive		85,665
F	Industry	6347	Hydropower plants	J	180,895
F1	Industry except thermal power plants	2198	Subtotal nonconsumptive		180,895
F2	Thermal power plants	4149	Total		266,560
G	Livestock	207			
H	Public urban	12,441			
I	Multiples	5566			
K	Commerce	0.1			
L	Others	0.5			
	Subtotal consumptive	85,665			
J	Hydropower plants	180,895			
	Subtotal nonconsumptive	180,895			
	Total	266,560			

Source: CONAGUA (2016)

2.4.3.3 Water Resources Stress

Water resource stress is defined as percentage of water consumed with regard to renewable water, and it was designed as an indicator of the water situation in a region. There are four categories: less than 10% is *NO STRESS*, 10–20% of renewable water consumption is a *LOW DEGREE*, the range from 20 to 40% the pressure is *MEDIUM*, the range from 40 to 100 is *HIGH*, and higher than 100% then of the pressure is *VERY HIGH*. Table 2.5 shows the values for each administrative region.

It can be seen that the most stressed area is RHA XIII *Valle of Mexico* that concentrates population and economic activities. It is followed by the desert regions and temperate regions heavily populated (Fig. 2.7).

Table 2.5 Stress over water resources indicator per Hydrologic Administrative Region (RHA)

Hydrologic Administrative Regions (RHA)	Total volume licensed (hm^3/year)	Renewable water (hm^3/year)	Stress over water resources (%)	Classification of stress
I	3958	4958	79.8	High
II	6730	8273	81.3	High
III	10,770	25,595	42.1	High
IV	10,798	21,678	49.8	High
V	1555	30,565	5.1	No stress
VI	9524	12,351	77.1	High
VII	3825	7905	48.4	High
VIII	15,724	35,079	44.8	High
IX	5742	28,124	20.4	Medium
X	5560	95,023	5.9	No stress
XI	2505	144,460	1.7	No stress
XII	4200	29,324	14.3	Low
XIII	4774	3442	138.7	Very high
Total	85,665	446,777	19.2	Low

Source: CONAGUA (2016)

2.4.4 Hydraulic Infrastructure

To supply water to 121 million people and support associated economic activities, an impressive hydraulic infrastructure has been built and continuously improved to attend the needs of a fast-growing economy. The largest water users are agricultural and municipal uses, but other activities need energy and water in their processes; therefore, the construction of dams to store water, produce energy, control floods, or other services has been a national priority. CONAGUA is responsible for the construction, operation, and maintenance of the larger hydraulic works, i.e., the storage and conveyance of water, the basis for the direct supply of water to households or agricultural lands. There are more than 5000 dams or reservoirs in the country, 180 have capacities over 20 hm^3. The potential water capacity of all the dams is estimated at 150,000 hm^3. A complete list and characteristics of the 180 dams is provided by CONAGUA (2016, p. 99).

There are over 3000 km of aqueducts that serve the largest cities. The discharge of those aqueducts is over 112 m^3/s. CONAGUA (2016, pp. 118–121) provides a very extensive list of the aqueducts and their characteristics. The extension of the aqueducts is a measure of local deficits, especially in relation to large cities. Most of the aqueducts are water transfers to serve large cities. The most outstanding examples are Mexico City and the Cutzamala System, Monterrey, Guadalajara, Ixtapa-Zihuatanejo, Tijuana, Ciudad Victoria, León, and Hermosillo, among others (CONAGUA 2016, p. 98). Most of those cities have populations above 1 million people. In most cases, water conflicts have aroused between the water source and the water delivery.

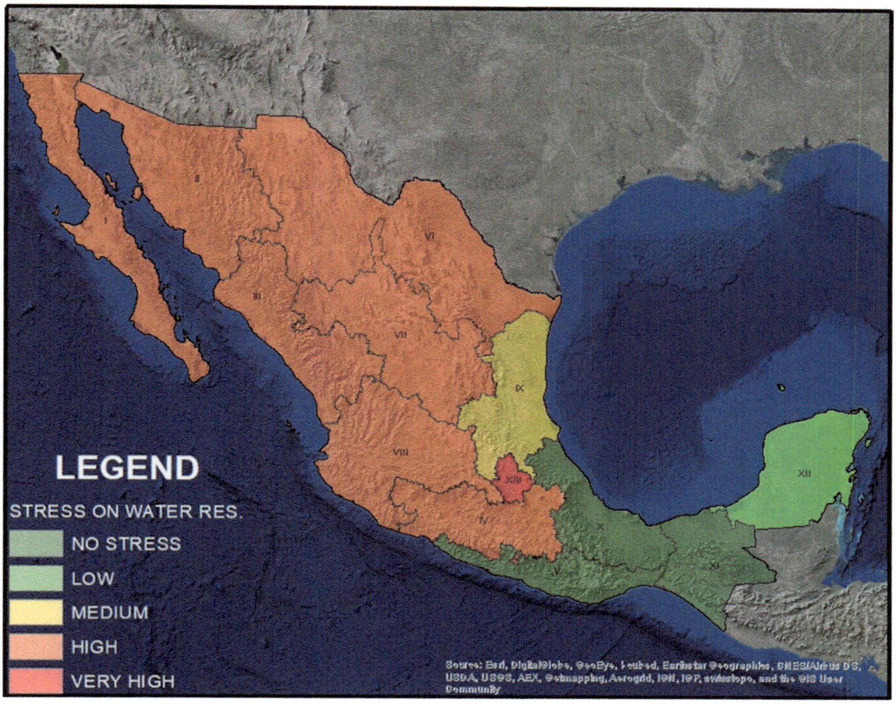

Fig. 2.7 Stress on water resources per Hydrologic Administrative Region (Source: modified from CONAGUA 2016)

2.4.4.1 Potable Water Supply and Sanitation

The water supply and sanitation services to the households are provided by municipalities, but CONAGUA funds, trains, and provides advice to the municipalities. Municipal services are straightforward in urban areas but in rural areas it is a big challenge, so it is important to analyze it separately. The water supply infrastructure consists of 874 drinking water treatment plants that yields 97.9 m^3/s to cover 94.4% of the population, 97.2% urban, and 83% rural (CONAGUA 2016, p. 112). Most importantly is the evolution of coverage of water supply and sanitation. Figure 2.8 shows the evolution of the potable water supply to households from 1990 to 2015, where it jumped from a 75.4% to 94.4% at the national level, but the coverage in the rural areas almost doubled from 46.5% to 83.0%.

Sanitation is provided by 2477 municipal wastewater treatment plants that treat 120.9 m^3/s and 2832 industrial wastewater treatment plants that discharge 70.5 m^3/s. The coverage of sanitation in 2015 was 91.4% national average with 96.6% in urban locations and 77.5% in rural locations. The evolution of the coverage of sanitation services for the population is shown in Fig. 2.9. It started from 58.6% to 91.4% in 25 years at a national level. Of course, the big difference was in the rural areas, where

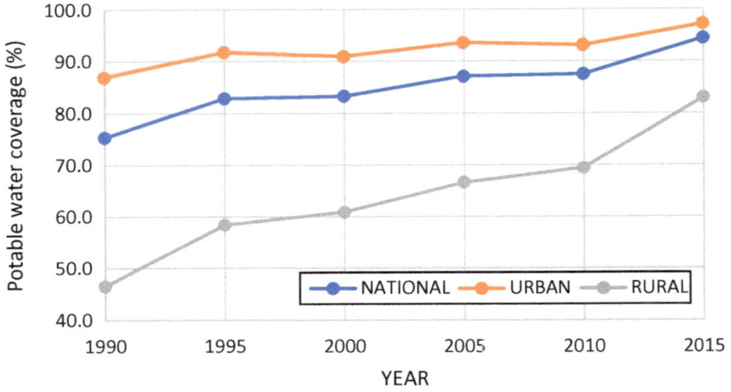

Fig. 2.8 Evolution of coverage of potable water and sanitation services in Mexico (modified from CONAGUA 2016, pp. 112–115)

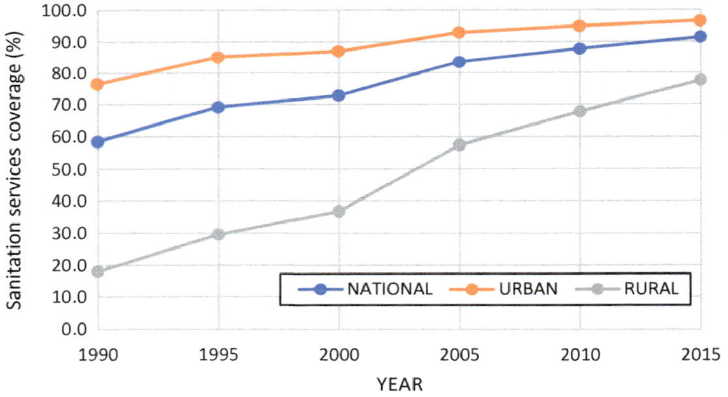

Fig. 2.9 Evolution of sanitation service, for the population in Mexico from 1990 to 2015 (CONAGUA 2016, pp. 114–116)

the coverage in 1990 was 18.1%, and it increased to 77.5%, still below of what is expected. The coverage in urban areas increased from 76.6% to 96.6%.

The major challenge in the future is to increase the water supply and sanitation to rural communities.

2.4.4.2 Hydro-agricultural Infrastructure

Although some of the hydro-agricultural infrastructure is shared with livestock and aquaculture, most of it is merely devoted to agriculture, the main water consumer.

Agriculture is conducted in about 26 Mha (million hectares), but some areas have been supported with infrastructure of two main types: irrigation and drainage work. To be eligible to financial aid, the farmers are organized in three different administrative units that will be called hydro-agricultural administrative units:

1. Irrigation districts. Since 1926, irrigation has been supported by the federal government in providing the means to store, divert, convey, and even apply irrigated water in the field in large areas or with many farmers.
2. Irrigation units. Smaller areas in size, or with few farmers, were organized in what are called *Unidades de Riego para el Desarrollo Rural* (URDERALES).
3. Drainage districts. In the 1970s, the federal government supported farmers in the humid tropics with the construction of drainage networks to control excess water in the field in what were called *Distritos de Drenaje y Control de Inundaciones* (Flood Control and Drainage Districts), and by the 1990s the program was regained and now called *Distritos de Temporal Tecnificado* (Technified Rainfed Districts).

The agricultural land in Mexico is about 26 Mha (million ha), but 7′774,435 ha with water concessions have irrigation works, and 3′375,898 ha have drainage networks. Drainage networks are not only in drainage districts but also in irrigation districts due to the presence of salt-affected soils in irrigation districts that require drainage to control salinity (Table 2.6). Although the area with water concessions of the three administrative units is comparable, irrigation units contribute more to the production and income of the agricultural sector, followed by the irrigation districts and the drainage districts.

A major challenge in the hydro-agricultural infrastructure is salinity in the irrigation districts and units. The salinity rate is in the order of over 10,000 ha/year and there is a potential 1.2 Mha to be salinized, especially in the largest irrigation districts in northwest Mexico (Arias et al. 2017). In the drainage districts, the major challenge is to increase the on-farm drainage to increase productivity and income (Arias Rojo and González 2017).

2.4.5 Water Governance

The ways societies rule their water resources and services have profound impacts on people's welfare and sustainable development. Access to water is a matter of daily survival or breaking the poverty vicious cycle. Therefore, water governance is a keystone to reduce poverty and inequality. Water governance does not depend on the mandate of institutions but the way decisions are made. The existence of democratic processes, the freedom to express opinions/concerns, equality in information sharing, and the right to organize are key elements to make equalitarian and efficient decisions. The 1992 water legislation opened the social participation to water management. There are two main areas where social participation in water governance was opened, at a river basin scale, through the

Table 2.6 Administrative units with collective irrigation and/or drainage infrastructure compared with rainfed agriculture

Administrative unit	Number	Users	Area with water concessions (ha)			National contribution (%)	
			Total	Irrigation	Drainage	Production	Income
Irrigation districts	85	431,602	2,529,510	2,529,510	1,490,709	25	24
Irrigation units	39,492	901,963	2,956,420	2,956,420	No data	40	30
Drainage districts	23	97,518	2,288,505	2,288,505	1,885,189	15	5
Rainfed	No data	No data	18,225,565	No data	No data	20	41
Total			**26,000,000**	**7,774,435**	**3,375,898**	**100**	**100**

Source: Modified from Arias et al. (2017)

Consejos de Cuenca (River Basin Councils), and at the hydro-agricultural administrative units, through the *Comité Directivos* (Executive Committees).

2.4.5.1 Basin Councils

In the 1992 water legislation, River Basin Councils were created with the idea of promoting social participation with an integrated river basin management approach. The social participation was open to elected representatives of the main water users. The main representatives of the Basin Councils are agriculture (representatives of irrigation districts, irrigation units, and/or drainage districts), municipal water service providers, private or municipal authorities, and industrial, energy, and municipal and state officials, all of them with the assistance of CONAGUA officials. The members of the Basin Councils are elected from each water sector, and the group is formalized before attorneys.

The Basin Council has frequent meetings, at least two annually, and they are also supported by other subsidiary representations. The most common subsidiary groups are the following:

- Tributary councils. If the basin is very large and/or there are several river systems, an equally conformed organization is created.
- COTAS (Technical Committee of Groundwater). When groundwater is extracted, each aquifer can be represented, too.

Despite the participation of water users' representatives in River Basin Councils, water conflicts are still growing. The lack of clear rules and responsibilities of the civil representatives, unequal information level, as well as lack of transparency in those projects might be associated with the low impact of these institutions on those conflicts.

2.4.5.2 Hydro-agricultural Administrative Units

Starting in 1991, farmer associations operate all hydro-agricultural administrative units – irrigation districts, irrigation units, and drainage districts – through their *Comité Directivos* (Executive Committees). Planning, operation, and maintenance works of the hydraulic infrastructure of each administrative unit are agreed among them. CONAGUA is still responsible for the major infrastructure, but timing and volume releases, as well as other activities, are discussed in those *fora*.

The frequency of their meetings is at least once a year, but generally, and depending on their size and complexity, they have at least four annual meetings. The more complex administrative units are large irrigation districts, since they are subdivided in *Módulos de Riego* (Irrigation Modules), where they are represented by the same procedure in terms of the selection and duration of the module authorities.

The selection process is through open elections among all users, whether a module or representatives of modules in a large irrigation district or drainage district, and the term is about 3 years. Although there is no information about the performance of the Comités Ejecutivos of the administrative units, it is clear that those committees that are transparent and allow free participation of stakeholders are in better conditions than those who are not transparent or do not allow free and open participation.

2.5 Future of the Water Sector

In general, the future of the water sector in Mexico looks promising. There are still several issues to be considered to reduce water conflicts and inequalities. Fortunately, Mexico has signed several international agreements where water is the strategic resource. The Millennium Development Goals and the Global Change Strategy are two of them. The commitments made have been added to the CONAGUA 2030 Water Agenda.

2.5.1 2030 Water Agenda

The 2030 Water Agenda identified four goals to attain.

2.5.1.1 Basins and Aquifers in Equilibrium

There are serious misbalances in some basins and aquifers due to either an excessive demand or short supply. On the average, the agricultural sector exerts the major demand, and therefore, activities are focused on a more efficient water use and/or reduced water consumption with different crops. Several hydraulic works have been designed to reduce water consumption either by improving the conveyance losses, applying more efficient irrigation systems, or using less water-demanding crops. In overexploited aquifers, water supply has been reduced and/or recharge techniques have been applied (CONAGUA 2011, pp. 13–15).

2.5.1.2 Clean Rivers

Currently, wastewaters reach 6.7 hm^3/year and it is expected to increase up to 9.7 h m^3. The main emphasis is on wastewater treatment to increase from 36.1% to 60%, though by 2012, it was established on 45.7%.

Sanitation is close to the goal established, 90%, but it needs a better monitoring system by the increase in the establishment of 4600 water quality monitoring sites to secure water quality.

2.5.1.3 Universal Coverage

This goal is related to improve water supply and sanitation in rural locations. Currently, the national water supply coverage is 94.4%, with 97.2% urban and 83% rural. The goal is 90% coverage, which means an extra effort in rural areas. As for sanitation, we have 91.4% coverage, 96.6% in urban and 77.5% in rural areas, which also reflects a major effort in the rural areas.

2.5.1.4 Security Before Catastrophic Events

After the identification and prioritization of critical areas, funds have been allocated to improve the infrastructure for flood control, as well as the relocation of human settlements in risk zones. The national weather service has improved their warning systems, and the level and cost of those events have been reduced, but not to zero.

2.5.1.5 Principles and Strategic Lines

Four principles and strategic lines have been devised to attain the following goals:

Sustainability The main idea is development without compromising future generations in three dimensions: environmental, by controlling overdrafted aquifers and establishing environmental flow and water pollution control, among others; economical, in the sense that natural resources have an economic value that needs to be addressed in the decision processes; and social, in relation to provide clean water and a safe environment to all citizens.

Integrated River Basin Management This approach is the most appropriate since whatever happens in the uplands of a river basin is reflected in the lowlands; therefore, measures must be made to analyze and operate at a river basin scale.

Water as a Finite and Strategic Resource With the current population growth and global warming, we need to consider water as a finite resource since the rate of degradation of water and the cost will increase in time.

Land Use Planning and Water Resource Allocation The more efficient way to work is to maximize land use to obtain the maximum gains, so efforts need to be taken to use land resources to its maximum capacity.

2.5.2 Global Warming Preparation

Although globally, climatological records have been collected in less than 150 years, indirect ways to estimate climatological variables show that they fluctuated so widely in the very last 1000 years. A human life is not enough to perceive the extremes that had already affected the earth; however, the impact of human activities with the generation of greenhouse gases undoubtedly is warming the surface of the earth. We need to identify what and to what extent global warming will affect us in order to develop strategies that will reduce or ease negative impacts.

The groups in charge of the global warming identification and prevention are getting information of unknown effects for the future; the question is, are we in time? And, do we have the resources to reduce or control those effects?

2.5.3 Water Governance

Water governance deals with the determination of who, how much, and how often water will be serviced. The representation of interesting parties in the decision-making process and the role of politics are important components of water governance. Finally, since the decisions are anchored in the interaction among government, civil society, and private sector, the implementation of good water governance implies the establishment of processes and rules of decisions that allow free and open participation of all sectors to solve water conflicts.

Water conflicts, like those for the water supply to cities, can be discussed and solved in the institutions created to promote participation. However, the role of Basin Councils has not made a difference, yet.

Therefore, good water governance is critical for the future of the water sector in Mexico.

Although the current water governance in Mexico has produced positive impacts, it is far from complete. Some of the areas that need improvements are discussed:

- *Decentralization of the decision-making process*. Although there is decentralization on the decision-making process, it is only in paper. There are remnants of central decisions over regional needs.
- *Participation of stakeholders*. The last legislation improved the composition of the River Basin Councils; however, there are deficiencies in the selection process of the representatives of water users and the percentage of representatives, and there is a lack of rules for conflict resolution, among others.
- *Integrated river basin management*. Although it is recognized in the new legislation, how is integrated river basin management conducted? And who is responsible to do it? Each river basin should have some way to represent the hydrologic processes and ways to compare or verify the interventions.

- *Clarification of roles and responsibilities*. What are the roles and responsibilities of the water user's representatives within the Council and what is the procedure to involve them in the decision-making processes? Regularly, CONAGUA officials oversee most of the organization, agenda, and recording of the meetings, with little participation of the nongovernment representatives.
- *Transparency in decision-making processes and information sharing*. The decision-making process is difficult when there is a lack of information and data collection and/or the information is not available to all stakeholders.

We can say that, although we are establishing the basic principles to good water governance, we need to do more.

2.6 Conclusions

Water resources in Mexico are so widely distributed due to its geographical location associated with climate. Northern Mexico is hot and dry, Central Mexico is temperate, and Southern Mexico is hot and humid. The location, combined with the relief, yields a wide variation in terms of surface and groundwater resources.

Although climate variation influenced on the pre-European cultures, water technology, institutions, and regulations were very similar. Changes in technology and changes in social conditions yielded new institutions and new regulations. They evolved to the current situation. What we can foresee is that new technology and/or new social conditions will continue evolving in terms of water legislation and institutional framework.

A description of the current situation of the water sector shows a higher stress on water resources in heavily populated areas, especially in arid or semiarid regions. It is possible to say that in those conditions, water efficiency is a key factor, since water transfers will only bring more conflicts and more stress on water resources.

Although Mexico is complying in terms of international commitments, like the Millennium Development Goals and Climate Change, water conflicts are getting more complicated. The best response is in the establishment of new water governance. Current water governance was introduced in the last legislation, but more work is to be done to reduce water conflicts, alleviate poverty, and improve water efficiency.

References

Arias, H., Craft, S., Baldenebro, C., Patron, R., Tarriba, J., González, A., Saucedo, H., Namuche, R., Garcia, N., & Fuentes, C. (2017, October 8–14). *Tile drainage: A strategy to cope with the food challenge facing Mexico*. 23rd International congress on irrigation and drainage (p. 402). Mexico City: International Commission on Irrigation and Drainage.

Arias Rojo, H. M., & González, A. (2017). *Agricultural drainage in Mexico*. Participation in the Sustainable Drainage Working Group. October 12, 2017. 23rd International congress on irrigation and drainage. International Commission on Irrigation and Drainage. Mexico City.

Braniff Cornejo, B. (2009). *Comercio e interrelaciones entre Mesoamérica y la Gran Chichimeca. Históricas Digital* (pp. 27–50). Coyoacán: Universidad Nacional Autónoma de México-Instituto de Investigaciones Históricas. http://www.historicas.unam.mx/publicaciones/publicadigital/libros/caminosymercados/mercados.html.

CILA. (1944). Tratado de Aguas México Estados Unidos. Comisión Internacional de Límites y Aguas. Diario Oficial de la Federación. Febrero de 1944. Mexico City.

CONAGUA. (2007). ACUERDO por el que se determina la circunscripción territorial de los organismos de cuenca de la Comisión Nacional del Agua. Comisión Nacional del Agua. Secretaría de Medio Ambiente y Recursos Naturales. Diario Oficial de la Federación. 12 de diciembre del 2007. México, DF.

CONAGUA. (2011). Agenda del Agua 2030. Continuidad y Seguimiento de la Agenda del Agua 2030. Comisión Nacional del Agua. Secretaría de Medio Ambiente y Recursos Naturales. Marzo del 2012. México, DF. 66 pp.

CONAGUA. (2016). *Estadísticas del Agua en México. Edición 2016. Comisión Nacional del Agua* (275 pp). México: Secretaría de Medio Ambiente y Recursos Naturales.

Hundley, N., Jr. (2000). *Las aguas divididas. Un siglo de controversia entre México y Estados Unidos* (283 pp). Mexicali: Universidad Autónoma de Baja California.

INEGI. (2000). *Diccionario de Datos Climáticos Escala 1:250,000 y 1:1,000,000 (Vectorial)* (57 pp). México.: Instituto Nacional de Estadística, Geografía e Informática.

Infante, J. A. (2011). La verdadera historia del origen de Los Mochis. En. http://mx.globedia.com/verdadera-historia-origen-mochis

Kroeber, C. B. (2009). Man, land, and water: Mexico's farmlands irrigation policies 1885–1911. *Journal of Latin American Studies, 17*(2), 473.

Luna Escalante, G. (2007). *Derechos, Usos y Gestión del Agua en el Territorio Yaqui* (156 pp). Hermosillo: Tesina Diplomado de Especialidad en Gestión Integrada de Cuencas Hidrográficas. Colegio de Sonora.

Molina Enríquez, A. (1909). *Los grandes problemas Nacionales*. México: A. Carranza e Hijos.

Moreno Vázquez, J. L. (2006). *Por Abajo del Agua. Sobrexplotación y Agotamiento del Acuífero de la Costa de Hermosillo, 1945–2005* (507 pp). Hermosillo: El Colegio de Sonora.

Palerm, A., & Wolf, E. (1972). *Agricultura y civilización en Mesoamérica. No. 32. Colección SEP/Setentas* (215 pp). México: Secretaría de Educación Pública.

Rojas Rabiela, T. (2009). *Las obras hidráulicas en las épocas prehispánica y colonial. En CONAGUA, 2009. Semblanza Histórica del Agua en México* (pp. 9–26). México: Comisión Nacional del Agua, Secretaría de Medio Ambiente y Recursos Naturales.

SEMARNAT. (2003). Norma Oficial Mexicana NOM-001-SEMARNAT-1996, que establece los límites máximos permisibles de contaminantes en la descarga de aguas residuales en aguas y bienes nacionales. Diario Oficial de la Federación. 23 de Abril del 2003. México, D.F.

Hector Manuel Arias-Rojo is External Consultant with over 60 publications on national and international magazines on watershed management, environmental planning and protection, soil and water conservation, and on-farm underground agricultural drainage. He has a Ph.D. in Watershed Management from the University of Arizona. He has a wide academic experience, in the federal and state Mexican government, as well as in the private sector. He worked in Colegio de Postgraduados, Universidad Autónoma Chapingo (UACH) and UNAM, being awarded Level II of the National Research System (Conacyt). As a federal government official, he worked in SARH (Secretary of Agriculture and Hydraulic Resources) as Director for Technical Norms on Soil Erosion Control; and as a Sonoran state government officer, he was Planning Director of Instituto del Medio Ambiente y Desarrollo Sustentable IMADES (Institute of Environment and Sustainable Development), and Comisión de Ecología y Desarrollo Sustentable CEDES (Ecology and

Sustainable Development Commission), where he planned, conducted, and decreed (i) Plan de Ordenamiento Ecológico Territorial de la Costa de Sonora (Environmental Zoning of the Sonoran Coast), decreed in 2009; and (ii) Plan de Ordenamiento Ecológico Territorial del Estado de Sonora (Environmental Zoning of the Sonora state), decreed in 2015. In the private sector, he led the World Wildlife Fund's Chihuahua Desert Ecoregion Program in the planning and instrumentation of the Conchos Watershed Management Program that contributed to the environmental flow determination, payment of hydrological environmental services, control of headwaters of Río Conchos, protection of endemic fishes, and support protected areas in the Chihuahua Desert. He led a nonprofit organization to design Watershed Management Plans in the Lower Mayo and Sonoran rivers. In Irridren de México, SA de CV, as Planning Director, he prepared a white book on a national strategy to face Mexico's food challenge with on-farm underground drainage in irrigation and drainage districts, and projects to support indigenous communities in irrigation districts of northwestern Mexico affected by saline soils.

Roberto Fernando Salmón-Castelo is a Mexican Commissioner for the International Boundary and Water Commission (Comisión Internacional de Limites y Aguas, CILA) and has a wide range of experience in hydraulic projects. He was Northwest Regional Manager of the National Water Commission (CONAGUA) from 2002 to 2006, and from then until 2008, he served as General Manager of the Northwest Basin Unit based in Hermosillo, Sonora. Additionally, Mr. Salmon was Planning and Special Projects Director for the Sonora Center for Research and Development in Natural Resources and has remained a partner in a variety of companies that carry out basin management projects and environmental studies. He has a Bachelor of Science degree from the University of Arizona, where he also received a Master of Science degree. Additionally, he is doing advanced studies in Water Resource Administration at the University of Arizona and is currently a candidate for a doctoral degree there.

Part II
Sectorial Issues

Chapter 3
Effects of Water Availability and Policy Changes for Irrigated Agriculture

Antonio Yunez-Naude and Patricia C. Aguilar-Mendez

Abstract Water conservation and its efficient use in agricultural production are a challenge that has to be addressed by the Mexican government. In this chapter we present our research results on two aspects of this problem related to water availability and water pricing. For our study we use a general equilibrium model applied to Mexico focused on agricultural and food production. Two policy options are evaluated based on the Mexican state's ownership and distribution of water resources: the reduction of water for irrigation and the increase of water prices paid for this use. Our results show that the water pricing option is superior because the negative effect of this option on crop production and food security is lower than reducing the water supply for irrigated agriculture. This conclusion is reinforced by the possibility that the reduction in water for irrigation may come from climate change and not from policy changes, i.e., pricing water for irrigation might promote a more efficient use of the same by farmers.

Keywords General equilibrium models · Irrigation · Agricultural policy · Water · Government policy

3.1 Introduction

In Mexico, water used by irrigated agriculture is inefficient and/or wasted mainly because the infrastructure for irrigation is inappropriate, and the use of irrigation water is practically free. This current situation is aggravated by climate change impacts that have reduced water availability especially in northern Mexico where most staple foods are produced by commercial farmers with access to irrigation. In

We want to thank the William and Flora Hewlett Foundation for its financial support and to Elisa Moreno for English editing.

A. Yunez-Naude (✉)
El Colegio de Mexico, A.C, Mexico City, Mexico
e-mail: ayunez@colmex.mx

P. C. Aguilar-Mendez
Independent Consultant, Mexico City, Mexico

H. R. Guerrero García Rojas (ed.), *Water Policy in Mexico*, Global Issues in Water Policy 20, https://doi.org/10.1007/978-3-319-76115-2_3

addition, urban and industrial growth is increasingly competing with agriculture for water. This scenario exemplifies the urgency for the Mexican government to design and implement policies that promote water conservation and its efficient use for agricultural production. In this paper we address some issues related to superficial water used by irrigated agriculture using a computable general equilibrium model applied to Mexico or CGEM.

The rest of the article is organized as follows. In the next part, we summarize the main features of the CGEM and the Social Accounting Matrix for Mexico or SAM to calibrate the model. In Sect. 3.3 we describe the Mexican economy and its agricultural sector according to the SAM. We dedicate Sect. 3.4 to present the CGEM simulations we performed to assess the probable effects of two water-related shocks: a decrease in water availability for irrigation and a rise in the price charged by the Mexican state to farmers with access to superficial water for irrigation. We conclude the article with a synthesis of our findings and with policy-oriented reflections.

3.2 The CGEM and the SAM

The application of a general equilibrium approach is appropriate for evaluating the possible effects of exogenous shocks in a particular economy. Among others, it captures the direct and indirect impacts of these sudden changes. To estimate the effects of shocks related to water availability and pricing for irrigated agriculture in Mexico, we have built a computable general equilibrium model applied to Mexico focused on its agricultural sector (CGEM from now on). To calibrate the CGEM, we use a SAM for the Mexican economy for 2002.[1] A SAM captures the complex interlinkages among domestic production, institutions, and the outside world of the economy under consideration. A SAM provides a starting point for economy-wide analysis; it is a useful expository device to portray the structure of economies and is commonly used as the basic data input for general equilibrium modeling. A SAM is a form of double-entry accounting flows of an economy during a period, generally a year. The rows of a SAM contain receipts and the columns expenditures; it presents the accounting entries in income and product accounts and in input-output production accounts as debit and credit entries in income balance sheets of institutions and activities. Activities may include rural and urban production (or any disaggregation of the two). Institutions in SAMs typically include different household groups, enterprises, government, and the rest of the world. In the context of economy-wide modeling, institutions are categories of economic agents.

Entries in a SAM have intermediate input demands between production sectors (the Leontief matrix is contained within the SAM): income (value added) paid by production sectors to different types of labor (e.g., skilled and unskilled), land, or

[1]When doing the research, 2002 was the last year where the data required to construct a SAM focused on agriculture was available.

capital; and water for the present study. The SAM also includes the distribution of value added across different private institutions (household groups and enterprises) and the distribution of these institutions' expenditures across consumption of domestically produced goods and services, imports, and savings. A government account collects income from activities and the private sector and redirects this income within the system (to government demand for goods and services, transfers to production activities, or household groups), saves, or uses its income to pay foreigners (for imported goods and services or repayment of debt).

A SAM is a square matrix, and a salient characteristic of a SAM, derived from double-entry accounting, is the equality between the sum of expenditures (column) and the sum of revenues (row) for every account in the system (Table 3.1 presents a scheme of the SAM used for the present study).

In order to conduct a general equilibrium impact analysis related to the water sector of the agriculture of Mexico, we built a SAM for 2002 (Yunez-Naude 2006). In this SAM, agricultural activities include six crops and/or groups of crops (Table 3.2). In addition, we distinguish crop production under two water regimes: irrigated and rain-fed production (so, the SAM has 12 crop sectors).

We also disaggregated crop production into five rural regions of Mexico: I, north; II, central; III, southwest; IV, southeast; and V, the Rio Bravo Basin (see Map 3.1).[2]

In the SAM, production uses intermediate inputs and factors of production: labor and physical capital for the nonagricultural activities plus land and irrigation water for agricultural activities. Urban labor includes professionals, white-collar, blue-collar, and non-skilled labor; and land is disaggregated by irrigated and rain-fed conditions.[3]

The SAM has two types of institutions: private and government. Private institutions consist of low-, medium-, and high-income households, enterprises, and Water Users Associations (WUAs) that collect water user fees for infrastructure and maintenance (see below, introduction to Sect. 3.4). Rural households are by region and distinguished from urban households. Government collects taxes, buys goods and services, conducts investments, and transfers part of its income to the private institutions. The remaining accounts of the SAM are the rest of the world and the savings-investment macro account.

The CGE used here is neoclassical in spirit, with agents responding to price changes and Walrasian, determining relative prices. As others, the CGEM is based on a set of simultaneous linear and nonlinear equations, which define the behavior of economic agents, as well as the economic environment in which these agents operate. This environment is described by a set of market equilibrium conditions and macroeconomic balances.

---

[2]The Rio Bravo Basin or RBB is located in part of the north and central regions of Mexico. We distinguished this region since it shares water of the Rio Bravo with the USA that has been subject to conflicts among the two countries and has remained so until the present.

[3]Details in Chap. 4 of Yunez-Naude (2006). Since water for irrigation is not valued, we estimated it using the fees paid by farmers to their Water Users Associations. In Aragon (2011) the estimation was improved by adding to the fees the implicit value of irrigated land rent.

Table 3.1 Scheme of a SAM for a country focused on agriculture

	Expenditures					
	1	2	3	4	5	
Receipts/incomes	Activities	Factors	Institutions	Capital	Rest of the world	*Totals*
1. Activities	Input-output matrix		Consumption (private and public institutions)	Investment	Exports	*Total sales (domestic demand)*
Agriculture						
Other sectors						
2. Factors	Value added in production				Factorial incomes from abroad	*Total value added or GDP at factor prices*
Labor						
Physical capital						
Land						
Water						
3. Institutions	Value-added taxes	Transfers to institutions accruing to factorial services			Foreign transfers	*Total institutional incomes*
Households			Family			
Rural			Government transfers			
Urban						
Firms						
Water Users Associations						
Government			Direct taxes			
4. Capital			Savings		Capital transfers	*Total savings*
5. Rest of the world	Imports					*Total imports*
Totals	*Total expenditure*	*Total payments to factors*	*Total expenditures from institutions*	*Total investment*	*Income from abroad*	*Totals*
						Income/expenditures

Source: Own

Table 3.2 Sectors/commodities in the SAM

Commodity accounts		
1. Maize	8. Dairy products	15. Intermediates
2. Wheat	9. Prepared fruits and vegetables	16. Consumer durables
3. Bean	10. Processed wheat	17. Capital goods
4. Other grains	11. Processed maize	18. Professional services
5. Fruits and vegetables	12. Sugar manufacturing	19. Other services
6. Other crops	13. Other processed foods	20. Construction
7. Livestock/forestry/fishery	14. Light manufacturing	21. Commerce and transportation

Source: SAM, from Yunez-Naude (2006)

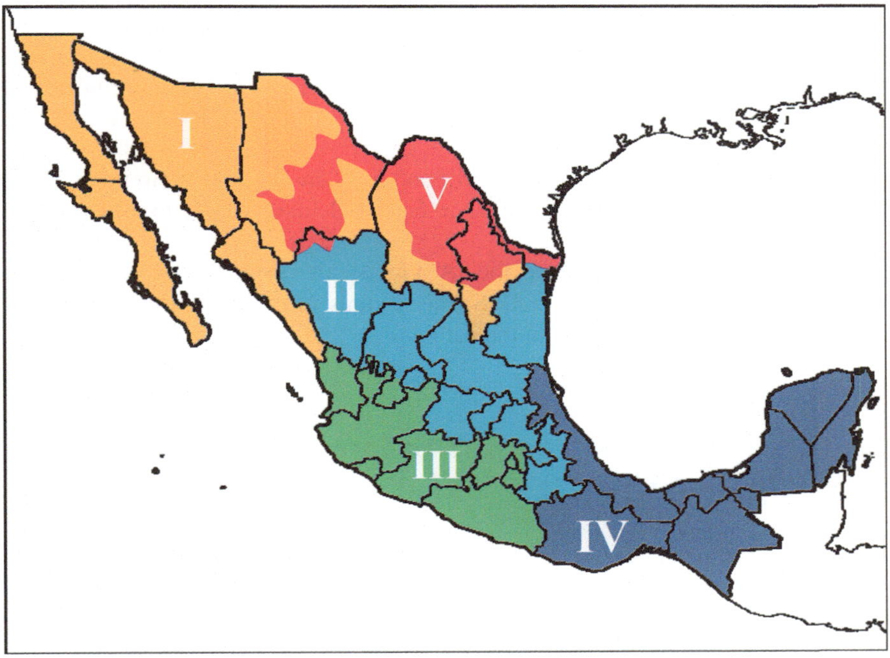

Map. 3.1 Location of the rural regions of Mexico (Source: Own elaboration, based on administrative divisions of Mexico and for the Rio Bravo basin, the hydrologic region number 24 Bravo-Conchos from Semarnat, at Diario Oficial de la Federación, June 2, 2011.). Note: I, north; II, central; III, southwest; IV, southeast; V, Rio Bravo Basin (RBB)

For production and consumption decisions, behavior is captured by nonlinear, first-order optimality conditions. That is, production and consumption decisions are driven by the maximization of profits and utility, respectively. The equations also include a set of constraints that must be satisfied by the system as a whole but not necessarily considered by any individual agent. These constraints cover markets (for factors and commodities) and macroeconomic aggregates (balances for the savings-investment account, the government, and the current account with the rest of the world).

There are alternatives for defining market-clearing conditions and model closures. Given the characteristics of the Mexican economy, we selected the following. (1) For each factor, the economy-wide wage is the market-clearing variable in a setting with perfect factor mobility across activities; (2) a flexible exchange rate clears the current account with the rest of the world; (3) flexible government savings clears the government account; and (4) the savings rates of domestic institutions are scaled to generate enough savings to finance exogenous investment quantities (i.e., we assume investment-driven savings). In line with the above clearing conditions, the model closures are as follows: foreign savings and import and export prices are fixed (the latter means that we assume that Mexico is a small open economy); taxes and subsidies are fixed, as well as the government's marginal propensity to save; we also assume flexible government savings.

Finally, we selected the consumer price index as the *numeraire*. So product and factor prices and the equilibrium exchange rate are defined relative to the consumer price index.

The CGEM was calibrated using the SAM for Mexico using the software *General Algebraic Modeling System* or GAMS (the features of the CGEM and the model code are available upon request to the authors).

3.3 The Economic and Social Structure of the Mexican Agricultural Sector

We present the main features of the agricultural sector and rural households of Mexico in order to have a framework for the discussion of our CGEM simulation results.[4]

Notwithstanding the expected effects of the North American Free Trade Agreement (NAFTA, Yunez-Naude and Hernandez-Solano, 2018), maize continues to be the single major crop produced in Mexico. According to the SAM, it accounted for more than 18% of total crop GDP. In terms of value of production, more than 50% of maize is cropped under rain-fed conditions, whereas most of wheat and fruits and

[4]The agriculture of Mexico has not experienced major changes during the present century (see Yunez and Solano 2017). So, the description presented in this section, based on the SAM used as the baseline for our CGEM, applies to the current structure of rural Mexico.

Table 3.3 Value of production (GDP), imports and exports of field crops, and participations

Crop(s)	GDP	%	Irrigated	%[a]	Imports	%	Exports	%
Maize	34.96	18.11	16.24	46.5	16.99	37.37	0.71	1.16
Wheat	4.46	2.31	3.86	86.5	1.86	4.09	0.32	0.52
Bean	10.02	5.19	2.94	29.3	1.16	2.55	3.6	5.88
Other grains	10.29	5.33	4.36	42.4	0.95	2.09	0.04	0.07
Fruits and vegetables	76.71	39.73	56.05	73.1	1.8	3.96	23.7	38.96
Other crops	56.65	29.34	33.54	59.2	22.7	49.93	32.89	53.69
Totals	193.09	100.00	116.99		45.46	100.00	61.26	100.00

Source: SAM, from Yunez-Naude (2006)
[a]Refers to the proportion of the gross value of irrigated crop(s)

Table 3.4 Distribution of field crops by region (%)

Crop\region	North (I)	Central (II)	Southwest (III)	Southeast (IV)	RBB (V)[a]	Totals
Maize	15.70	26.90	42.60	14.66	0.15	100
Wheat	77.54	11.96	9.86	0.42	0.22	100
Bean	27.62	50.53	13.33	8.39	0.13	100
Other grains	15.85	51.34	26.19	1.64	4.98	100
Fruits and vegetables	34.84	20.46	26.35	17.52	0.83	100
Other crops	12.72	27.28	32.96	26.44	0.59	100

Source: SAM, from Yunez-Naude (2006)
[a]Rio Bravo Basin

vegetables are cropped with irrigation. Maize is also the most imported crop, whereas fruits and vegetables are the major crops exported by Mexico (Table 3.3).

A high proportion of maize comes from the southwest (region III), under rain-fed conditions; it is followed by the central region (II), where it is produced under both irrigated and rain-fed conditions; the north (I) is also an important maize production region, mostly under irrigation, whereas in the southeast (IV), this grain is cropped under a rain-fed water regime (their participation in gross domestic production of maize is 42.6%, 26.9%, 15.7%, and 14.7%, respectively, Table 3.4). Most of the wheat produced in Mexico comes from its northern rural region which is also a major producer of vegetables, both groups of crops being produced under irrigation. The central region is the major producer of beans and "other grains" under rain-fed and irrigated conditions (barley among them), and the Rio Bravo Basin (RBB) is also a relevant producer of other grains under irrigation.

An important feature of the rural economy of Mexico is the heterogeneity of its households. This is shown not only by the sharp differences in total incomes per household group but also in the weight of their income sources (Table 3.5). For

Table 3.5 Total income of rural households by income group and source and participations

Source	Low income	Medium income	High income
	(Billions of pesos)		
Labor	326.20	668.48	844.64
Land	11.68	26.86	24.86
Capital	319.44	826.84	1,250.34
Government transfers	91.33	222.70	220.77
World transfers	40.66	52.56	0.00
Totals	**789.32**	**1797.44**	**2,340.61**
	(%)		
Labor	41.33	37.19	36.09
Land	1.48	1.49	1.06
Capital	40.47	46.00	53.42
Government transfers	11.57	12.39	9.43
World transfers	5.15	2.92	0.00
Totals	**100.00**	**100.00**	**100.00**

Source: SAM, from Yunez-Naude (2006)

example, the share of labor on total income decreases as income level rises: from 41.33% for poor rural households to 36.09% for the richest households, and the opposite applies to the capital.[5]

3.4 Simulations and Results

The economy-wide nature of the CGEM gives results on the possible direct and indirect effects of exogenous shocks on productive factor and output prices, production and consumption, household incomes, trade, and macro-variables (exchange rate, savings, investment, etc.). Based on a CGEM approach, we conducted several experiments to quantify the possible effect of natural shocks, trade liberalization, and agricultural policy changes on Mexico's agriculture (Yunez-Naude 2006 and Dyer et al. 2005).

Our report shows the CGEM effects on the Mexican economy and its agricultural sector by region of a decrease in water availability and an increase of water prices for superficial irrigation.

A water supply shock for agricultural production may come from nature, the process of growth and/or from policy changes. One consequence of climate change

[5]The value of land rents is similar for medium- and high-income households due to the process of normalization of irrigated to rain-fed lands that we applied to value water used for irrigation (see above).

is the modifications of water availability for agricultural activities. The reduction of irrigation water supply in some regions can be coupled by an increased demand for other uses due to industrial and urban growth.[6] Taking into consideration that the Mexican state has property rights on water and its distribution, these scenarios may lead to a policy aimed to reduce the irrigation water supply.

With respect to pricing, the Mexican government has the option to increase water prices for irrigation as a means to encourage farmers valuate this resource in their economic decision, to enhance irrigation water efficiency and to sustain water demands by other agents. In addition, and as we will discuss below, irrigation water pricing would augment the WUA's income that can be invested in water infrastructure and technology resulting in improvements in operation and maintenance and leading to a more efficient irrigation water use.[7]

Before presenting our simulations and results of the two water-related shocks in superficial irrigation water, it is convenient to summarize the relevant features of the water sector institutional framework, as well as water costs and subsidies for irrigation in Mexico.

As said, the Mexican state is the owner of the country's water resources. The Federal Government is the agent that distributes water and/or gives concessions or permits for its use and is carried out mainly through the National Water Commission (Comision Nacional de Agua – CONAGUA). Water users for agricultural and forestry activities are treated differently than other agents as they are not charged for their use of the liquid. The only cost irrigation farmers pay is the maintenance and operation fee to their respective WUA. However, an additional fee must be paid if the amount of water used by a farmer exceeds the volume of water authorized (see details in Guerrero 2008). Notwithstanding this, farmers exceeding the conceded volume pay much less than other water users; for example, in 2010, farmers paid less than 24% of the fee charged to residential consumers for a cubic meter of water. Another way to measure the implicit subsidy received by irrigation farmers is by using the fee paid by farmers exceeding the volume of water conceded and applying it to the volume of the water allowance, i.e., in 2009, 11.73 cents per cubic meter multiplied by the 61.8 billion of kilometers of water authorized to farmers with superficial irrigation amounts to 7.3 billion pesos (Yunez-Naude and Aguilar 2012). To this we must add that agriculture is the major water consumer in Mexico: for in between 1996 and 2005, it consumed more than 75% of the total water used in Mexico (ibid.).

[6]This is illustrated by water depletion in central Mexico, due to the increasing demand of water in the cities of Mexico, Cuernavaca, and Toluca (see Lopez Morales 2012).

[7]The reported results are in terms of the percent changes caused by the simulated shock with respect to the base, i.e., to the calibrated SAM data.

3.4.1 Water Availability Shock

To conduct a realistic simulation of a water shock for the agriculture in Mexico is a challenge since a myriad of processes are involved. With respect to climate change, fluctuations in the variability of precipitation, reductions in the irrigation water supply in arid regions, and an increase in precipitation in the tropics are just some of the phenomena involved. In fact, climate change produces increasing temperatures and extreme weather conditions, and its variability directly affects agricultural production. So, modelling changes in water supply for irrigation is challenging as it requires selecting alternative data to be used in the simulation (for example, changes in precipitation during an historical period, variations in precipitation by region, rainfall predictions, etc.). Among all the alternative experiments related to a water availability shock, we decided to report the results of a simple one consisting in a 20% reduction of irrigation water availability across all rural regions of Mexico as a means to glance at the possible direct and indirect consequences of this shock.[8]

A 20% drop in irrigation water causes an increase in agricultural prices and in the price of wheat in particular (2.2%, Table 3.6). The reduction of water supply causes a decrease in irrigated agricultural production, reduces food exports, and increases their imports. However, these changes are insufficient to maintain domestic sales with respect to the baseline: domestic sales decrease between 29.5% (wheat) and 8.15 (maize, Table 3.7).

Table 3.6 Effects of a 20% reduction of irrigation water on prices of composite commodities (% change from base)

Commodity		Commodity	
Maize	0.90%	Light manufacturing	0.00%
Wheat	2.20%	Intermediate goods	0.00%
Bean	1.90%	Capital goods	−0.10%
Other grains	1.00%	Consumer durables	0.00%
Fruits and vegetables	1.60%	Construction	−0.10%
Other crops	0.70%	Professional services	−0.10%
Livestock	0.10%	Other services	0.00%
Dairy products	0.00%	Commerce and trade	−0.10%
Fruits and veg prepared	0.40%		
Processed wheat	0.50%		
Processed maize	0.10%		
Sugar manufacturing	0.30%		
Other foods	0.00%		

Source: Own estimations

[8]Based on the droughts experienced by Mexico during the present century Cook et al. (2007) and Stahle W. et al. (2009), the level of the simulated water supply shock is not unreasonably high. Notwithstanding this, we discuss its results with another simulation that approximates regional water shock effects of El Niño.

Table 3.7 Effects of a 20% reduction of irrigation water on domestic sales (% change from base)

Commodity		Commodity	
Maize	−8.15%	Light manufacturing	−0.03%
Wheat	−29.50%	Intermediate goods	0.37%
Bean	−10.32%	Capital goods	1.45%
Other grains	−18.58%	Consumer durables	0.26%
Fruits and vegetables	−11.94%	Construction	0.20%
Other crops	−10.34%	Professional services	−0.01%
Livestock	−1.01%	Other services	0.02%
Dairy products	−0.68%	Commerce and trade	−0.01%
Fruits and veg prepared	−4.81%		
Processed wheat	−4.65%		
Processed maize	−1.36%		
Sugar manufacturing	−2.95%		
Other foods	−1.13%		

Source: Own estimations

Whereas domestic production of irrigated crops decreases (−5.96%), domestic supply of crops under rain-fed conditions increases (1.28%). This later rise is not sufficient to maintain domestic agricultural supply: it drops by 2.38% (Table 3.8). The reduction of irrigated crop production is similar among the five rural regions; however, if we consider that around 60% of irrigated crops are in the northern region, we can conclude that it is the most affected by the water supply reduction. With respect to particular crops, it is convenient to notice that rain-fed maize and wheat production increases more with respect to other rain-fed crops; however, wheat production in Mexico is mostly an irrigated crop (see Table 3.8).

General equilibrium impacts on production of the water shock are reflected on the changes in households' real incomes since rural households located in the north would suffer the most: all northern households experience higher income reductions with respect to households in the central, southwest, and southeast irrespective of their baseline income level (Table 3.9). In addition to households, real income of the remaining institutions decreases. This is particularly true for the WUAs, whose income decreased by 14.15%. This result is worrisome because it means a considerable drop in WUAs resources used for irrigation infrastructure maintenance.[9]

[9]The signs of the results for the northern two agricultural regions previously discussed are the same to those of a previous simulation based on Tiscareño et al. (2003) estimations on changes on seasonal precipitations by Mexico agroecological region caused by El Niño with respect to neutral years. They consisted in simulating a 48.7% reduction of water availability for the north and the RBB regions, by 4% for the central, and by 6.5% for the southeast, as well as a rise of 32.6% in water supply for the southwest (see Yunez-Naude 2006).

Table 3.8 Effects of a 20% reduction of irrigation water on domestic production (% change from base)

All crops by region		All regions by crop	
Region		Crop	
Rainfed			
North	3.07%	Maize	2.85%
Central	1.26%	Wheat	6.09%
Southwest	1.53%	Bean	1.15%
Southeast	0.58%	Other grains	0.26%
RBB[a]	0.98%	Fruits and veg	1.87%
Subtotal	***1.28%***	Other crops	−0.59%
		Subtotal	***1.28%***
Irrigated			
North	−5.93%	Maize	−14.62%
Central	−5.05%	Wheat	−6.70%
Southwest	−6.57%	Bean	−21.91%
Southeast	−6.49%	Other grains	−15.59%
RBB[a]	−8.56%	Fruits and veg	−4.38%
Subtotal	***−5.96%***	Other crops	−4.43%
		Subtotal	***−5.96%***
Totals			
Field crops		*−2.38%*	
Livestock		*−0.16%*	
Processed food		*−0.27%*	
Nonagriculture		*0.05%*	
Gross domestic product: Mexico		*−0.03%*	

Source: Own estimations
[a]Río Bravo Basin

3.4.2 Water Price Shock

In the SAM we used the estimated water value added for irrigated agriculture based on the difference between irrigated and rain-fed land rents (see Aragón, *op. cit.*). As previously mentioned, in Mexico, water is a resource owned by the state, and a possible policy option is to use irrigation water pricing to promote water efficiency in agriculture. Based on the low fees charged by the WUA, the simulation consists of doubling the implicit price of water resulting from the calibration of the CGEM. As expected, one indirect effect of the simulated water price shock is an increase of crop prices, and the price of processed maize and wheat also rises (Table 3.10).[10]

[10]The price drop of "other crops," sugar, and "other processed crops" is due to other general equilibrium effects of the water price shock, such as the changes in the composition of agricultural production.

Table 3.9 Effects of a 20% reduction of irrigation water on households' real income (% change from base)

Urban households		*Rural households*	
Low income	−0.38%	Medium income	
Medium income	−0.39%	North	−1.12%
High income	−0.42%	Central	−0.25%
Enterprises	−0.40%	Southwest	−0.10%
Water Users Associations	−14.15%	Southeast	−0.14%
Rural households		RBB	−0.46%
Low income		High income	
North	−1.69%	North	−8.91%
Central	−0.83%	Central	−3.70%
Southwest	−0.35%	Southwest	−1.73%
Southeast	−0.28%	Southeast	−1.98%
RBB	−0.62%	RBB	−3.82%
Government income			−0.28%

Source: Own estimations

Table 3.10 Effects of doubling irrigation water pricing on prices of composite commodities (% change from base)

Commodity		Commodity	
Maize	1.40%	Light manufacturing	0.00%
Wheat	3.70%	Intermediate goods	0.00%
Bean	2.60%	Capital goods	−0.10%
Other grains	1.60%	Consumer durables	0.00%
Fruits and vegetables	0.90%	Construction	−0.10%
Other crops	−0.20%	Professional services	−0.10%
Livestock	−0.20%	Other services	0.00%
Dairy products	−0.20%	Commerce and trade	−0.10%
Fruits and veg prepared	0.20%		
Processed wheat	0.90%		
Processed maize	0.20%		
Sugar manufacturing	−0.10%		
Other foods	−0.10%		

Source: Own estimations

One of the consequences of the increase in food prices is the drop of their domestic sales. The most affected commodities are grains and beans, followed by processed wheat (Table 3.11).

Doubling the price of irrigation water would reduce irrigated crop production, as well as production of processed foodstuffs, and increase production under rain-fed conditions. However, the overall general equilibrium effect would be a drop in field crop production and a slight reduction of processed foods domestic supply (Table 3.12). In part, the rise in crop production under rain-fed conditions is

Table 3.11 Effects of doubling irrigation water pricing on domestic sales (% change from base)

Commodity		Commodity	
Maize	−2.32%	Light manufacturing	0.07%
Wheat	−7.33%	Intermediate manufacturing	0.24%
Beans	−3.22%	Capital goods	0.47%
Other grains	−4.58%	Consumer durables	0.12%
Fruits and vegetables	−0.98%	Construction	0.67%
Other crops	0.20%	Professional services	0.00%
Livestock	0.00%	Other services	0.00%
Dairy	0.02%	Trade	0.01%
Processed fruits and vegetables	−0.41%		
Processed wheat	−1.03%		
Processed maize	−0.50%		
Sugar	−0.24%		
Other processed foods	−0.13%		

Source: Own estimations

explained by the drop of rural labor wages coupled with the fact that rain-fed production is more labor intensive than irrigated crop production (Table 3.13). However, the overall effect of the water price rise in crop production is negative, reducing its gross domestic production by 1.33% with respect to the baseline. Notwithstanding, the Mexican economy is not negatively affected by this drop since its GDP slightly rises by 0.07% (Table 3.12).

The water price shock leads to a decrease in rural labor wages, ranging from 8.69 to 1.73 in the north and in the Rio Bravo Basin (RBB), respectively; it also causes a considerable reduction of irrigated land rents in all regions and a rise in rain-fed land rents, especially in the central and southwest rural regions. In addition, payments for water increase in all regions (Table 3.13).[11] These factor price movements explain in part the divergence in the general equilibrium impacts on sectorial GDP.

The simulated water price shock negatively affects real income of all institutions. The exceptions are government and WUA incomes. For household groups, the general equilibrium impact of the shock is stronger in high-income rural households in the north, RBB, and in the centre with respect to poor- and medium-income households (Table 3.14).

A relevant result is that WUA income considerably rises with the water price shock (48%). This revenue increase could be used for irrigation infrastructure improvements and for the adoption of water-saving technologies.[12]

[11]The general equilibrium effect on water prices is below the simulated price shock (between 52.29% and 32.16% vs 100%). This result suggests that the estimated price of irrigation water is below its marginal product, i.e., that we underestimated the value added of water for irrigated agricultural production.

[12]We conducted a stylized experiment based on simulating the investment of WUA incomes to promote irrigated crops production under a water drop scenario. The results show the positive

Table 3.12 Effects of doubling irrigation water pricing on gross domestic production (% change from base)

All crops by region		All regions by crop	
Region		Crop	
Rain-fed Agriculture			
North	7.00%	Maize	5.95%
Central	3.58%	Wheat	12.33%
Southwest	3.75%	Beans	3.52%
Southeast	1.28%	Other grains	0.75%
RBB	2.94%	Fruits and vegetables	3.26%
Subtotal	*3.21%*	Other crops	0.87%
		Subtotal	*3.21%*
Irrigated Agriculture			
North	−6.63%	Maize	−26.31%
Central	−3.47%	Wheat	−11.11%
Southwest	−6.37%	Beans	−36.36%
Southeast	−6.95%	Other grains	−23.50%
RBB[a]	−3.36%	Fruits and vegetables	−3.63%
Subtotal	*−5.76%*	Other crops	0.24%
		Subtotal	*−5.76%*
Totals			
Field crops		*−1.33%*	
Livestock		*0.02%*	
Processed food		*−0.29%*	
Non agriculture		*0.14%*	
Gross domestic product: Mexico		*0.07%*	

Source: Own estimations
[a]Rio Bravo Basin

3.5 Final Reflections

A reduction of water availability for irrigation (either by a policy change or global warming) would negatively affect domestic crop and food production. Since more imports of these goods would not maintain the baseline level of agricultural and food sales (in part because of the increase in the prices of these goods), food security in Mexico would worsen, and extreme poverty could rise under this scenario.

The negative effects of a water shock, in both economic and social terms, show the importance of government intervention to promote water-efficient use in agricultural production. Two policy options available are to increase the price of

effects of this type of investment in reducing the negative effects on production and incomes due to a reduction in irrigation water availability (Yúnez-Naude 2006).

Table 3.13 Effects of doubling irrigation water pricing on factor prices (% change from base)

Labor		Land		Water and capital	
Urban Labor		*Dry*		*Water for irrigation*	
Professional	−0.02	North	0.04	North	48.91
White collar	−0.02	Central	2.07	Central	46.44
Blue collar	0.07	Southwest	2.51	Southwest	47.91
Unskilled	−0.02	Southeast	0.88	Southeast	52.59
		RBB	0.61	RBB	39.16
Rural labor		*Irrigated*		*Physical capital*	
North	−8.68	North	−20.29	Capital	−0.07
Central	−4.01	Central	−19.75		
Southwest	−5.23	Southwest	−20.21		
Southeast	−2.13	Southeast	−21.87		
RBB[a]	−1.73	RBB	−26.13		
Physical capital		−0.07			

Source: Own estimations
[a]Rio Bravo Basin

irrigation water and to invest in enhancing the use of water-efficient technologies and water infrastructure maintenance (Lopez-Morales and Duchin 2011).

It is recognized that valuating natural resources or natural capital is a fundamental incentive for economic agents to make appropriate decisions leading to an efficient use of these resources.[13] Worldwide this principle is not generally applied to irrigated agriculture since the common practice is to subsidize water used by farmers. However, we are convinced that for Mexico, irrigation water pricing is appropriate due to the state practice of charging, through the Water Users Associations, a very low fee for the water assigned to farmers for irrigation. Furthermore, pricing water for agricultural production is a necessary incentive for farmers to adapt to the effects of climate change on food production (see, e.g., Cook et al. 2007; Cortez Vazquez 2006; Mendelsohn et al. 2009 and Stahle et al. 2009).

The results obtained by simulating a 100% increase in the price of irrigation water indicate that it will reduce irrigated crop production and households' income. However, another general equilibrium effect of this price shock is a considerable rise of the income received by the WUAs as well as an increase of government receipts. These additional earnings could be used for investments leading to the adoption of water-saving technologies in agriculture..[14]

[13]A recent study incorporates natural resources and the environment in a SAM for Mexico is in Govea Vargas (2017). The SAM is for 2012, and it indicates that policy challenges to promote a more efficient use of water by farmers remain. Based on the SAM built by Govea, a construction of a GC for impact analysis is in progress.

[14]In this respect, a note of caution is necessary, though. As Ward and Pulido (2012) argue from their study of the Upper Rio Grande Basin of North America (2008): "Adoption of more efficient irrigation technologies reduces valuable return flows and limits aquifer recharge. Policies aimed at reducing water applications can actually increase water depletions. Achieving real water savings

Table 3.14 Effects of doubling irrigation water pricing on real incomes (% change from base)

Urban households		*Rural households*	
Low income	−0.04%	Medium income	
Medium income	−0.04%	North	−0.54%
High income	−0.05%	Central	−0.08%
Enterprises	−0.08%	Southwest	−0.05%
WUAs[a]	48.03%	Southeast	−0.04%
Rural households		RBB	−0.07%
Low income		High income	
North	−0.85%	North	−5.36%
Central	−0.41%	Central	−1.96%
Southwest	−0.18%	Southwest	−0.85%
Southeast	−0.09%	Southeast	−0.96%
RBB	−0.15%	RBB	−2.07%
Government income		0.05%	

Source: Own estimations
[a]Water Users Associations

If we consider as alternative policies the reduction of water supply for irrigation and the –still pending-- increase of water fees to farmers with irrigated crops, our results show that the latter is superior for several reasons. As said, water pricing would increase WUAs and government receipts, whereas the water supply shock would reduce both of them. In addition, the negative effects on total agricultural production and households' income due to the water pricing policy option would be lower with respect to the water restriction policy (compare figures in last panel of Tables 3.8 and 3.12 and Tables 3.9 and 3.14). Hence, the reduction of food security would be lower under the water pricing scenario. However, irrigated crop production would decrease more under this scenario with respect to the reduction of the water supply shock. This means that the lower negative impact on crop production of pricing water for irrigation would be based on a higher increase in crop production under rain-fed conditions with respect to the water supply shock (compare figures of the first two panels of Tables 3.8 and 3.12). An increase of crop production under rain-fed conditions requires policy interventions such as investments to facilitate rain-fed farmers to have access to the relevant markets, a condition that is not captured in the CGEM since in the model we rule out transaction costs (see Solano and Yunez-Naude 2016; Dyer and Taylor 2011).

To enhance rain-fed agriculture and the marketing of the crops produced by it is precisely one major policy of the current federal administration (2013–2018) through a new strategy called National Crusade against Hunger (Cruzada Nacional Contra el Hambre – CNCH) that combines agricultural productive policies of small poor farmers with social policies to reduce poverty (*Diario Oficial de la Federación* (DOF) Jan. 2013 and Berdegue et al. 2015). Among others MasAgro, a component

---

requires designing institutional, technical, and accounting measures that accurately track and economically reward reduced water depletions. Conservation programs that target reduced water diversions or applications provide no guarantee of saving water" p. 18215.

of the CNCH and a project of the International Center for the Improvement of Maize and Wheat (CIMMYT), promotes the use of high yields maize varieties by small traditional rain-fed farmers (http://masagro.mx/index.php/es/).

As said, the negative effects on households' income of the water pricing policy scenario are lower than the ones obtained under the reduction of irrigation water supply. In addition, under both scenarios the poorest households are the ones less hit, particularly so under the price shock experiment. The later difference means that the increase in the incidence of poverty could be lower if water pricing is the adopted policy.

Another important result of the general equilibrium effects of the two simulations is that the rural households of southern Mexico would be less negatively affected by them, especially so under the water price shock scenario (Tables 3.9 and 3.14). The result is partially explained by the fact that agricultural production in the south is generally done under rain-fed conditions and is relevant because the southern regions are the poorest and less developed in Mexico. If we add to this, the fact that proper water pricing for agriculture could lead to a more efficient use of this resource, we can argue that increasing the price of irrigation water could be an advisable state policy.

With respect to actual agricultural water policies in Mexico, the current federal administration does not have an explicit purpose to reduce irrigation water supply; instead, it stresses the promotion of technology adoption for a more efficient use of water in agriculture, as well as the increase of public investment in irrigation infrastructure.[15] By contrast, irrigation water pricing is explicitly referred to only once in the Ministry of Agriculture strategies for 2013–2018.[16] In addition, the ministry makes no mention in the document to the Water Users Associations and their potential role in promoting investments and water-efficient technology adoption. In fact, the Federal Rights Law (*Ley Federal de Derechos*) approved by Congress in 2013 maintains the same rates charged to farmers in 2011–2013 for their water use (*DOF*, December 11, 2013).

We hope that the results discussed in this chapter contribute to a better understanding of the policy options that could lead to a more efficient use of water for agricultural production with no major negative effects on food security in Mexico.

[15]However, the implementation of these purposes were not fulfilled, especially so for the last 3 years, in part because of the budget restrictions faced by the Federal Government.

[16]DOF, December 13, 2013. In this official document, the only mention to water pricing is in one of the Ministry of Agriculture strategies. In strategy 1.4, the Ministry states the purpose to promote the efficient use of natural resources, among others, by establishing in coordination with the National Commission of Water (CONAGUA), prices and tariffs of water that reflect its economic cost and promote its efficient use and conservation, p. 98. On December 27, 2013, we searched for details in CONAGUA's website (http://www.conagua.gob.mx/) and found no explicit mention of current plans for water pricing in general or specific to water for irrigation.

References

Aragon Castaño, G. (2011). *Un estudio sobre el uso del agua en México para la irrigación.* MA thesis in Economics, El Colegio de Mexico.

Berdegué, J., et al. (2015). *Territorios Productivos: Un Programa Articulador para Reducir la Pobreza Rural a través del Incremento de la Productividad, la Producción y los Ingresos.* RIMISP/Centro Latinoamericano para el Desarrollo Rural. Documento de Trabajo No. 131, 24 de march 2015, en. http://www.rimisp.org/proyectos/publicaciones-y-documentos/?p=programa-cohesion-territorial-para-el-desarrollo&c=documentos-de-trabajo

Cook, E. R., et al. (2007). North American drought: Reconstructions, causes, and consequences. *Earth Science Reviews, 81*, 93–134.

Cortez Vazquez, J. (2006). Mexico. In K. A. Shein (Ed.), *State of the climate in 2005*, special supplement, *Bulletin of the American Meteorological Society, 87*(6), S66–S67.

Diario Oficial de la Federación. (2013a, January 22). Decreto por el que se establece el Sistema Nacional para la Cruzada contra el Hambre, Secretaría de Desarrollo Social (Ministry for Social Development).

Diario Oficial de la Federación. (2013b, December 11). Several decrees of the Ministry of Finance, 2nd Section, pp. 30–33.

Diario Oficial de la Federación. (2013c, December 13). Vol. DCCXXIII no. 10, *Approval of the Ministry of Agriculture, Livestock. Fisheries and Rural Development Program, 2013–2018*, 50–112.

Dyer, G., & Taylor, J. E. (2011). The corn price surge: Impacts in rural Mexico. *World Development, 39*, 1878–1887.

Dyer, G., Taylor, & Yunez-Naude. (2005). Disaggregated rural economy-wide models for policy analysis. *World Development, 33*(10), 1671–1688.

Govea Vargas, Y. (2017, May) *Impactos de las políticas para el desarrollo sustentable de México: un enfoque multisectorial.* MA thesis in economics, El Colegio de México.

Guerrero, H. (2008). El Agua de México: Consecuencias de las Políticas de Intervención en el Sector. *El Trimestre Económico,* Lecturas No. 100, México, Fondo de Cultura Económica: 31–58 and 93–114.

Hernández Solano, A., & Yúnez Naude, A. *Impactos del cambio climático en la economía rural de México: un enfoque de equilibrio general.* CEE, El Colegio de México, Documento de Trabajo 2016-6. http://cee.colmex.mx/2-principal/336-documentos-de-trabajo-2016

Lopez-Morales, C. (2012). "Valoración de servicios hidrológicos por costo de reemplazo: Análisis de escenarios para el Bosque de Agua", Documento de trabajo de la Dirección General de Investigación en Ordenamiento Ecológico y Conservación de Ecosistemas. México: Instituto Nacional de Ecología. http://ine.gob.mx/descargas/dgioece/doc_bosque_de_agua.pdf

Lopez-Morales, C., & Duchin F. (2011, December). Policies and technologies for a sustainable use of water in Mexico: A scenario analysis. *Economic Systems Research, 23* (4), 387–407.

Mendelsohn, R., et al. (2009). A Ricardian analysis of Mexican farms. In *Environment and development economics* (Vol. 15, pp. 153–171). Cambridge: Cambridge University Press.

Stahle, W., et al. (2009). Early 21st-century drought in Mexico. *EOS, 90*(11), 89–90.

Tiscareño, M., et al. (2003). Modeling el Niño southern oscilation climate impact on Mexican agriculture. *Geofísica Inernacional, 42*(003), 331–339.

Ward, F. A., & Pulido M. (2012, November 25). Water conservation in irrigation can increase water use. In P. S. Dasgupta (Ed.), *Proceedings of the National Academy of Sciences of the United States of America (PNAS)* (Vol. 105, 4, pp. 18215–18220), University of Cambridge, Cambridge. http://www.pnas.org/content/105/47/18215.full

Yunez-Naude, A. (2006). *Economic assessment of policy interventions in the water sector*, Final report for the World Bank Project, Mexico Economic Sector Work (ESW).

Yunez-Naude, A., & Aguilar, P. (2012). Análisis sectorial para una economía verde, final report for the project MX-GESS: Inter-American Development Bank, National Institute of Ecology and United Nations Program for the Environment.

Yunez-Naude, A., & Hernandez-Solano, A. (2018). The Mexican agricultural sector two decades after NAFTA: Expectations, facts and policy challenges. In A. M. Buainain, Z. Navarro & M. Rocha de Sousa (Eds.), *Globalization and agriculture: Redefining unequal development* (pp. 71–92). Lanham: Lexington Books.

Antonio Yunez-Naude is a Professor at the Center of Economic Studies, El Colegio de Mexico. He is a member of the Mexican Academy of Sciences, has the highest level as National Researcher from the National Council of Science and Technology, and has a Ph.D. in Economics from the London School of Economics and Political Science, University of London. His research interests are economic development, agricultural, rural and resource economics and trade. Prof. Yunez-Naude recent research includes the estimation of microeconomic general equilibrium effects of climate change on Mexico's rural households.

Patricia C. Aguilar Mendez is an economist majoring in agricultural economics and evaluation of rural public policies. She has served in the public sector as well as an independent consultant. At the Federal Government of Mexico, she has held diverse positions at the Secretariat of Agriculture, and in the Congress, she collaborated at the Center of Studies on Sustainable Rural Development for the Security and Food Sovereignty. Currently, she is Manager of Investigation at the previously named Rural Financial. She has a master degree in Economy and International Politics (CIDE) and was a Hubert H. Humphrey fellow hosted by Cornell University.

Chapter 4
Mexico, the Water Stress: Challenges and Opportunities in Wastewater Treatment and Reuse

Luis Alberto Seguí Amórtegui, Gabriela Moeller-Chávez, and Andrés De Andrés Mosquera

Abstract A proper treatment of wastewater is important to avoid the consequences of releasing untreated wastewater into water bodies and soils, which causes harmful effects on human health and on the environment and also leads to negative social and economic consequences. Due to this, regeneration and reuse activities are of paramount importance. Many countries practice the regeneration and reuse with successful results. In developing countries because of technological and economical deficiencies, the use of strong and efficient systems of lesser cost and sophistication is required. The current situation in Mexico is described by mentioning the need of a model to make these activities as sustainable as possible and to achieve the United Nations Sustainable Development Goals (SDG). The need to reinforce the reuse of regenerated water regarding direct or indirect potable reuse is discussed. A model of recycled water management to assure a continuous water supply is proposed from the technical, health, economical, legal, and environmental perspective to preserve quality and increase stock in the region.

Keywords Reclaimed water · Sustainable water treatment · Reuse · Efficiency production · Scarcity

4.1 Introduction

Most human activities that use water produce undoubtedly wastewater. As the overall demand for water grows, the availability of this precious resource diminishes and the quantity of wastewater produced and its overall pollution load are continuously increasing worldwide. The consequences of releasing untreated or

L. A. Seguí Amórtegui (✉) · A. De Andrés Mosquera
EAE Business School, Barcelona, España

G. Moeller-Chávez
Universidad Politécnica del Estado de Morelos, Progreso, Mexico
e-mail: gmoeller@upemor.edu.mx

H. R. Guerrero García Rojas (ed.), *Water Policy in Mexico*, Global Issues in Water Policy 20, https://doi.org/10.1007/978-3-319-76115-2_4

inadequately treated wastewater can be classified into three groups, (a) harmful effects on human health, (b) negative environmental impacts, and (c) adverse repercussions on economic activities (UNESCO 2017), and then, a proper treatment of wastewater is obligated, making this resource appropriate for different types of reuse.

There are many countries conducting different types of wastewater regeneration and reuse. Several studies justify and support this practice. In industrialized countries, many projects and research for the regeneration of wastewater have been developed, obtaining, besides the reuse of reclaimed water to meet specific water demands, additional benefits of environmental protection and prevention of health risk. In developing countries, it is also necessary to cover these aspects, but, as a result of economic and technological deficiencies, the use of strong and efficient systems of lesser cost and sophistication is required (Soderberg 2016).

When considering those areas where rainfall is less than 200 mm per year and based on the premise that, "Reuse of wastewater occurs primarily in those places where water availability is low," there are basically three situations where the practice of wastewater regeneration and reuse is accentuated: (a) those continental zones where rainfall is scarce, (b) those continental zones where topography makes the site inaccessible, and (c) the islands, where probably, due to a lack of infrastructure for collecting water, a reuse practice turns into an option for water supply. The only documented cases where direct potable reuse is performed are in Windhoek, Namibia, and Orange County, California, and other demonstrative projects in the USA (Denver, Colorado, Cloudcroft, New Mexico, and Big Springs Texas). Also, other projects are implemented to practice indirect potable reuse (Hispanhol 2014; Moeller et al. 2016). For all the other purposes, there are evidences of the reuse of regenerated water in several parts of the world and for different uses. It is important to note that all over the world there are places where the regeneration and reuse of wastewater is a reality and it is conducted to a greater or lesser degree.

Clearly, regeneration and reuse of wastewater is increasingly important, particularly in areas where the possibility of access to other sources of supply is increasingly complicated. Good water management, with an exchange of uses that allows replacing regenerated water with water from conventional sources, will enable the release of water volumes for priority uses, like human consumption, since the reuse of water directly or indirectly regenerated for human consumption is very limited due to health reasons and especially to the negative perception of consumers (Moeller et al. 2016).

Technological advances and cheaper technology increasingly allow the impulse of wastewater regeneration and reuse systems (WRRS). There is well-documented information about regeneration and reuse costs. In some areas, it begins to be less than water transportation and/or supply from conventional sources over long distances.

Wastewater regeneration and reuse have found a place in countries with scarce water resources. In some countries of Africa, the Middle East, and Asia, they have become a source of supply of great relevance, and an exponential increase can be noted, especially with the implementation of cutting-edge technologies such as ultra and RO membranes. Countries such as Australia, South Africa, Mexico, and the USA are planning to develop, or increase, the reuse of water in the short, medium, and long term, because water reuse for targeted potable and non-potable applications could reduce stress on existing drinking water supplies and deliver energy benefits (NSF 2015). Also to fulfill requirements of objective 6 of SDG, reuse is strongly recommended (United Nations Organization 2015).

4.2 Current Situation in Mexico

Mexico is a country with an extraordinary wealth of natural resources that presents a serious and huge inequality in the distribution of economic wealth. It has a population of approximately 119,530,753 inhabitants (INEGI 2016); 76.8% (86.29 million of inhabitants) of the population lives in urban areas located on 2% of the national territory. Rural population represents 23.2% (29.99 million of inhabitants). This situation promotes the existence of two different countries: urban, which in some cases is highly developed and enjoys all the technological advances, and rural, where not even the most basic needs are fulfilled. The availability of water in the country has an average annual volume of 3692 cubic meters per capita, a figure that places it as a country with no water problems; however, upon regional balance highly marked deficit areas appear. Table 4.1 contains data from gross extraction of water for various uses. Note that agriculture and industry pose enough demand to absorb regenerated water derived from municipal effluents.

The use of aquifers represents 38.6% of the total extraction, and it is worth noting that irrigation is the sector that uses it the most of groundwater (76.3%). Usually, this type of water is considered to be of very high quality; consequently preserving it for human consumption is preferred. Here a question emerges regarding the use of high-quality water for irrigation, with low efficiencies (around 50%), being able to replace it in some areas with regenerated water. With this replacement, not only the benefit of water is obtained, but also, using regenerated water from household sources is helpful for crops by providing them with organic matter, nitrogen, and phosphorus.

Table 4.1 Gross extraction of water for diverse uses in Mexico in 2015

Use	Origin		Total volume (km^3)	Extraction percentage
	Superficial (km^3)	Underground (km^3)		
Agriculture	41.89	23.47	65.36	76.30
Public supply	5.16	7.32	12.48	14.60
Self-contained industry	1.61	2.07	3.68	4.30
Electric power[a]	3.70	0.45	4.15	4.80
Total	**52.35**	**33.31**	**85.66**	**100.00**

Source: CONAGUA (2016a)
[a]Excluding hydropower

In spite of the water consumed in irrigation through conventional sources of supply, wastewater irrigation is extensive in Mexico; of the 86 existing irrigation districts, there are about 40 irrigation districts that use wastewater, irrigating approximately 330.000 ha (CONAGUA 2016a). It is important to emphasize that Mexican regulation permits the irrigation of tall stem crops with treated wastewater (criterion established by the WHO – World Health Organization), which is called "restricted crop irrigation"; officials responsible for the irrigation districts monitor that these irrigation practices with partially treated wastewater comply with the irrigation only in such tall stem crops, as corn, wheat, or barley.

It is estimated that in Mexico, 229.1 m^3/s of urban wastewater are produced, of which 212 m^3/s are collected in sewerage and 120.9 m^3/s are treated before being discharged. Treatment infrastructure at the national level corresponds to 2477 plants in formal operation, treating 120.9 m^3/s of regenerated water susceptible of being reused. Treatment processes for municipal effluents are various, although stabilization ponds and activated sludge systems predominate (CONAGUA 2016b). Through this scenario, it is possible to understand how WRRS in Mexico, which may be compared with the global forefront and at the same time present considerable setbacks in potable water, sewerage, and sanitation coverage, can coexist. An example of this situation is the Metropolitan Zone of the Valley of Mexico (ZMVM) with a population of 8,918,653 inhabitants in 2015 (INEGI 2016); it ranked as the fourth most populous city in the world, being exceeded only by Tokyo, Delhi and Shanghai (United Nations Organization 2014). The city requires approximately 32 cubic meters of water per second to cover the demands of drinking water.

Table 4.2 presents various sources of supply that deliver water to Mexico City: 55% of the supply comes from underground sources based on springs and deep wells located at strategic points in the city, which have already reached 450 m in depth, while the other 50% is transferred from neighboring basins. Once used, wastewater is discharged to the unitary sewage system consisting of 13,000 km of primary sewage system and secondary sewage system. Eighty-seven percent of municipal wastewater from the ZMVM remains untreated and is sent outside the Valley of Mexico basin through the main drainage system (OCAVM 2016).

Untreated wastewater is used to irrigate 90,000 ha of crops in the Valle del Mezquital, state of Hidalgo. Returning runoffs from irrigation drain into Panuco River tributaries, which empty into the Gulf of Mexico. Recently, the largest wastewater treatment plant in the country was constructed and started operation in

Table 4.2 Supply sources for Mexico City

Source	Percentage	Flow rate (m^3/s)
Valley of Mexico wells	55	14.908
Springs	3	0.792
Sistema Lerma	12	3.832
Sistema Cutzamala	30	9.575
Total	**100**	**29.107**

Source: OCAVM (2016)

the municipality of *Atotonilco de Tula*, Hidalgo. It has a treatment capacity of 23 cubic meters per second during the dry season (through a conventional process) and an additional module (through a physical-chemical process) to treat 12 cubic meters per second during the rainy season. Completion of this plant was in the year 2017.

4.3 Problems of Water Regeneration and Reuse

Current technological developments enable us to obtain regenerated or treated municipal wastewater with different characteristics of quality depending on the use to which it is destined. These technologies allow us to reach any level of quality no matter how strict the requirement is. Obviously as the water quality requirements are more demanding, the process of treatment becomes more complex and therefore more expensive. Likewise, it is important to note that any process of regeneration requires consideration of the (byline products) residual matter obtained as well as its ultimate disposal. Thanks to advances in technology, there is a broad spectrum of treatment systems that allow us to obtain suitable grades in quality no matter how strict it is, for any type of reuse of regenerated water.

Research and development of new technologies for water regeneration and reuse in the international arena has increased in recent years mainly in the field of "membrane technologies." One of the problems we face with the use of membrane technologies is that these technologies require a very consistent and specific water quality in order to function. In the case of wastewater with industrial input, this is a real problem, which leads to severe difficulties in exploitation. Membrane technologies require sound pretreatments, for example, osmosis requires a treatment that avoids process jamming; when ultrafiltration is set, several times it gets jammed not in the osmosis but in the ultrafiltration. It is simply transferring the problem from one unit to another. Apart from the deficiency and legal void for water collected in sewerage systems monitoring, there is a high risk in wastewater treatment plants (WWTPs). They receive pollutants that produce failures in the processes, mainly in the biological ones, which have an impact on unavoidable problems for tertiary treatments.

This situation creates a production of regenerated water that is not continuously guaranteed, situation that promotes a very high risk of not supplying water, when designing a system, with a view to produce on demand. It is therefore important to study and establish treatment processes that are flexible and robust, because in wastewater with industrial contribution, there is a constant risk of the presence of compounds that can damage membrane technologies at any given moment. On the other hand, the use of membrane technologies requires highly skilled personnel; this leads to the difficulty of personnel rotations and therefore a cost of regular staff throughout the year, whether it occurs or not, which significantly increases the total cost of the cubic meter when the regenerated water production is low.

In short, the problems that we face nowadays are, on the one hand, to ensure water supply despite the potential episodes of contamination that may damage the biological treatment of the WWTP and, on the other hand, a reduction in production costs.

4.4 Proposal for a New Management Model

As a result of the abovementioned problem, it is necessary to establish a new management model. The current model is a result of a gradual scheme depending on the demands, providing storage in basins to cover consumption peaks; that model works, if and only if there is a guarantee in production and continuous water quality that meet national and/or international criteria. The possibility of increasing the storage of basins is nonviable, due to high investment and maintenance costs.

The concept of sustainability involves an economic and social development that meets the needs of today without compromising future generations. Although the definition of sustainable development is not very precise regarding its operational range, undoubtedly there must be an implicit planning to obtain immediate effective solutions to a short-term problem that is not more costly in the long run.

The cycle that follows both surface water and groundwater through the different uses to which it is intended has as a consequence is the deterioration of the water resource, as well as of the environment; however, when integrating regeneration, recycling and reuse technologies are a part of the hydrological cycle; proper water quality can be achieved to allow the balance of ecosystems and the provision of the resource within the concept of sustainability. Still the practice of wastewater regeneration and reuse is the best empirical example of a sustainable development that has to continue in practice.

The new management model considers using the aquifer as storage, allowing production throughout the year, including those times in which there is no demand on the part of users. This allows us to close the water cycle and link the handling and management of surface and underground resources. At the international level, this model is already being used in some places, having as a representative case at the global level the one in Dan, Israel (Juanico and Friedler 1999), and Orange County, California (OCWD 2003). Although Mexico is an international benchmark in the field of wastewater regeneration and reuse, the practice of recharging aquifers with regenerated water is incipient.

Next, some basic arguments to advance and demystify the recharge of aquifers with residual water or regenerated water are presented, these arguments arise from technical, legal, health, environmental, social, and economic perspectives.

4.4.1 Technical Perspective

Recharging aquifers with regenerated water is imposed, from the standpoint of California, that is, through the use of the most advanced technologies, primarily those related to membranes, since these technologies produce a water of high quality (Leverenz et al. 2011). However, as of today we are more concerned about getting the best quality than considering the minimum requirements necessary for the use of regenerated water, as well as the treatment provided by the same aquifer as regenerated water polishing.

4.4.2 Legal Perspective

In this area, Mexico has a comprehensive and modern legal framework, which provides the possibility for recharging aquifers with regenerated water. Basically, there are two components to be taken into account from a legal point of view for the reuse of regenerated water in the recharge of aquifers, regarding the acquisition of the right to the regenerated water and regarding the quality of regenerated water that will be stored in the aquifer, because of health issues (Moeller et al. 2016).

4.4.2.1 Acquisition of the Right to the Regenerated Water Reuse

As stated in the legal framework, it is necessary for the authority in the field of water to assume water trading to ensure private sector investment in the production of regenerated water for its later use. Water management is the responsibility of the water authority; that is why we consider that the paradigm of reuse simply passes through treated water grants, whether or not the same user regenerates them. This will allow at any time to maintain the management of the resource, incorporating the so-called new sources of supply to water management.

4.4.2.2 Criteria for the Quality of Regenerated Water

The progress of the planned regeneration and reuse of wastewater depends not only on technological advances but also on the existence of a robust legal framework that establishes guidelines for the reuse that entails no risk to beneficiaries. Legislation on wastewater reuse on a global scale is a complicated issue, because while there are countries with regulations of a legal nature, others just give recommendations, each with their own parameters and indicators. Asano (Asano 1998, 2001, 2004) presented a historical overview of the regulations for reuse in the USA and a discussion on the implementation of water quality criteria. An AIDIS compilation presents a review on the Latin American perspective presented by several authors (AIDIS-UNESCO 2016).

In this way, one of the first actions to allow the reuse of regenerated water will be to establish the necessary criteria that these waters should meet in order to be reused in multiple applications depending on the level of treatment performed. There are basically two trends regarding the criteria for wastewater reuse. One is established by the World Health Organization (WHO 1989, 2006, 2017); and the other is submitted by the State of California in the USA and if it is for agricultural reuse by the Food and Agricultural Organization and the United States Environmental Protection Agency (Pescod 1992; Bastian and Murray 2012).

However as of today, the criteria required for reusing regenerated water in the recharge of aquifers reflect more the concerns over the possible contamination of the aquifer and the health risk (WHO 2017; Moeller et al. 2016) than the study of substance diffusion into the aquifer. This is a clear expression of the lack of multidisciplinary work among areas such as hydrogeology, environmental engineering, and public health. In this sense, there are many empirical works that support the low risk that entails the recharge of aquifers with regenerated water; an example of this are the studies of Delovitch et al. (2003), Jarusutthirak et al. (2003), and Okun (2002).

Mexico has been a pioneer in defining and establishing a standard that clearly defines the criteria and "rules" to artificially recharge an aquifer. Mexican Official Standard NOM-014-CONAGUA-2003 (DOF 2009) is an example of the importance in the implementation of regenerated wastewater reuse in the country.

4.4.3 Health Perspective

Pathogen organisms are considered hazardous pollutants for indirect wastewater reuse for human consumption for the magnitude and speed of their effects and toxic chemicals for their long-term effects (Jiménez et al. 2012; Cortés Muñoz et al. 2013). Wastewater facilities must assure a safe treated wastewater removing all these noxious pollutants and microorganisms. Also there is a growing awareness of the presence of pollutants such as hormones, antibiotics, steroids, and endocrine disruptors in wastewater that poses a new set of challenges as their impact on the environment and health has yet to be fully understood (WHO 2017; Moeller and Buelna 2012).

The World Health Organization published the first guidelines for the safe use of wastewater and excreta in agriculture and aquaculture (WHO 1989). Since that time, more experience has been gained and more evidence is being produced. The new WHO guidelines on water reuse in agriculture, recently published (WHO 2017), provide all the basic information on the health risks and how to set health-based targets, by quantifying the risk and developing pathogen reduction targets.

4.4.4 Environmental Perspective

Regeneration and reuse of wastewater have a huge potential for the recovery of water resources and the reduction of pollution. Anderson (2003) carefully examines the environmental benefits of the reuse of regenerated waters; United Nations establishes wastewater as the untapped resource (UNESCO 2017). The upcoming World Water Forum 2018 launches the same (WWF 2018). The reuse of wastewater, either untreated or regenerated, has been a very common practice in the vast majority of countries considered as "dry." Nevertheless, over these past years, the planned reuse of reclaimed water has taken on a significant boom, basically because of the shortage of water resources; the primary importance of this planned reuse is not to lose the control and ownership of these waters.

Recharging an aquifer with regenerated water presents as the highest environmental contribution, increasing water stocks within the water unit, releasing volumes of water that can be recovered within water management for the preservation of aquatic ecosystems of the management unit; this concept is named indirect potable reuse (Moeller et al. 2016).

4.4.5 Social Perspective

Recharging an aquifer with regenerated water entails several social advantages: (1) restoring or maintaining groundwater levels and subsequent acceptance by groundwater users, (2) water availability with no need of investing in pipeline for its distribution, and (3) the conceptual change of regenerated water for groundwater, which promotes a positive social perception.

4.4.6 Economic Perspective

Recharging an aquifer with regenerated residual water would allow a steady output of regenerated water, independent of the current demands of users. In the year 2009, the cost of importing water to the Metropolitan Area of the Valley of Mexico was 8.48 pesos/m^3, of which 7.45 pesos/m^3 (87.8%) corresponded to energy costs. On the other hand, the cost of extracting water in the ZMVM was 5.24 pesos/m^3 of which 88.7% (4.65 pesos/m^3) corresponded to energy costs (CONAGUA 2010). The cost of water regeneration (including all costs, investment, exploitation, and maintenance) from residual water up to the level of drinking water oscillates at the international level, between 10 and 19 pesos/m^3 (Seguí et al. 2014), 0.70–1 USD/m ^{3} (Moeller-Chavez et al. 2004). It is important to recognize that the difference in costs is not so significant, if we consider that in the case of the current supply, only exploitation and maintenance costs are being considered.

The total cost obtained as of today can be considered high when compared with the total costs of conventional sources of water (although the fact that the cost of conventional sources does not usually include investments should be kept in mind). However, what should be really considered is how much the user is willing to pay for regenerated water, especially in this case study where the user has no other supply alternative.

In addition, it is necessary to incorporate into the economic analysis not only the internal or private impacts understood as those that are directly linked to the production process for regenerated water and its subsequent reuse but to incorporate the economic analysis of external impacts (e.g., pollution control, increase in water availability, or guarantee of supply), which, while they may be more difficult to measure, their importance is not minor, since an impact of these characteristics may virtually cause the censorship of the project or the economic viability of the same. In the work of Seguí et al. (2014, 2005, 2009), the economic techniques for the determination of income and costs of the abovementioned impacts are described in detail.

4.5 Conclusions

The success of the proposed management model is subject to the involvement of the authority in the field of water, which will be responsible for managing surplus resources stored in the aquifer. Under this assumption, water management in the hands of the authority will be responsible for receiving, on the one hand, the volumes of regenerated water in the aquifer, recording the delivery, and on the other hand, granting concessions of surplus volumes to those users who require it. The aquifer will always maintain a volume that will ensure the current producers of regenerated water its supply.

The proposed model of management will respond to the assurance in water supply, causing users who choose for wastewater regeneration and reuse to consider their investment safe and at minimum risk.

Within an integrated water management, the results obtained in these studies open the possibility to increase water resources in the region. The authority responsible for the management of the water in this area can promote, through this model, private participation in wastewater regeneration and reuse, without ever losing its property.

Finally, it is fundamentally a very deep paradigm shift, in which we grant more importance to water quality and not to the origin of the water; only in this way, can we continue without any problems to the scenario of indirect and direct reuse for potable reuse. In this way, challenges will be fulfilled, risks will be minimized, and opportunities will be greater.

References

AIDIS-UNESCO. (2016). Uso seguro del agua para el reúso. In P. Tello, P. Mijaylova & R. Chamy (Eds.), sp. ISBN: 978-85-93571-00-8.

Anderson, J. (2003). The environmental benefits of water recycling and reuse. *Water Supply, 3*(4), 1–10.

Asano, T. (Ed.). (1998). *Wastewater reclamation and reuse, Water Asano Quality Management Library* (Vol. 10). Lancaster: Technomic Publishing Inc.

Asano, T. (2001). Water from (waste) water – The dependable water resource the dependable water resource (The 2001 Stockholm Water Prize Laureate Lecture). *Water Science & Technology, 45* (8), 23–33.

Asano, T. (2004). Water and wastewater reuse, An environmentally sound approach for Sustainable Urban Water Management. http://www.unep/

Bastian, R. & D. Murray. 2012 *Guidelines for water reuse.* US EPA Office of Research and Development, Washington, DC, EPA/600/R-12/618.

CONAGUA. (2010). Compendio del Agua de la Región Hidrológico-Administrativa XIII. Edición 2010. Organismo de Cuenca Aguas del Valle de México. Comisión Nacional del Agua. www.conagua.gob.mx Diciembre de 2010.

CONAGUA. (2016a). Estadísticas del Agua en México 2016. Comisión Nacional del Agua. http://www.conagua.gob.mx

CONAGUA. (2016b). Status of potable water, sewerage, and sanitation sub-sector in sp. (Situación del subsector de Agua Potable, Alcantarillado y Saneamiento. Comisión Nacional del Agua. http://www.conagua.gob.mx

Cortés Muñoz, J., Calderón Mólgora, C. G., Martín Domínguez, A., Espino de la O, E. E., Gelover Santiago, S. L., & Hernández Martínez, C. L., Moeller Chávez, G. E. (2013). Endocrine disruptors in water sources: Human health risks and EDs removal from water through nanofiltration. https://doi.org/10.5772/54482

Delovitch, E. I., Icekson-Tal, N., Avraham, O., & Michail, M. (2003). The long-term performance of soil aquifer treatment (SAT) for effluent reuse. *Water Supply, 3*(4), 239–246.

DOF. (2009). Diario Oficial de la Federación de 18 de agosto de 2009. NORMA Oficial Mexicana NOM-014-CONAGUA-2003, Requisitos para la recarga artificial de acuíferos con agua residual tratada.

Hispanhol, I. (2014). Reuso Potavel Direto, In P. Tello, P. Mijaylova, & G. Ruiz (Eds.), *Temas de Ingeniería Sanitaria y Ambiental in Sp, AIDIS.* ISBN 978-607-02-5853-4.

INEGI. (2016). Instituto Nacional de Estadística, Geografía e Informática. Encuesta Intercensal 2015. http://www.beta.inegi.org.mx/proyectos/enchogares/especiales/intercensal/

Jarusutthirak C., Amy G. y Foss D. (2003). Potable reuse of wastewater effluent through an integrated soil aquifer treatment (SAT) – Membrane system. Water Supply 3 3 25–33.

Jiménez, B., Chavez, A., Gibson, R., & Maya, C. (2012). Unplanned aquifer recharge in El Mezquital/Tula Valley, México. In C. Kazner, T. Wintgens, & P. Dillon (Eds.), *Water reclamation technologies for safe managed aquifer recharge* (pp. 141–143). London: IWA Publishing.

Juanico, M., & Friedler, E. (1999). Wastewater reuse for river recovery in semi-arid Israel. *Water Science and Technology, 40*(4–5), 43–50.

Leverenz, H. L., Tchobanoglous, G., & Asano, T.. (2011). Direct potable reuse: A future imperative. *Journal of Water Reuse and Desalinisation.* doi:https://doi.org/10.2166/wrd.

Moeller, G, & Buelna, G. (2012). Contaminantes emergentes: Su importancia, retos y perspectivas sobre la medición. El tratamiento y la reglamentación in sp. ¡a. Edición. IMTA, INRS, CRIQ AIDIS 6 RETAC. ISBN: 978-607-7563-53-2. Mexico.

Moeller, G., Guillén, R., Treviño L., & Lizama, C. (2016). Reúso de aguas residuales tratadas como fuente directa e indirecta de agua potable, in Sp. (pp 91–107). AIDIS. En Uso seguro del agua para el reúso. Eds. Pilar Tello, Petia Mijaylova & Rolando Chamy. ISBN: 978-85-93571-00-8.

Moeller-Chavez, G., Seguí, L., Alfranca, O., Escalante, V., Pozo, F., & Rivas, A. (2004). Water reuse in the Apatlaco River basin (Mexico): A feasibility study. *Water Science & Technology, 50*(2), 329–337.

NSF (National Science Foundation), U.S. DOE (U.S. Department of Energy), and U.S. EPA (U.S. Environmental Protection Agency). (2015, April 28–29). Energy – Positive Water Resource Recovery Workshop Report.

OCAVM. (2016). In sp. Organismo de Cuencas Aguas del Valle de México. www.gob.mx/conagua

OCWD. (2003). Orange County Water District. http://www.ocwd.com/gwrs/

Okun, D. A. (2002). Water reuse introduces the need to integrate both water supply and wastewater management at local and regulatory levels. *Water Science & Technology, 46*(6–7), 273–280.

Pescod, M. B. (1992). Wastewater treatment and use in agriculture. FAO irrigation and drainage paper 47. ISBN 92-5-103135-5. Food and Agricultural Organization of the United Nations Organization. Rome.

Seguí, L., Alfranca, O., Guerrero García-Rojas, H., & Moeller, G. (2005). Price for water reclamation, how can it be established? A Mexican case study. *Water Supply, 5*(2), 87–96.

Seguí, L., Alfranca, O., & García J. (2009). Techno-economical evaluation of water reuse for wetland restoration: A case study in a natural park of northeastearn Spain. *Desalination, 246*(1), 179–189. ISSN: 0011-9164.

Seguí, L., Alfranca, O., & Moeller G. (2014). Metodología para el análisis técnico-económico de los sistemas de regeneración y reutilización de las aguas residuales. *Tecnología y Ciencias del Agua*, V(2), 55–70. ISSN 0187-8336.

Soderberg, C. A. (2016). Panorama mundial del reuso de aguas residuales tratadas. (pp. 31–42). En uso Seguro del agua para el reúso AIDIS, 2016. ISBN: 978-85-93571-00-8.

UNESCO. (2017). *The United Nations world water development report WWDR: "Wastewater: The untapped resource", United Nations world water assessment programme*. Paris: UNESCO.

United Nations Organization. (2014). United Nations Organization (UN). http://www.un.org/es/index.html

United Nations Organization. (2015). United Nations Sustainable Development Goals. Goal 6: Ensure access to water and sanitation for all. (SDG 6). http://www.un.org/sustainabledevelopment/water-and-sanitation

WHO. (1989). *Health guideline for the use of wastewater in agriculture and aquaculture*. Technical Report Series, No. 778. Geneva: World Health Organization.

WHO. (2006). *Guidelines for the safe use of wastewater, excreta and greywater*, Vol. 1. Geneva 27: World Health Organization.

WHO. (2017). *Potable reuse: Guidance for producing safe drinking-water*. Geneva: World Health Organization. Licence: CC BY-NC-SA 3.0 IGO.

WWF. (2018). 8th world water forum, Brasilia, 2018: Theme: People (Water, Sanitation and Health).

Luis Alberto Seguí Amórtegui holds a Ph.D. from the Universidad Politecnica de Cataluña (UPC) and a Master's Degree in Economic Engineering from Universidad Nacional Autonoma de Mexico. Dr. Seguí has devoted his 25 years of professional experience to private and public sectors in various technical and executive positions. Currently, he is a consultant for management and environmental issues and a professor in EAE Business School and UPC. The research he has developed has resulted in the production of eight publications in international refereed journals, with anonymous evaluation and impact factor. In addition, he has published 5 chapters in books, 10 works in several media (journals, videos, lectures, and Internet), and more than 30 papers in national and international congresses. As a researcher, he has collaborated in four competitive process research projects.

Gabriela Moeller-Chávez has a background in Chemistry and holds a Master's Degree and Ph.D. in Environmental Engineering from Universidad Nacional Autonoma de Mexico. Her main areas of interest are focused in industrial and municipal wastewater treatment issues, especially regarding biological processes, the reuse of treated wastewater, handling, treatment and disposal of sewage sludge, water microbiology, and training and education on those issues. In her professional and academic activity, she has worked in chemical and microbiological quality of water, as well as in the treatment of municipal and industrial wastewaters. She has held various academic and administrative positions at the Universidad Nacional Autonoma de Mexico in the Division of Graduate Studies of the School of Engineering as well as in the Mexican Institute of Water Technology, IMTA: Subcoordinator of Waste Water Treatment, Full time researcher in the area of Waste Water Treatment and Professional development; Coordinator of Water Treatment and Quality. National Consultant for the UN Program on UN Development and World Weather Organization (OMM) and the National Association of the Chemical Industry (AMIQ). Guest tutor at graduate program in Engineering at UNAM and in Water Sciences at IMTA. Faculty member of the Biotechnology Master Degree at UPEMOR and Ph.D. studies at Universidad Autonoma del Estado de Morelos, UAEM. She is currently Academic Director in Environmental Engineering and Biotechnology Engineering at the Universidad Politécnica del Estado de Morelos. Accreditations: National Researcher, Level 2. Member of the Mexican Academy of Sciences (AMC) and the Engineering Academy (AI). She received the Academic Merit Award for her professional career from the National Association for Engineering Faculties and Schools (ANFEI) (June, 2016).

Andrés de Andrés Mosquera has a Major in Economic and Entrepreneurial Sciences and holds a Master degree in Applied Economics. He is a Professor of Social and Business Sciences at EAE Business School, affiliated center of the Universidad Politecnica de Cataluña, Spain. He is Academic Director of Business Administration at EAE Business School. As a researcher, he is a member of the EAE Research Group (GRIT_EAE) and of the Teacher Innovation Group (GID-EAE). Outside the labor sphere, he is a member of the board and treasurer of "Dit i Fet", civil association that tries to alleviate situations of extreme marginality by hosting elder men with scarce resources.

Chapter 5
Industrial Water Use in Mexico: Analysis of Efficiencies Using Water Price Elasticity

Hilda R. Guerrero García Rojas, Faustino Gomez-Santiz, and Erandi Maldonado-Villalpando

Abstract Efficient use of water in Mexico is determined by a tariff structure, which is defined according to the type of user and a categorized classification of water availability zones which arises from the scarcity principle, assuming that the water policy for an efficient water use in Mexico follows the economic principle of supply and demand. In this chapter an analysis of changes in water demand for the industrial sector is performed by hydrological regions in relation to variations in tariff structure. Analyzed data take into account a 14-year period where irregular changes can be seen both in tariff and in demand for water in hydrological regions. According to the result of the price elasticity of demand, there is no direct relationship between an increase in tariff and water demand. This implies that, if there is an efficient water use policy, at least with analyzed data, both the pricing by availability zones and the tariff structure by type of user do not behave as an adequate economic instrument, given that there is a significant variation between applying an increase to the water tariff and the expected response of its demand.

Keywords Industrial water use · Water price elasticity · Water availability zone · Efficiency analysis · Water consumption in the industry

5.1 Introduction

Article 20 of the National Water Law (LAN) states that exploitation, use, and catchment of national waters by natural or legal persons shall be conducted through a concession granted by the National Water Commission (CONAGUA). When the

H. R. Guerrero García Rojas (✉) · E. Maldonado-Villalpando
Facultad de Economía, Universidad Michoacana de San Nicolas de Hidalgo (UMSNH), Morelia, Michoacan, Mexico

F. Gomez-Santiz
Independent Consultant, Morelia, Mexico

H. R. Guerrero García Rojas (ed.), *Water Policy in Mexico*, Global Issues in Water Policy 20, https://doi.org/10.1007/978-3-319-76115-2_5

exploitation purpose of agencies and decentralized organisms is to provide water services in an urban or domestic public character, including those processes that these services comprise, national water catchment will be performed through granted allocation. In other words, water use and exploitation are granted by allocation in decentralized organisms (most of them allocated to municipalities and, these in turn, to third parties); meanwhile, concession refers to the issue of titles for water use or exploitation for natural or legal persons with specific economic activities, such as agricultural or industrial activities. It is important to differentiate this latter aspect. Decentralized organisms that locally manage water provide water not just to the domestic sector; they also supply this resource to the industrial sector, so in terms of the National Water Law, these industries might be classified as "supplied industries," whose rights and obligations come directly from the local entity. Unlike allocations, titles issued by concessions for specific water uses, in this case industrial use, are classified as concessions for self-supplied industry, which is the one that has its own catchment to obtain water, independently from the public supply network; therefore its rights and obligations derive solely from procedures of CONAGUA.

Hence, water consumption pricing in the industrial sector, self-supplied industry, is established by CONAGUA; while water pricing in the supplied industry may change according to current ordinances of each local organism. Likewise, in both cases, the pricing level is expressed in a pricing structure, most of the times differentiated by types of user (domestic, commercial, and industrial, among others). As a part of public policies to encourage economic activities, pricing structure contains certain mechanism for redistributing costs through subsidies. This is true at least in the agriculture sector, which benefits from this mechanism.

Generally, the guiding axis of water policy in Mexico can be distinguished in two types: (i) offer management and (ii) demand management. On the one hand, offer management of water is conducted through infrastructure projects for the distribution of the resource and other programs for water supply. On the other hand, the management of water demand is conducted mainly through mechanisms for setting rates. This latter aspect is regulated by the National Water Law and is established through agreements in the Federal Law of Rights (CONAGUA 2016a). The main economic instrument of fiscal nature to establish rates is through the creation of availability zones with the objective of regulating demand as a function of its availability, as an incentive toward a rational and efficient use, by internalizing social and environmental costs.

Taking into account that the pricing structure of water is composed of types of users and availability zones, how does this disposition affect water consumption in the industrial sector? In this chapter we will analyze the use of water in Mexico in self-supplied industry as a first approach toward the construction of efficiency indicators for water resource in the industrial sector.

5.2 Water Used by the Industrial Sector

According to the Public Registry of Water Duties (REPDA) data, the total volume of water allocated to economic sectors in the country reached more than 85,664 hm^3 in 2015. From that, 38.89% were concessions for underground water extraction, while 61.11% were concession for superficial water use. The water volume allocated to self-supplied industry represented[1] 4.3% of the total. Regarding those more than 3683 hm^3 allocated, 56.25% come from superficial sources and 43.75% from underground sources.

The total water allocated in the industry is distributed in 5500 titles, and just 8.3% correspond to titles for superficial water catchment. From the total of issued titles, about 709 economic units are covered with more than one extraction site in their respective title. For instance, in the case of Federal Electricity Commission, this parastatal company has one allocated title registered in the Mexicali Municipality, whose total volume granted, the highest of all records in the industry, is 83.57 hm^3 with an extraction distributed in 113 catchment sites (Table 5.1).

In general, most of the economic units with larger appropriation concessions are companies related to parastatal companies or economic activities licensed to private companies, as in the case of Minera Peñasquito S.A. de C.V., a mining company which owns five titles with a total water volume granted of 25.70 hm^3.

Regarding other mainly water-consuming industries, these are located within a range of concession of less than 4 hm^3 per year. Eighty-five percent of economic units with a single catchment site, and therefore a single title, concentrate 57% of the total volume licensed in the sector and represent 85% of economic units (Table 5.1).

According to data extracted from the National Water Information System (SINA), the trend of water volumes licensed to self-supplied industry[2] showed a significant variation during the period 2001–2014 (Fig. 5.1). In 2001, the total volume licensed in the industry was 6619 hm^3, while in 2005 this figure shifted to 7084 hm^3, that is, an increase of more than 7% in the aforementioned period. Nevertheless, a 58% decrease in the concession volumes was registered from 2005 to 2006.

As of 2006, concession volumes tend to be almost uniform with no major increase, but the granting of rights for consumption followed an average percentage growth of 20.26% during the period 1995–2013 and 15.40% in 2000–2013 period; that is, despite a reduction in the concession of the total water volume, the number of titles increased, leading to an increase in the number of users.

The chart on the right of Fig. 5.1 shows that from 2011 to 2013, the number of titles presented a 144% increase, from 10,755 to 26,223 in the concessions. If water volumes licensed at the regional level, that is, at the hydrological region level, are reviewed, in 2001–2014 period, the hydrological regions with a continuous growth

---

[1]Self-supplied industry includes concession to commercial, services, agroindustry, and industrial sectors per se.

[2]Referred data from self-supplied industry in Fig. 5.1 includes commercial, services, agroindustry, and industrial sectors per se.

Table 5.1 Licensed volume and economic units from underground source

Underground sources that protect the title	Total economic units	Concessions (volume hm^3)	% hm^3	Underground sources that protect the title	Total economic units	Concessions (volume hm^3)	% hm^3
1	4082	787.94	56.81	16	1	1.08	0.08
2	455	172.62	12.45	17	2	1.31	0.09
3	125	60.65	4.37	20	2	16.87	1.22
4	59	35.55	2.56	21	1	1.37	0.10
5	29	23.47	1.69	23	1	16.22	1.17
6	16	14.51	1.05	24	1	21.56	1.55
7	9	9.46	0.68	25	1	1.46	0.11
8	2	2.29	0.16	29	1	60.61	4.37
9	3	2.40	0.17	32	1	30.85	2.22
10	1	1.24	0.09	35	1	22.44	1.62
11	2	2.48	0.18	36	1	1.74	0.13
12	1	0.02	0.00	48	1	1.74	0.13
13	3	3.78	0.27	108	1	2.22	0.16
14	2	2.52	0.18	113	1	83.57	6.03
15	2	4.87	0.35	**Total**	**4807**	**1387**	**100**

Source: CONAGUA (2016b)

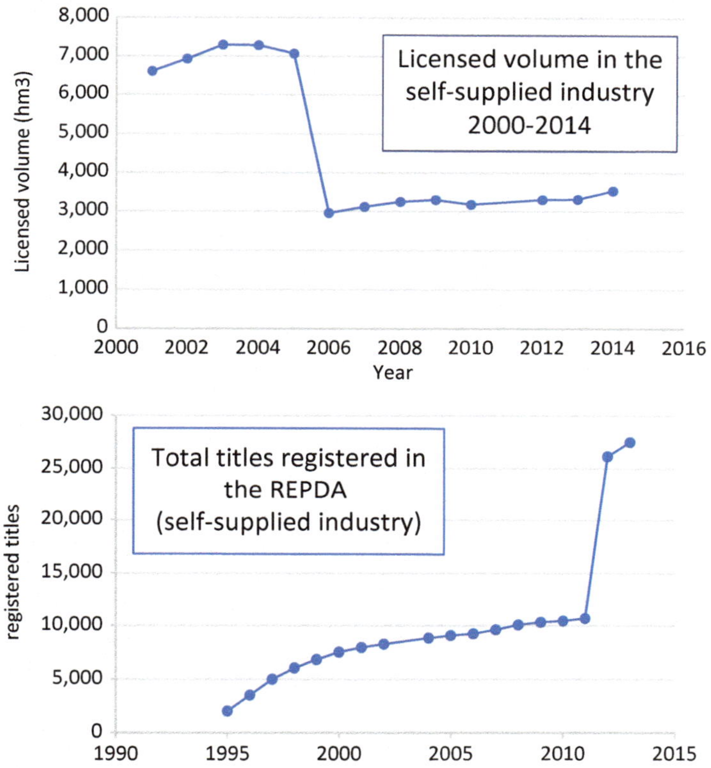

Fig. 5.1 Licensed volume for industry (Source: CONAGUA 2016b)

in water volume licensed for self-supplied industry were Northeast, Southern Pacific, Lerma-Santiago-Pacific, and Yucatan Peninsula. On the other hand, in the Northern Gulf region, the volumes of concession remained almost constant, while regions like California Peninsula, Central Gulf, and Waters of the Valley of Mexico have shown a trend toward the reduction of licensed volumes (Table 5.2).

According to data from CONAGUA, Mexico receives around 1.5 million hm^3 of rainfall water per year of which 71.6% is estimated to perform evapotranspiration and return to the atmosphere (CONAGUA 2012); 22.2% drains to rivers and streams, where additional inputs and outputs from neighboring countries flow, and the remaining 6.2% infiltrates recharging aquifers. Therefore, Mexico has about 471,500 hm^3 of renewable fresh water per year, which is also known as mean natural availability (DNM). From that, CONAGUA has licensed more than 266,559 hm^3 to all economic sectors in the country, where 85,644 are intended for consumptive use (CONAGUA 2016c).

Based on the DNM values, concession titles are distributed or assigned, and the tariff structure of water is formed. If, as an example, water balance in an aquifer is favorable or in surplus, CONAGUA can still grant concession titles. Of the

Table 5.2 Licensed volumes (hm^3) by hydrological regions, self-supplied industry

Region	2001	2002	2003	2004	2005	2006	2007	2008	2009	2010	2012	2013	2014
I Baja California Peninsula	217	281	282	283	284	85	88	91	95	95	96	96	96
II Northwest	32	31	39	41	52	61	79	91	91	91	117	121	125
III Northern Pacific	68	64	65	60	60	61	61	58	58	56	57	57	59
IV Balsas	3406	3403	3402	3401	3423	266	269	227	220	217	219	221	304
V Southern Pacific	13	13	13	14	16	16	21	18	20	21	26	26	23
VI Rio Bravo	277	282	287	288	317	202	203	205	213	211	218	212	215
VII Central Basins of the North	106	106	104	105	108	73	58	61	64	81	84	85	99
VIII Lerma-Santiago-Pacific	331	344	352	367	393	386	402	411	446	450	492	505	504
IX Northern Gulf	203	221	539	528	532	462	461	467	466	470	471	473	469
X Central Gulf	1446	1673	1675	1677	1364	876	877	876	878	725	722	691	782
XI Southern Border	84	80	77	61	48	48	95	103	104	101	106	105	114
XII Yucatan Peninsula	152	157	165	171	194	247	320	445	503	517	551	573	599
XIII Waters of the Valley of Mexico	284	285	308	302	295	187	199	211	162	168	166	173	172
Total	**6619**	**6940**	**7308**	**7298**	**7084**	**2971**	**3133**	**3264**	**3320**	**3202**	**3325**	**3338**	**3562**

Source: CONAGUA (2016c)

Table 5.3 Annual average runoff and recharge by hydrological region

HAR	Total surface mean natural runoff (hm^3/year)	Total recharge of aquifers (hm^3/year)	Renewable water (hm^3/year)
Waters of the Valley of Mexico	1112	2357	3469
Baja California Peninsula	3341	1658	4999
Central Basins of the North	5745	2320	8065
Northwest	5073	3251	8324
Rio Bravo	6857	5900	12,757
Balsas	17,057	5842	22,899
Northern Pacific	22,650	3290	25,940
Northern Gulf	24,146	3969	28,115
Yucatan Peninsula	4541	25,316	29,857
Southern Pacific	30,800	1551	32,351
Lerma-Santiago-Pacific	26,005	9749	35,754
Central Gulf	90,419	4705	95,124
Southern Border	141,128	22,718	163,846
Total	**378,874**	**92,625**	**471,499**

Source: CONAGUA (2016c)

13 hydrological regions, Waters of the Valley of Mexico region is the one with less mean annual availability (renewable water; see Table 5.3), which added to water demand is the one with the greater pressure.

5.3 Declared Volumes for Payment of Rights for Water Extraction, Use, and Catchment per Hydrological Region

Declared volumes are those that users of natural waters or owners of a concession title reported in a given period. Consistent with the extracted volumes and geographical location of users, differentiated tariffs are set for water availability zones per municipality.[3]

According to data from CONAGUA (2016d), the volumes of water declared by the self-supplied industry during the period 1999–2015 had no significant variations,

[3]Reforms to the National Water Law in 2014 established new availability zones repealing the existing ones; therefore zoning is established in availability zones per aquifer or basin.

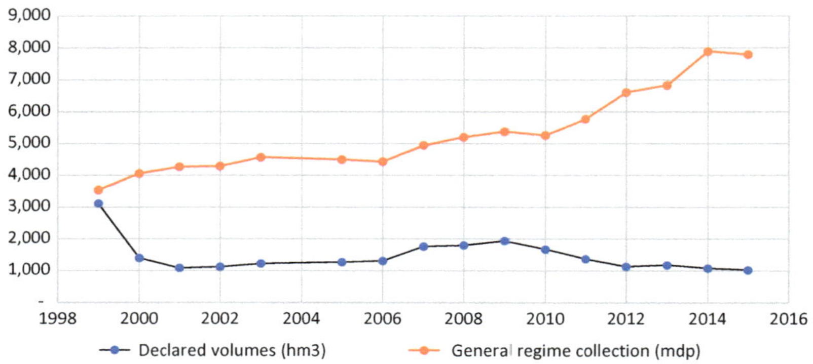

Fig. 5.2 Declared volume and collection amount for rights payment of the self-supplied industry (Source: Own elaboration based on CONAGUA (2016a, d)

with an almost horizontal behavior, except for data registered in 1999 when the declared volume was the highest. If the declared volume of water data is compared to data from water consumption collection in self-supplied industry, it can be noted in Fig. 5.2 that during the considered period, collection increased 121%, while the declared volume decreased in 67%. In other words, declared volumes decreased 4.80% per year, and collection increased 5.6% per year.

In percentage terms, the declared volume in 2005 in the industrial sector represented 1.02% of the nationally declared total volume; in 2010 this amount was 1.18%, and in 2014 it was 0.76%. The evolution of declared volumes per hydrological region shows that Central Gulf, Río Bravo, and Lerma-Santiago-Pacific regions present more frequently larger volumes of declared or consumed water. Meanwhile, the main regions with larger collection are Rio Bravo, Lerma-Santiago-Pacific, and Waters of the Valley of Mexico. Total collection of the three regions represented in 1999, 62% of the total; in 2005 this amount was 57% and in 2014 it was 54%.

5.4 Availability Zones and Rates

Availability zones determine water pricing in Mexico. Until 2013 it was divided into nine areas defined by municipal boundaries. Guerrero García Rojas et al. (2015) present an analysis of the pricing structure for water in nine availability zones in Mexico. In 2014, with the amendment of the National Water Law (CONAGUA, 2014), the availability zones were constituted in four zones whose limits are defined in two types: (1) rates for the extraction of superficial waters are structured according to availability levels in each hydrological basin; (2) rates for exploitation of underground waters are defined by the existing water availability in each aquifer. Guerrero García Rojas et al. (2018) in Chap. 7 describe the topic of water pricing as a policy and economic instrument for the management of water resources, addressing the current structure of water price related to availability water zones.

5.5 Water Sustainability Indicator in the Industrial Sector

Data from the National Institute of Statistics and Geography (INEGI, 2009a) registered, according to the 2009 Economic Census, more than 3.7 million economic units that reported water consumption volumes in their production processes.[4] Economic Census in 2014[5] presented updated values where the number of economic units registered is 4,230,745. Note that figures presented by INEGI regarding water consumption are measured as the relation of the monetary value by its consumption. The monetary value that each economic unit pays for water consumption in cubic meters depends of the area where it is situated according to a regionalization by availability zones. This limits the analysis since, for instance, several economic units of the beverage industry are located throughout the country where rates for water consumption are diverse. This could lead to a mistaken analysis in the sense that not necessarily the one who pays more is the one with a larger consumption.

According to INEGI (2009b), the value of water consumption for all economic activities in 2009 was more than 25,283 million pesos. In 2014, this value represented more than 35,182 million pesos, a 39.2% increase compared to 2009.

Analyzing the value of water consumption in the eight industrial sectors considered by Guerrero García Rojas (2005), note that the food industry (which includes the sugar industry)[6] is the one that presents the highest value of water consumption, which in 2009 was more than 1759 million pesos, and in 2014 this value reached more than 2161 million pesos, a 23% increase compared to the previous period. Nationwide, the value of water consumed in this sector represented 6.14% in 2014 and 6.95% in 2009 regarding the total value. Taking 2009 data and dividing the value of the total water consumption in the food industry with the total number of economic units, the mean value for water consumption is 12 thousand pesos per economic unit; in 2014 this figure reached 23 thousand pesos per economic unit.

The chemical industry is the second largest water consumption industry. In 2009, this value exceeded 1094 million pesos (4.32% related to the value of water consumption of all national economic activity), and in 2014 it was more than 1712 million pesos (4.86% regarding the national total). The average value of water

[4]It does not mean that water was used as input in the final product of the economic unit.

[5]Economic Censuses in Mexico are conducted every 5 years, so for this publication, 2014 Census is the most recent one; therefore the analysis in this chapter includes a period of study up to 2014. The previous Economic Census is from 2009, and the following will be published in 2019.

[6]For the integration of data from 2010 Economic Census, the sugar industry is added to the food industry. The latter limits the water consumption analysis in this sector.

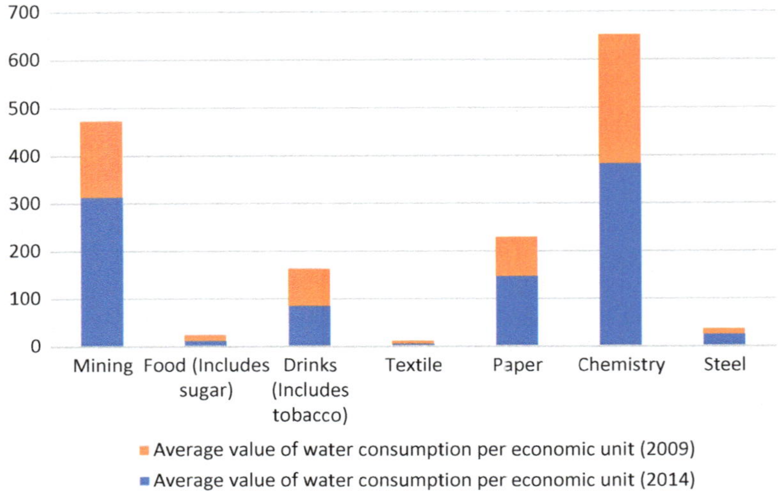

Fig. 5.3 Average value of water consumption ($ in pesos) per economic unit and by sector (2009–2014) (Source: Own elaboration based on INEGI 2014)

consumption per economic unit was 268 thousand pesos in 2009, and in 2014 this value rose to 384 thousand pesos per economic unit.

Regarding the value of water consumption in the steel industry, in 2009 it represented more than 735 million pesos which equals 2.9% of the water consumed value of all economic activities in the country. In 2014, this value was 1,809.78 million pesos, which equals 5.14% of the total value of the water consumed by the economic activity. When calculating the average value of water consumption per economic unit, we found that in 2009 this value was 11 thousand pesos per economic unit and in 2014 it was 26 thousand pesos.

Figure 5.3 shows that the economic units with a larger value of water consumption average are those related to chemical, mining, paper, and beverage industries. The average value of water consumption in the chemical industry was 384 thousand pesos per economic unit; in the mining industry, this value was 314 thousand pesos; in the paper industry, this value was 147 thousand pesos per economic unit, and in the beverage industry this value was 86 thousand pesos per economic unit.

Unlike Fig. 5.3 that represents the average value of water consumption per economic unit, Fig. 5.4 shows the average value of water consumption per industrial sector with respect to the national total, where the food sector presents the main value and more than 171 thousand economic units are located, 137% more than in the textile sector.

Regarding water uses directly used as input, the sector in which water use contributed in a greater amount to the generated added value was the textile industry with 2.66%. In the steel sector, this value was 2.19%, and in the paper industry, it was 1.62% (Table 5.4). In other words, water contributed on average 1.44% to the total production value in the considered industries. Seemingly, in the chain of

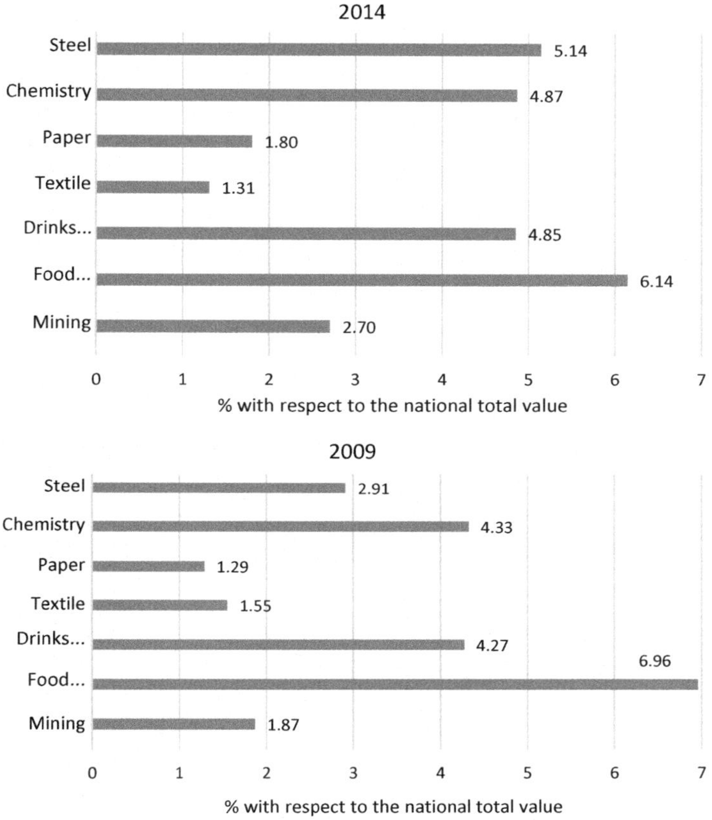

Fig. 5.4 Percentage water consumption by industry with respect to the national total (Source Own elaboration based on INEGI 2014)

production, the value of the water resource is not significant since both in terms of cost (per water consumption) and added value (the value that water adds to the product) are minimal.

5.6 Water Consumption in Five Economic Sectors

The water consumption analysis per volume in the industries is more interesting in the sense that it provides more precise information regarding actual water consumption. Even though data presented in Table 5.5 does not include a total of analyzed industries or a total of the activity branches of the included industries, values can be

Table 5.4 Value of water consumption in industry, main economic activities, 2009–2014

Industry	2014 Economic units	Water consumption (millions of pesos)	Water consumption at added value generated (%)	2009 Economic units	Water consumption (millions of pesos)	Water consumption at added value generated (%)
Mining	2,940.00	933.89	0.79	2,832.00	466.00	0.88
Food (includes sugar)	171,369.00	2,161.47	0.67	144,104.00	1,759.12	0.79
Drinks (includes tobacco)	19,937.00	1,707.18	1.40	13,888.00	1,080.78	0.96
Textile	72,357.00	460.34	2.66	66,555.00	392.46	2.35
Paper	4,302.00	633.41	1.62	3,966.00	326.22	0.89
Chemistry	4,460.00	1,712.38	0.74	4,084.00	1,094.23	0.42
Steel	70,380.00	1,809.78	2.19	65,019.00	735.94	0.89

Source: Own elaboration based on INEGI (2014)

Table 5.5 Volume of consumed water from five industrial activities (1999–2004–2009)

Industry	1999			2004			2009		
	Economic units	Water consumption (hm^3)	Production value (millions of pesos)	Economic units	Water consumption (hm^3)	Production value (millions of pesos)	Economic units	Water consumption (hm^3)	Production value (millions of pesos)
Food	2027	7.14	44,663.80	70,093	24.18	173,102.07	85,577	7.78	257,538.41
Drinks	13,416	202.17	48,705.81	17,657	89.97	155,337.26	21,768	377.25	223,687.13
Chemistry	431	0.32	19,221.23	546	5.74	41,518.27	516	0.05	51,116.29
Paper	282	17.99	22,802.39	124	49.14	31,330.54	340	1.95	46,951.17
Mining	10,116	8.57	8,547.31	23,030	6.16	84,927.16	28,082	1.68	61,335.83
Total	**26,272**	**236.19**	**143,940.54**	**111,450**	**175.19**	**486,215.30**	**136,283.00**	**388.71**	**640,628.83**

Source: SEMARNAT (2015)

useful for the analysis of efficiency in water consumption. This table contains data from five industries, which were integrated from a total of 29 types of economic activity out of 112 that compose these 5 industries.

Regarding consumed water volumes, it is noted that the beverage industry is the main water consumer. In 1999, its consumption represented more than 202 hm^3, which is equivalent to 7% of the total consumption of the industry in 1999 and 18% if it is compared to the consumption in 2014.[7] In that year, the value of the production of the industry was 48,705.81 million pesos, which was the production of 13,416 economic units, where the average water consumption per economic unit was 15,069 cubic meters.

In 1999, the volume of the total water consumed in the economic units referred to in Table 5.5 was 236.19 hm which is equivalent to 7.60% of the total water volume declared by the industries (INEGI, 1999). In that same year, the sum of the value of the production of those industries was more than 143,940 million pesos. The beverage industry contributed with more than 33%, while the food industry contribution was 31%. In the latter, the consumed volume was 7.14 hm^3. In 2009, the water consumption of the beverage industry was more than 377 hm^3, which is equivalent to 19.5% of the total water declared by industries in that same year.

Values in the table show significant variations regarding water consumption. For instance, the beverage industry reported in 1999 more than 202 hm^3 of consumed water, while in 2004 this value was 89.97 hm^3, 124% less than in the previous period. However, the following consumption period rose to 377 hm^3, 319% more than the preceding period.

5.7 Water Demand in the Industrial Sector: An Approximate Analysis

In order to establish an approximate analysis on water rates per hydrological region, the rates applied in 2013 will be considered as baseline year. Likewise, the median shall be taken as a central trend measure, because applied rates in the baseline year present a normal distribution. In order to proceed with the calculation of water consumption in the industrial sector of the seven economic activities with the largest consumption rate shown in Table 5.4, it is considered as an assumption that the total value of water consumption is the product of water consumption according to current rates per cubic meter in the corresponding period. Hence, its inverse would be the result of water consumed in that very same period. The main assumed identity is that the collection or the total value of water consumption is a function of the rate per square meter by the total declared or consumed volume:

[7]Total volume consumed is the declared volumes for the payment for extraction, use, or catchment rights.

Table 5.6 Approximate volumes of water consumption in industry nationwide (hm^3)

Industry	2014 Economic units	2014 Water consumption (hm^3)	2014 Water consumption per employed person (m^3)	2009 Economic units	2009 Water consumption (hm^3)	2009 Water consumption per employed person (m^3)
Textile	72,357	52	338	66,555	44	346
Paper	4,302	71	653	3,966	37	363
Mining	2,940	105	1,120	2,832	52	761
Drinks (includes tobacco)	19,937	192	1,140	13,888	122	742
Chemistry	4,460	193	745	4,084	123	528
Steel	70,380	204	1,063	65,019	83	606
Food (includes sugar)	171,369	243	277	144,104	198	238
Total	**345,745**	**1060**		**300,448**	**659**	

Source: Own elaboration based on INEGI (2014)

$$\textit{Value of water consumption} = \text{rate per m}^3 * \text{total volume consumed}$$

Therefore, the water volume consumed by the industry was about 1060 hm^3 in 2014. In that year, food industry reported greater water consumption with 243 hm^3, followed by the steel industry with 204 hm^3. In fact, 78.5% of water volumes consumed corresponded to food, steel, chemical, and beverage industries (Table 5.6).

On the other hand, in 2009 the total volume consumed was 659 hm^3, that is, 61% less than in 2014. In general, this would mean annual increases in water consumption.

Nevertheless, comparing Table 5.6, Fig. 5.2, and Table 5.5, there are some discrepancies regarding total volumes consumed in those compared years. For instance, data reported in Fig. 5.2 indicate that the water volume consumed in the industry in 2014 was 1082 hm^3, while data in Table 5.6 report a consumption of 1062 hm^3, a slightly significant difference. However, comparing data from 2009, the total volume of water consumed as indicated in Fig. 5.2 shows 1939 hm^3 consumed, while data from Table 5.6 present a consumption of 659 hm^3, A 1280 hm^3 difference. Also, we will find discrepancies by comparing the abovementioned data with those from Table 5.5, where, obviously, these differences are explained by incomplete data. Likewise, these differences could be explained as the result of data aggregation, that is, industries classified by CONAGUA as self-supplied industries are significantly different from the classifications of INEGI. On the other hand, it is important to note that these differences in the aggregated values could be the result of variations in fees among availability zones, where the consumed and collected amounts that resulted from water consumption in each availability zone are uncertain. It is possible to have a zone with a high demand and low rate and vice versa. However, these values give a rough idea about the functioning of water demand management with a rate structure policy based upon availability zones.

5.8 Water Rates Based on Hydrological Regions

Considering data that resulted from the collection for the payment of water consumption rights according to hydrological regions, Table 5.7 was created, and it shows the average price of water per hydrological region. According to this data, in 1999 the hydrological regions with the highest price per cubic meter for water consumption were the Baja California Peninsula region (I), 6 pesos per cubic meter, and Central Basins of the North (VII), 6.7 pesos per cubic meter.

In 2003, the hydrological region Central Basins of the North (VII) registered the highest average in water fees. Upon reviewing the total average of prices, a linear trend is noted with certain variations between 2005 and 2009 (Fig. 5.5).

Data registered in 2014 show that Lerma-Santiago-Pacific region presents the highest price per cubic meter (11 Mx pesos/m^3). The trend line of rates per hydrological region is presented in Fig. 5.6, where the black and red dots show the highest and lowest rates, respectively, during the 1999–2014 period.

Table 5.7 Water average tariff per m^3 per Hydrological Administrative Region (Mexican pesos)

HAR	1999	2000	2001	2002	2003	2005	2006	2007	2008	2009	2010	2011	2012	2013	2014
I	6.00	6.30	6.20	8.20	3.90	5.70	4.20	1.20	7.00	1.20	8.20	10.50	10.60	10.10	9.80
II	2.10	2.20	2.50	2.60	2.40	2.70	2.00	0.90	4.70	6.80	5.90	4.30	6.70	8.80	8.10
III	3.10	3.40	3.90	3.80	2.60	3.70	2.70	2.70	0.70	5.40	5.90	3.60	6.40	7.00	8.20
IV	1.20	1.00	2.60	3.30	2.60	2.90	3.60	1.80	1.00	2.80	3.00	4.30	3.10	2.90	4.10
V	2.60	2.10	7.30	5.50	3.90	5.60	5.50	4.80	2.00	2.30	8.00	6.90	6.70	5.60	7.00
VI	5.50	5.40	5.10	6.60	6.80	5.90	5.30	4.60	5.00	1.90	3.00	1.40	9.60	8.00	9.20
VII	6.70	8.10	8.80	9.70	16.60	7.80	9.20	5.30	5.30	9.40	7.20	8.10	8.50	11.20	5.30
VIII	5.20	5.70	5.10	4.60	5.00	5.90	6.30	4.80	3.50	8.80	7.80	9.10	5.60	10.00	11.00
IX	1.70	2.10	2.50	2.30	3.00	2.40	2.30	1.80	2.30	4.80	3.50	3.50	3.40	3.20	3.60
X	1.60	0.90	1.60	1.30	1.10	1.10	1.10	0.90	1.00	0.50	0.70	2.70	1.90	2.20	2.50
XI	3.40	3.30	3.80	4.60	3.70	2.10	2.10	1.20	0.70	3.90	1.20	1.80	1.50	1.50	1.70
XII	1.20	2.10	1.50	2.10	1.20	1.20	0.90	0.50	2.80	3.20	3.90	4.20	4.10	3.30	4.60
XIII	0.30	5.20	6.10	6.60	7.50	7.30	6.70	6.00	7.40	10.80	8.20	7.90	8.70	6.60	10.20
Average tariff	3.12	3.68	4.38	4.71	4.64	4.18	3.99	2.81	3.34	4.75	5.12	5.25	5.91	6.18	6.56

Source: Own elaboration

Average water price per hydrological region is the result of dividing the collection from extraction, use, or catchment of national waters and the declared volumes for the payment for extraction, use, or catchment rights. It is considered as a price because it is the monetary amount charged for the consumption of cubic meter of water. The water cost would be the amount that users pay for the consumption of certain amount of water given the prices per cubic meter

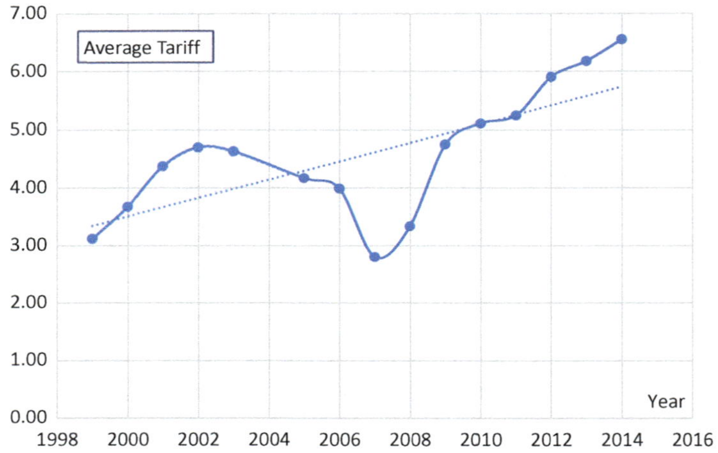

Fig. 5.5 Evolution of the average tariff per cubic meter of water (Source: Own elaboration)

Fig. 5.6 Average trend line of water tariff 1999–2014 per Hydrological Administrative Region (Source: Own elaboration)

Hydrological Administrative Region		Trend line
I	Baja California Peninsula	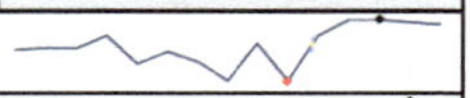
II	Northwest	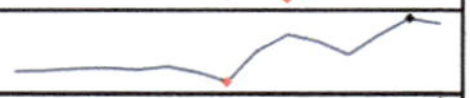
III	Northern Pacific	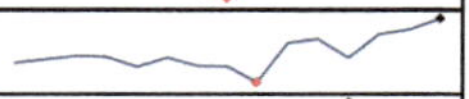
IV	Balsas	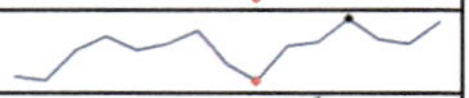
V	Southern Pacific	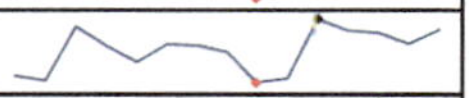
VI	Rio Bravo	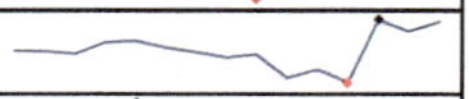
VII	Central Basins of the North	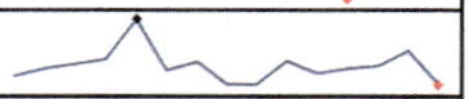
VIII	Lerma-Santiago-Pacific	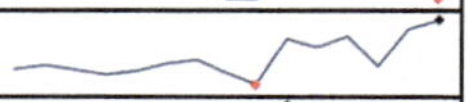
IX	Northern Gulf	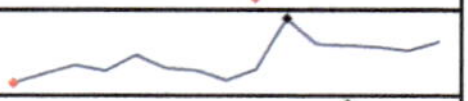
X	Central Gulf	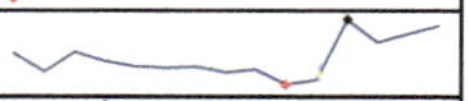
XI	Southern Border	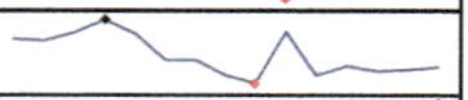
XII	Yucatan Peninsula	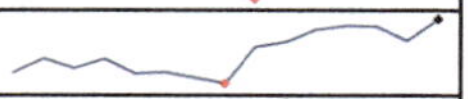
XIII	Waters of the Valley of Mexico	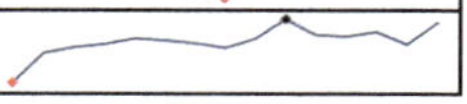

Logically, regions with higher water pressure would present increasing rates, as the hydrological region of the Metropolitan Zone of the Valley of Mexico, but the region of the Central Basins of the North shows greater variations downward even though it is considered as a high water stress region. Obviously, the effects of averaging the added values of collection per hydrological region should be considered, since rates vary according to availability zones, in this case, defined according to municipalities until 2014. If we consider these figures in a lower territorial scale, as an example, municipalities, the map of water rates or prices would present a more detailed picture regarding water consumption or demand and its relation with the existing prices in that area.

With the abovementioned data of rates and water consumption volumes, the following sections will analyze if the water policy around the management demand generates positive effects of efficiency of water consumption in the industrial sector. That is, to determine as far as possible, whether the rates structuring per availability zones make water use more efficient.

5.9 Price Elasticity of Water Demand

In order to determine the variations of demand regarding prices, Table 5.8 presents data of elasticity price of water demand.[8] Values considered for the table take into account percentage changes of water prices and percentage changes of declared water volumes during the period 1999–2014.

Observing the elasticity price of demand of users located in the Baja California Peninsula hydrological zone (I), we realize that in 2001, the elasticity price value was $3.10 > 1$; this would mean a change in demand toward price changes. From 1990 to 2000, the average water price goes from 6 to 6.3 Mx pesos per cubic meter; in that same period, the demand (declared volume) shifts from 5.9 to 6.7 hm^3, which responded to an 11% increase in demand in spite of a 4% increase in price. We could really conclude that there was no significant variation in price as well as in demand.

In fact, the comprehensive data analysis seems to indicate that a change in price per cubic meter of water apparently does not affect water demand, or rather the demand acts in a multivariate manner. Perfectly elastic or inelastic demands may be found. It is elastic in the sense that region or zone rates would force economic units in the industrial sector to move toward those with a lesser price. Or, it is inelastic when as a result of the conditions or the type of activities conducted the industry is not able to easily move toward other lesser price sectors. Or, in other cases, elasticity

[8] According to the definition of elasticity price of the demand, inelastic or relatively inelastic demand is when the elasticity price of the demand is between zero and -1; in this case, the amount varies, but in a relatively slight manner; it is an elastic or relatively elastic demand, when the elasticity price of the demand is between -1 and $-\infty$. In absolute terms, when elasticity price of the demand is more than 1, the latter would mean that the demand strongly reacts toward a change in price; on the other hand, it is said that the demand is perfectly elastic when the demand is horizontal; an infinitesimal change in price would cause an infinitely large change in the demanded amount; the perfectly inelastic demand is when the demanded amount remains as a constant even with a price change and the elasticity price of demand equals zero, and it is said that the good has a perfectly inelastic demand.

Table 5.8 Price elasticity of demand

HAR	2000	2001	2002	2003	2005	2006	2007	2008	2009	2010	2011	2012	2013	2014
I	3.10	2.78	–0.29	–0.79	–0.87	–1.00	–1.00	–0.90	0.18	0.20	0.10	–19.46	–0.32	–4.91
II	2.01	–0.29	1.67	–2.71	0.91	–1.00	–0.77	–0.49	–1.75	–0.31	–1.24	0.27	0.04	–0.60
III	0.94	–1.15	1.77	–0.61	–0.50	–1.00	–21.39	–1.04	–0.93	–1.81	–1.20	–1.22	–0.08	–0.85
IV	–1.83	–1.01	–0.79	–1.54	–0.79	–1.00	–0.80	–1.32	–1.22	4.24	–0.42	–0.76	–2.61	–0.91
V	–2.96	–0.20	0.10	–1.32	–1.72	–1.00	–1.12	–1.05	0.09	–0.96	–0.42	–2.54	–0.59	0.15
VI	1.87	0.94	–1.16	–1.37	0.39	–1.00	–2.15	0.25	–1.07	–0.96	–0.85	–0.92	–1.69	–0.81
VII	0.53	–0.90	–1.71	–0.65	–0.95	–1.00	–0.76	–12.12	–1.20	–1.66	–2.27	1.94	–0.36	–0.95
VIII	0.40	–1.90	–1.14	0.62	–1.20	–1.00	–0.99	–1.36	–0.95	–1.61	–0.92	–0.98	–0.76	–0.59
IX	0.74	–0.55	–0.74	–1.32	–2.17	–1.00	–1.00	–0.83	–1.03	–1.16	–3.62	–3.41	–1.51	–1.20
X	–1.09	–1.18	–1.48	–0.92	–0.66	–1.00	–0.85	–3.04	–1.59	–1.19	–0.97	–0.62	0.25	–0.59
XI	–18.29	–0.18	–0.05	–0.01	–0.23	–1.00	–0.53	–0.45	–0.39	–0.15	–0.13	0.03	–0.21	0.03
XII	–0.07	–1.99	–0.84	–1.11	0.68	–1.00	–1.00	–0.98	–2.19	0.08	–2.68	–3.59	–1.18	–1.01
XIII	–0.98	–0.36	–1.02	–0.40	–3.02	–1.00	–0.78	–0.69	–0.85	–0.95	3.39	0.96	–1.43	–0.84
Total	**–0.87**	**–0.84**	**–1.17**	**–3.44**	**–0.68**	**–1.00**	**–0.95**	**0.20**	**–2.91**	**–0.81**	**–1.12**	**–0.68**	**0.45**	**–0.72**

Source: Own elaboration

equals the unit, which would mean constant changes, in the same proportion, along the demand curve.

A simple linear regression model, which considers time series data from 1999 to 2014 period, was defined to analyze changes in the demand from the previously described identity, where:

$$Value\ of\ water\ consumption\ (PT) = rate\ per\ m^3(p) * total\ volume\ consumed\ (D)$$

Defining the following identity, where:

$$Total\ volume\ consumed\ (D) = \frac{Value\ of\ water\ consumption\ (PT)}{Rate\ per\ m^3\ (p)}$$

So, the water demand model per region is defined as a function of the rate or unitary prices per cubic meter of water:

$$D = p$$

Table 5.9 was created from above, where variables show a negative slope, which indicates that as prices increase, the amount of the demand tends to decrease. Another aspect to consider is the values of the coefficient, which has a positive value that reflects the amount of water demand when the price is zero. Even when presenting consistent coefficient values, the adjustment level in most of the models is minimal, which indicates a low D and p correlation.

Only the cases where the adjustment is above 50%, that is, hydrological regions IV, VI, VII, X, XI, and XIII, precisely the ones with a higher degree of water pressure, have been considered for analysis purposes.

If we consider the demand function in region IV, it is defined as:

$$f\ (D) = -43.48p + 240.77$$

Then:

$$f^{-1}(D) = 5.5 - \frac{1}{43.5}D$$

From the inverse function, we graphically obtain (Fig. 5.7):

According to the aforementioned data, in the case of Balsas region (IV), at a zero tariff, the demand would be above 240 hm^3 of water consumed, while the demand would be zero if the rate would be set to a price of \$5.5 or more per cubic meter. This same trend is noted in the Central Basins of the North, Central Gulf, and Waters of the Valley of Mexico regions, which are the areas where larger rates for water consumption are registered.

The demand function for the hydrological region Waters of the Valley of Mexico is:

Table 5.9 Linear regression models of water demand

HAR	Variable/constant	Coefficient	*t*	R-square	Standard deviation
I	Variable (*p*)	−1.0622	−1.99	0.2200	0.5345
	Constant	16.7108	4.21		3.9695
II	Variable (*p*)	−0.9576	−0.35	0.0088	2.7134
	Constant	63.1185	4.57		13.8242
III	Variable (*p*)	−5.6764	−2.39	0.2903	2.3719
	Constant	47.2769	4.05		11.6760
IV	Variable (*p*)	−43.4889	−4.43	0.5839	9.8120
	Constant	240.7723	8.31		28.9653
V	Variable (*p*)	−3.7221	−3.20	0.4219	1.1645
	Constant	43.7244	6.79		6.4425
VI	Variable (*p*)	−38.4864	−4.71	0.6135	8.1648
	Constant	393.3994	7.71		51.0012
VII	Variable (*p*)	−5.5164	−4.64	0.6064	1.1877
	Constant	103.2018	9.96		10.3668
VIII	Variable (*p*)	−14.5911	−3.22	0.4252	4.5337
	Constant	290.8476	8.88		32.7706
IX	Variable (*p*)	−18.4571	−3.46	0.4615	5.3287
	Constant	151.9204	9.55		15.9128
X	Variable (*p*)	−212.6974	−4.17	0.5534	51.0655
	Constant	667.7225	8.07		82.7750
XI	Variable (*p*)	−8.0549	−3.87	0.5173	2.0794
	Constant	76.1435	13.80		5.5177
XII	Variable (*p*)	−21.8219	−3.73	0.4983	5.8519
	Constant	109.2455	6.37		17.1487
XIII	Variable (*p*)	−158.5277	−4.48	0.5887	35.4145
	Constant	1398.7340	5.19		269.5214
Total	**Variable (p)**	**−284.4131**	**−4.57**	**0.5990**	**62.1883**
	Constant	**2550.3640**	**10.15**		**251.2094**

Source: Own elaboration

$$D = -159p + 1399$$

where $f^{-1}(D)$ is:

$$p = \frac{D - 1399}{-159}$$

Figure 5.8 shows that the demand reaches its peak around 1400 hm^3, when the price equals zero and the maximum price that makes the demand reduce to zero equals \$8.79/m^3. In both graphs (5.7 y 5.8), we can understand that, facing minor price adjustments, both of them would show large variations in the amounts of water

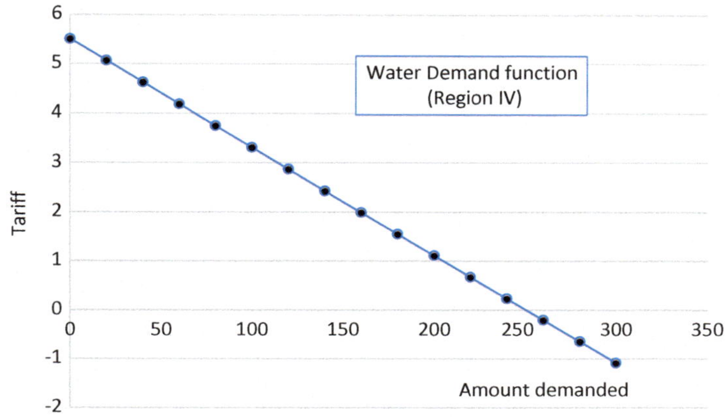

Fig. 5.7 Demand function of region IV (Source: Own elaboration)

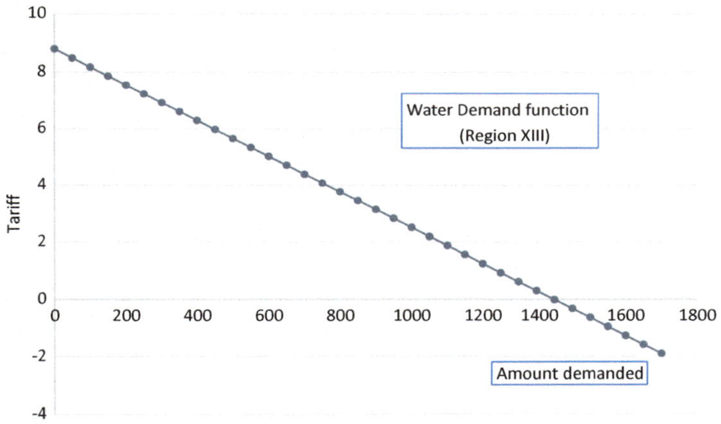

Fig. 5.8 Demand function of the region Waters of the Valley of Mexico (Source: Own elaboration)

demanded. However, considering that the goodness of fit is fairly acceptable, we could not establish clear conclusions around the dynamics of tariff and water consumption in the region. What we can assert is that we found no clear relation between tariff and price. Therefore, if we take into account the information given in the previous sections, it seems that the tariffs do not reveal a direct or inverse influence on water consumption.

In summary, considering the analogy regarding the elasticity price of demand along with the model of water demand per hydrological region, it may seem that tariffs according to availability zones are not the most appropriate policies for an efficient water demand management, because a higher price not necessarily means a reduction in the demand, much less a rational use of the water resource.

5.10 Conclusion

The limited statistical information on sectorial water consumption in the industry restricts to some extent the particular conclusions about water demand per hydrological zone and availability zone, due to differences in the aggregation of data regarding the industrial water consumption between INEGI (that publishes the value of water consumption by industry and sector of the economic activity classified according to the North American Industry Classification System) and CONAGUA (that publishes data from revenues and declared volumes of self-supplied industry which is classified in the industry per se, agricultural sector, and service and commerce sector).

On the other hand, if the inspection and monitoring of water policies work properly, then the total declared water volumes would represent around 35% regarding licensed volumes. The above would mean that there is a 65% water surplus, and in some regions, this surplus could reach over 90%.

Finally, if the prices or rates differentiated by availability zones are used as allocation mechanisms to make water use more efficient in the industry, they do not show significant variations in the demand for changes in the price or rate. This can be explained, in part, because the expense for water consumption does not represent a significant variable within its cost function.

The price of water, even though it is defined by an economic principle of scarcity by the water availability zones, in fact does not impact on consumption modifications, since, as can be seen in the exercise, the price elasticity of the water demand estimated by hydrological region varies in very wide ranges, which is not reflected in low water consumption, limiting the efficient use of the water resource, not because of the economic instrument (elasticity) but as a result of the management instrument (tariff).

References

CONAGUA. (2012) Atlas Digital del Agua México.

CONAGUA. (2014). Ley de aguas nacionales. México.

CONAGUA. (2016a). Ley Federal de Derechos. México.

CONAGUA. (2016b). Registro Público de Derechos de Agua. México.

CONAGUA. (2016c). Estadísticas del agua en México. México.

CONAGUA. (2016d). NUMERAGUA. México.

Guerrero García Rojas, H. (2005). Industrial water demand in Mexico: Econometric analysis and implications for water management policy. Tesis de Doctorado. Universite de Toulouse 1 – Sciences Sociales.

Guerrero García Rojas, H., Gómez Santiz, F. & Rodríguez, J. R. (2015). Water pricing in Mexico: Pricing structures and implications. In A. Dinar, V. Pochat & J. Albiac Murillo (Eds.), *Water pricing experiences and innovations* (Global Issues in Water Policy 9). DOI 10.1007/978-3-319-16465-6_2. Springer International Publishing Switzerland.

Guerrero García Rojas, H., García Vega, D., & Herrera Torres, H. (2018). Chapter 7: Water price policy, Its institutional role as an economic instrument for water management. In H. Guerrero-

Garcia-Rojas (Ed.), *Water policy in Mexico: Economic, Institutional and environmental considerations* (pp. XXX--YYY). Cham: Springer.
INEGI. (1999). Censos Económicos 1999, Industria Manufacturera, Materias Primas y Productos.
INEGI. (2009a). Censos Económicos 2009. Aguascalientes.
INEGI. (2009b). Panorama censal de los organismos operadores de agua en México. Censos Económicos 2009. Aguascalientes.
INEGI. (2014). Censos Económicos 2014. Aguascalientes.
SEMARNAT. (2015). Compendio de Estadisticas ambientales. México.

Hilda R. Guerrero García Rojas is a Professor of Economics and Environment in the Facultad de Economia, Universidad Michoacana de San Nicolas de Hidalgo (UMSNH), Morelia, Michoacán, Mexico. She holds a PhD in Economics from the Toulouse School of Economics. She is among the leading water policy experts in Mexico and contributed to policy work in Mexico. Her PhD thesis focused on "economic assessment of policy interventions in the water sector; industrial water demand in Mexico: econometric analysis and implications for water management policy." She worked for Instituto Mexicano de Tecnología del Agua (IMTA), and as a consultant, she has been collaborating with the World Bank, Colegio de México (COLMEX), OECD, and with the Inter-American Development Bank (IDB). Publication topics involving Guerrero include "Water pricing reforms in Mexico: The case of manufacturing sector"; "Applying water demand: Case for industrial use on Mexico"; "Water in Mexico: Impact of policy interventions in the sector"; and "Water pricing reforms in Mexico: The case of the manufacturing sector", among others. Guerrero's areas of interest are natural resources and environmental economics, integrated water management (IWM), economic instruments applied to natural resources, water price and its reforms, applied economics, environmental economic valuation, micro-econometrics, climate change in water sector, and economics of climate change.

Faustino Gomez-Santiz is an Economist at the University of Michoacan and has done his Master's in Human Geography at el El Colegio de Michoacán (Colmich). Currently, he is studying his PhD in Environmental and Urban Studies at El Colegio de México. His areas of research are environmental economics assessment, ecological economics, common property resources, and neoinstitutional economy. He has published and disseminated scientific articles at international conferences on topics such as economics assessment of water, payment for environmental services (hydrologic), and community management mechanisms. He is an independent consultant for public and private companies.

Erandi Maldonado-Villalpando is an Economist from the Universidad Michoacana de San Nicolás de Hidalgo (UMSNH). She is currently a Master in Public Management of Sustainability and is in the third semester of the Master of Science in Local Development (PNPC Program – CONACYT), both programs at the Facultad de Economia, UMSNH. She has worked on issues of social and solidarity economy, as well as those of ecological economy, environmental economics, and sustainable management in natural protected areas. Before joining the master's program, she worked for the State Council of Science, Technology, and Innovation (CECTI) of the State of Michoacán, where she aroused her interest in the study of innovation, directing her research on issues related to public policies from the perspective of social innovation and social innovation networks for local territorial development.

Chapter 6
Water Utilities: Is Their Sustained Financial Efficiency Achievable? – The Mexican Case

Ricardo Sandoval-Minero

Abstract Mexico reached the Millennium Development Goals (MDGs) in terms of access to water and sanitation infrastructure, but the quality of service provision is still unsatisfactory. In face of the new Sustainable Development Goals (SDGs), whose reach is much wider and ambitious, the country faces a huge financial and technical challenge. Besides facing growing problems due to the scarcity and bad quality of water sources, the gap between demand and supply also widens because of rapid and messy urban growth in metropolitan areas, a lack of investment in infrastructure rehabilitation and extension, and limited capacities for managing and operating the water and sanitation systems. The sector's infrastructure development has relied on a centralized financial system where federal subsidies have been allocated according to operating rules, designed to induce water utilities' performance improvement, but in the end, without any linkage to improvement commitments or ex post performance assessment. From a general analysis, it appears that institutional dysfunctions and instability have impeded the sustainable operation of these services, despite the financial and technical efforts deployed; in a way, the continuous and discretional allocations of subsidies seem to reinforce negative loops of service deterioration, users' unwillingness to pay or collaborate, political interference, and external "rescue" of failing systems. In the present context of financial scarcity, climate variability, impacts of global commercial trends on local resources, and social inequality, Mexico could finally be forced to undertake a profound institutional reform to give water services the stability, professionalism, resilience, and creditworthiness they need, to attract alternative financial resources and ensure a more sustainable operation. Such a reform faces huge cultural and legal obstacles, but the stakes leave no space for inaction.

Keywords Sustainable development goals · Urban water · Water utilities · Financial performance

---

R. Sandoval-Minero (✉)
Sextante Consultores SC., Mexico City, Mexico
e-mail: rsandoval@sextanteconsultores.mx

H. R. Guerrero García Rojas (ed.), *Water Policy in Mexico*, Global Issues in Water Policy 20, https://doi.org/10.1007/978-3-319-76115-2_6

6.1 Introduction

In Mexico, almost eight of every ten people live in urban localities (INEGI 2010), while 73% of the total gross production[1] is generated in 56 metropolitan areas (ONU HABITAT 2011). Despite an apparent high water and sanitation coverage, Mexican cities face growing problems for providing potable water with proper continuity and quality, replacing aging infrastructure and supporting urban expansion.

Federal, state, and municipal authorities have set up different measures in order to reinforce the resource's preservation, build new infrastructure, improve technical and economic efficiencies, create public awareness, and set up accountability and transparency frameworks. In a context of climate variability, growing water scarcity and financial limitations, a lot remains to be done, especially to set up better institutional arrangements which could lead urban systems toward higher quality and performance instances, attract alternative financial sources, and ensure a more sustainable and resilient operation.

In this chapter, an overall assessment of the performance of urban water and sanitation systems is presented; a call for setting up a new model is made, consistent with the need to guarantee equal access to water to urban and rural population while preserving the environment, to reach the Sustainable Development Goals (SDGs).

6.2 Overall Assessment: Urban Water and Sanitation Problems and Capabilities in Mexico

6.2.1 State of the Problem

Mexico is a federal republic, with an extension of 1964 million square kilometers; 76.8% of its population is considered "urban" (living in towns with more than 2500 inhabitants). Water and sanitation coverage in urban areas is relatively high, and a huge effort has been made to build wastewater treatment plants and to support municipal utilities with financial and technical resources to develop new water sources, improve technical and commercial efficiencies, and extend the services coverage. Some of the utilities serving bigger cities perform with acceptable indicators and are based on best international practices. Nevertheless, there are big challenges still to be resolved:

- In 30 years, the pollutant load from point sources has increased in 42%, while groundwater, which accounts for 75% of water supply for public uses, is not

[1]Total gross production means the value of all the goods and services commercialized by the economic units of a municipality. Gross domestic product is not reported for the municipal or city scale (ONU HABITAT 2011, p. 39).

being properly protected against salt intrusion, overexploitation, wastewater infiltration, and other impacts (Aboites et al. 2008).

- Water availability per person has diminished because of the population growth, but in Mexico's case, this is worsened by the concentration of most of the urban, industrial, and agricultural zones of the country in the central and northern regions, where less water is available (Saltiel 2008).
- It has been reported that in Mexico, 95.7% of households have access to water supply infrastructure in urban zones (CONAGUA 2016); nevertheless, only 80% of them have the service inside the house, only 75% have reported receiving water every day (INEGI 2016), merely 62.1% have constant supply, and only 25.3% believe they can drink from the tap without getting ill (INEGI 2015).
- In urban localities, only 51.7% of the respondents of a survey considered water service to be "satisfactory" (INEGI 2015).
- At the country level, 91.0% of households have sewage or sanitation infrastructure, 96.6% in urban zones and 74.2% in rural localities (CONAGUA 2016). Still, 8% of households share their sanitary installations with other households in rural areas, and 5% do it in urban ones (INEGI 2016).
- Wastewater treatment capacity was doubled in 15 years; it is estimated that 91.5% of sewage is collected, and 57% of that volume can be treated with the existing installed capacity of 57% (CONAGUA 2016). Nevertheless, in a study performed by the Engineering Institute of UNAM in 2015, only 54% of the wastewater treatment plants (WWTPs) smaller than 100 l/s are working, but less than a half work properly. From bigger WWTPs, 82% are working, but only 41% work right. Almost 50% of WWTPs in the country are classified in the range of "awful" or "bad" performance (Morgan 2016). Among the causes of this deficiencies, the study proposes public expenditure with no proper planning, leading to hasty investment decisions, as well as design and construction mistakes, lack of resources for operation and maintenance, and excessive rotation of the personnel.

  It's estimated that more than 70% of water bodies in the country show some degree of pollution, while economic activity and public supply in numerous cities depend on overexploited aquifers and degrade environmental flows (UN-Water 2013).
- Although Mexico has one of the largest irrigated areas in the world, close to 6.5 million hectares, a large part of this infrastructure continues to operate under irrigation systems whose efficiency could be significantly improved in addition to the lack of maintenance that has damaged its capacity (CONAGUA 2015, p. 77). Almost 77% of the cultivable area is not irrigated; since the 1990s, the expansion of the area under irrigation has been very limited, since the country has concentrated on improving water productivity. At this point, the economic feasibility of new irrigation schemes depends to a large extent on the capacity of the country to achieve improvements in agronomic practices (UN-Water 2013).

All these figures suggest a common explanation: while there has been a huge financial and human effort, deficiencies in planning, resource allocation, construction and operation of the infrastructure, and distribution of responsibilities and accountability have hindered the reduction of coverage gaps and caused the loss of installed capacities for obsolescence, lack of maintenance or improper operation.

Clearly, the problem for achieving sustainable urban water services in Mexico is more complex than having budget to build water or sewage connections or to import water from other basins; urban areas require to be served from farther sources and to collect and treat wastewater in more extended areas; aquifer recharge areas are being paved and riversides invaded, while climatic variability increases the risks of flooding with more intense precipitation events.

6.2.2 Institutional Framework

In Mexico there are three governmental "orders" (with no hierarchical relationship between them): federal, state, and municipal. While the Constitution enacts federal government, through the National Water Commission (CONAGUA by its Spanish acronym), to take charge of the administration of the national waters within the country, the National Waters Law also gives CONAGUA relevant capacities in at least two important issues: first, to support financially and technically the extension of services coverage through budgetary allocation and the definition of execution rules and norms and, second, to promote efficient practices, in physical and financial processes, among public water suppliers and productive users. Within this framework, CONAGUA can also build and operate water supply and sanitation infrastructure in coordination with states and municipalities. In the other hand, since 1982, Article 115 of the Constitution gives the municipality the responsibility for the provision of water and sanitation services, with the assistance of the state if needed. This has given place to a very varied set of institutional arrangements, with some states having a wider intervention (even a direct participation as water suppliers) and others setting aside of predominantly municipal utilities. Before 1982, bigger cities had been managed as federal organisms or had a more important federal participation, since they needed bigger infrastructure works which were normally financed and built by the national government since the late nineteenth century. There is no formal mechanism for economic regulation of water services (formal and technically sound mechanisms to ensure a proper balance between price and quality of water and sanitation services), with the exception of private participation schemes. In most cases of public utilities, there is a board in charge of the government system and a group of directors in charge of the operation. The board presents a tariff proposal to the municipality (city council), which then presents a revised proposal to the State Congress; the common practice is to set an upper limit according to general inflation rate, but without analyzing the utilities' cost and technical structures, and disconnect from the investment programs. In the absence of a formal supervision, it is common that utilities operate inefficiently and lack of resources for sustaining appropriate levels of operation and maintenance, transferring the costs of this inefficacy through resource overexploitation and capital improvement deferral. In most cases, political interference in staffing, administration, public works bidding processes, and even in operational decisions blocks even the most committed staff from getting higher sustainable performances. It can be stated that along with the lack of a planned and effective control of land use, institutional instability is a key root problem, leading to unsustainable provision of water and sanitation services.

6.2.3 The Performance of Urban Water and Sanitation Systems in Mexico

There are different systems in place to collect and analyze performance data from water and sanitation utilities; none of them, though, is based on audited data but on the information provided by the operators themselves.

The Citizens Council for Water, an autonomous consultative organization set by the National Waters Law at the national level, organized in 2009 and 2010 a survey and classification effort, to try out a method of prioritization while promoting a benchmarking experience. Fifty cities were surveyed and their data analyzed and classified. Here we review some of the major findings (CCA 2011), in spite of some criticisms and possible improvements:

- Services coverage, measured in terms of household connections which make possible the supply of potable water and the discharge of wastewater, appears to be high, as reflected in national statistics; nevertheless, when it comes to continuity, numbers vary a lot, with less than 40% of the utilities reporting 100% of connections with continuous service. Sewage coverage does not match either the capacity of wastewater treatment.
- There are noticeable differences in terms of labor productivity and commercial efficiencies. It seems to be a geographical discrimination, where cities located in the northern part of the country show better indicators, with the exception of metered connections, where more dispersion is present.
- Although some utilities declare to have relatively high technical efficiencies, that is, low leakage ratios, it must be noticed that with a limited metering capacity, this figure should be questioned. The lack of appropriate measurement and invoicing of flows also hits the financial health of utilities or affects the users economy, since many of the cities not having adequate metering ratios usually charge fixed charges per household, which discourages efficient use at the home level and can cause a decrease in the available income of the families, without relationship to the cost structure of the utility and to the quality of service.
- An outstanding issue addressed by the CCA survey was a set of questions related to the existence of seven good governance practices: long-term planning, governing board, users' participation in the board, audited annual reports, management autonomy, independent tariff setting, and customer service systems. Once more, the best performing utilities seem to have also the best practices, and they are mainly located in the central and northern parts of the country.

The report concluded, from the weighted addition of all the indicators (a method partly contested by the utilities themselves), that the cities of Leon, Saltillo, Monterrey, Aguascalientes, and Cancun ranked the best five utilities in Mexico. The authors try to point out how those utilities with private sector participation seem to show consistently better indicators, but a caveat should be stated in this direction: first, most of them already had good indicators and a local tradition inclined toward efficiency and, second, the availability of a long-term arrangement including relevant federal subsidies, along with agreed tariff formulae, has provided these utilities with an income stability not usual in the sector. Nevertheless, it does seem to be a relationship between good governance arrangements and good performance, though further research is desirable.

Beyond these findings, it seems worrying that even among the best urban utilities in the country, there are important challenges in terms of their internal efficiency and functioning, worsened by the growing need of funding to bring water from new sources, protect existing ones, and collect, treat, and dispose of wastewater, with limited management and financial autonomy. Many factors which worsen the scene haven't been considered in CCA's assessment, such as the existence of impoverished neighborhoods and sections within and around some of the best performing cities, in which many users wouldn't afford much higher tariffs. The stakes for the future development of the country are huge, considering the relevance of urban economies for national wealth.

6.2.4 Public Policy Responses

As it was stated before, Mexican Constitution gives the state order the capacity to intervene in the regulation, operation, and control of municipal water supply and sanitation services where municipalities lack the capacity to assure adequate levels of service. Historically, though, the states have only taken part of public works planning and construction programs, and only recently there have been some states whose water commissions or similar organizations set up capacity building programs. There is still no formal economic regulation, although some state organisms have a role in analyzing and supporting the approval of water tariffs; states set up investment programs, along with federal ones, in one hand to support the extension of services coverage and in the other to promote the improvement of water and energy use efficiencies, as well as managerial development and training. The lack of continuity in the utilities' managerial staff due to political interference, along with a wider lack of accountability (where a municipal government can receive a well-working utility and leave a broken one to the next administration, and no one is called to respond), has systematically blocked any long-term development program in most of the cities. State programs usually mirror federal ones, which has given CONAGUA's criteria a central role in defining actual policy orientations, because of its budgetary and political strength, accentuated by frequent negligence at the state level in taking a wider role in water and sanitation regulation and capacity building, beyond public works execution. This distance between the operator and the government sphere where policies are defined has important consequences, as it is discussed later in this chapter.

CONAGUA, whose central role is *resource regulation*, should therefore concentrate on guarding the integrity of national waters in terms of quality and availability, by measuring, registering, and controlling water extraction and discharge permits and rights. Nevertheless, it is also entitled to promote the expansion of water and sanitation services and their efficient operation, as it was mentioned before. While results in terms of resource regulation are far from being effective (with growing problems of water pollution and overexploitation), substantial resources are being dedicated to the development of big infrastructure projects called "strategic projects."

6.2.5 Water and Sanitation Capital Finance in Mexico

Federal policy investment on water and sanitation infrastructure seeks to act as a catalyst of the state and local or private funding, achieving a wider reach by raising these funds in face of the lack of public funds available (CONAGUA 2012), even when these local funds usually come from federal fiscal collection, which is redistributed to states and municipalities through funds or programs. CONAGUA assigns federal funds by means of investment programs, each one with operating rules that usually vary slightly from one year to the other and must adjust to the rules defined also by state authorities.

Most of the funds are allocated by transferring the resource to state or local authorities, under a coordination agreement where executors commit to follow federal rules; in some states, local programs are set up to support specific goals and targets. Most of the municipalities take part only with 25% or less of the funds requested.

A salient characteristic of water and sanitation services capital financing in Mexico is the concentration of resources and decision-making at the federal level. In the period 2007–2011, CONAGUA contributed nearly half of the total budget on water and sanitation, although it participates also defining rules and through incentives in 80% of the total investment (World Bank 2016); the rest of the investment comes from other areas of the government and the private sector. Between 2010 and 2015, resources exceeded the needed amount to cover gaps in coverage (Campanaro and Rodriguez 2014), but due to the economic situation of the country, between 2015 and 2017, such federal resources were drastically reduced; in 2017, the reduction was greater than 70% (Montoya 2017).

Financing capital investments in urban water and sanitation sector in Mexico rely mostly on federal subsidies, which suffer from some deficiencies:

- *Fairness* – States or cities that have projects and resources, to capture federal subsidies in a timely manner, end up receiving more support; money does not always benefit those that most require the help and gets allocated without reference to any performance commitments.

- *Efficiency* – Allocating resources to municipalities and states around the country involves a complex job which does not always guarantee the best allocation of resources and the achievement of greater social and economic benefits, despite the existence of mechanisms for performing ex ante evaluation of bigger federal investments. Hasty bidding processes and the lack of information contribute to an inefficient allocation of resources.
- *Stability* – Infrastructure investments come from resources that are passed annually by the Congress, without a multi-year scheme or guaranteed medium-term continuity. An operator is unable to know, at the time of preparing its tariff proposal, if there will be resources for capital investments next year, with which the utility could improve its efficiencies and lower the burden on the user. Similarly, when local resources are required to get federal funds, operators must revise their financial planning and sometimes use operating income for investment purposes as matching funds, deferring maintenance, or tolerating gaps in service.
- *Sufficiency* – Investments in the sector represent a minimum proportion of the GDP, compared with countries that have similar coverage but invest more (World Bank 2016). As noted before, federal subsidies don't cover the estimated minimum amount to correct the gaps. Many new connections are financed by the users themselves directly through connection rights or by real estate companies, transferring the cost to users.

A relevant feature of the institutional framework in the water and sanitation sector is the absence of formal mechanisms of economic and performance regulation. As subsidies are granted without performance improvement commitments, water utilities do not develop the administrative and financial expertise that they would, if they were competing for funding on capital markets. In terms of operating income, the rates are approved in municipal councils normally without bonding the income plan with the investment plan.

This establishes three conditions that constitute the core institutional problem of urban water and sanitation services in Mexico:

- *Virtual impunity*. There is no formal commitments or responsibilities associated with performance; bad operation or investment decisions can occur without consequences on public servants who operate an organism at a given moment; the controlling or auditing mechanisms focus in reviewing the compliance with regulations in the field of administration and expenditure, but not on qualifying the performance nor the impact of the services. Many utilities, however, have achieved sustained improvements, but their capacities can be dismantled in no time. This has led to the municipal authorities to use operators as instances to pay favors with political jobs, contracts, and benefits to individuals and private groups. Despite the presence of remarkable achievements in some urban water and sanitation systems, instability and vulnerability are the rule in the industry.

- *Bailouts*. The possibility of obtaining federal resources through lobbying or political pressure by partisan affinities, in the absence of clear allocation criteria and thanks to the flexibility to transfer resources between beneficiaries and programs, reinforces the tendency to operate the systems without a long-term strategy for improvement. An operator that falls in "bankruptcy" but has sufficient political or economic relevance can expect successive events of "bailout" by the state and federal government. And agencies with more capabilities for generating projects and resources capture more funds. The better performance of big cities comes in part from these conditions, as well as from the availability of more qualified human resources, diversified economies, and higher incomes.
- *Interferences*. Operating costs are recovered through tariffs, which are usually calculated based on the historical record of operating costs plus the general inflation rate. Water utilities are usually affected by political interference in their technical, administrative, and legal decisions. Different timings in resource availability and the lack of predictability of resources for capital expenditures provoke constant interferences and disruptions in the operation of the utility.

The allocation of funding has sought to be a lever to induce efficiency in service through rules of operation in the subsidy programs, but it cannot avoid interfering with the dynamics of the operation. The municipality, which should verify the consistency between goals and resources of the utility, neglects to play this role, partly because of the specialized nature of the sector but also because they can get political earnings without committing municipal resources, nor being accountable. There is a lack of clarity in the boundaries of roles and responsibilities.

In short, lack of incentives and trained personnel, coupled with corruption and improvisation in the public service (presence of more political than technical managers), causes deficiencies in the administration, operation, and management of the assets. Therefore, the quality of the service worsens and infrastructure gets inadequate and obsolete. This translates into indifference or even social rejection (users refuse to fulfill their obligations because they can't exercise their rights easily either), which leads to a high vulnerability and instability of organizations (Lentini and Ferro 2014). Even though tariffs are showing an increasing trend and new governmental accounting rules are being implemented, guaranteeing the constitutional right to water is still difficult in the present context. With incomes insufficient to sustain operational costs and an acute lack and intermittence of funds to finance the expansion of coverage, the weakness of the operators cannot guarantee the rights fulfillment of users, nor the achievement of SDGs.

It is clear that reaching a sustainable operation of water and sanitation services is more than a cashflow problem. It requires timely, efficient, equitable, and stable financing. This will not happen under the present institutional framework.

On the other hand, international trends also show an inability of the actors of the sector to attract alternative sources and to realize the benefits of investments in the long term. In a context of growing financial scarcity and competition, this becomes a major threat to the development of sustainable water and sanitation services.

6.2.6 Private Investment in Water and Sanitation in Mexico

Possibly as a reflection of the lack of conditions to attract private capital in a stable and sustainable way, private investment in water and sanitation in Mexico has had an erratic behavior, usually linked to major projects by special vehicles and financial structures tailored to each case; the participation of private capital in projects occurs in many different proportions and through varying mechanisms.

The main financial vehicle to support private investment in the sector is the National Fund for Infrastructure (FONADIN), but it has only contributed about 5% of its funds to water infrastructure projects. Other options that have been used are investment and property trusts, public-private partnerships, and contracts for the provision of services.

Private participation in the operation of water systems has also undergone different stages of development, stagnation, and relapse. The first systems operated under a concession agreement (Aguascalientes, Cancun, Navojoa) were followed by a successful experience of a public-private joint venture in Saltillo and later by its extension to Ramos Arizpe; more recently, concession agreements in Puebla and Medellin took place, but public management was restored in Navojoa and Ramos Arizpe. There are several BOT (Build-Operate-Transfer) and similar contracts for the construction of water and sewage treatment plants, as well as some contracts for the provision of services to outsource specific functions, as it is the case of commercial operations in Mexico City. Another scheme that has been sought to implement is the "integral improvement of the management" or MIG, for its Spanish acronym, in San Luis Potosí, which has remained limited by administrative difficulties and is in the process of redesign.

It should be noted that at the municipal level, part of the distribution and connection infrastructure is often being projected and built by real estate developers and individual users.

Visibly, there is no unified public policy to promote and structure private involvement in financing water and sanitation infrastructure. Schemes are decided case to case, using vehicles and structures ad hoc, sometimes without sufficient transparency and often in response to initiatives from the companies themselves, rather than following a national comprehensive strategy.

6.2.7 An Assessment of the Current Model

Undoubtedly, important goals in terms of coverage extension and efficiency improvements have been and will be achieved by means of the implementation of the federal programs for the expansion of infrastructure and the improvement of efficiency in the water and sanitation sector.

Nevertheless, there are some aspects that deserve to be revised:

- Most funds are allocated disregarding the outputs, performance, and evolution of water utilities; those utilities having stronger administrative and financial capacities are better placed to get funding from federal programs, leading to a somehow "regressive" scheme, where weaker utilities receive less support than those already stronger.
- In some cases, the origin and destination of the funds are not consistent, which leads to a lack of accountability and economic rationality. For instance, in the Valley of Mexico, the rights paid for the wholesale water supply are derived into a trust fund, where they are used to build flood control infrastructure, while operating funds for assuring proper maintenance of the system come from general budget and have to compete against many other projects throughout the country. Another example comes from the PRODDER program (Program for the Refund of Water Rights), in which water extraction rights are returned to utilities if they present an investment plan and matching funds, on a yearly basis. This poses two problems. First, water extraction rights have been designed to promote efficiency and prevent overexploitation, but now they are seen as a means for financing infrastructure expansion, instead of perhaps being used to ensure the protection of the integrity of national waters in quality and availability. Second, the refunds depend on the behavior of tax collection and national policies set by the Ministry of Finance, which has often led to late, incomplete, and partial refunds, causing the delay or cancellation of the programmed water works.
- In general, funding through existing programs is subject to yearly modifications in the operating rules and to delays in the actual transfer of the resources. It is especially difficult for the utilities to implement midterm financial planning when they are unaware of how much money they will get and when and under which conditions; sometimes funds are allocated very late in the fiscal year, leading to hasty bidding processes.
- A better framework for the allocation of existing funds is needed, in which bigger support to the weaker utilities would be provided, rewarding verifiable improvements with financial incentives. The participation of the state authorities must be effectively encouraged with concrete incentives and penalties, not only as a part of the funding sources but most importantly as the catalysts for achieving better institutional arrangements in the municipal settings.

In a way, the current model considers that, since most of the urban water utilities show low economic and technical performances, it is necessary for the federal government to subsidize infrastructure financing and performance improvement. This approach seems to be correct or unavoidable in the short term, but in the long term, it appears to be perpetuating a vicious cycle, where municipal authorities rely on federal and state funds to periodically "rescue" their systems, thus refusing to give the utilities the institutional freedom, stability, and accountability they need for a sustainable operation.

6.2.8 SDGs: More Ambitious Goals in a Complex Context

In the year 2000, the "Millennium Goals" were agreed at the United Nations, with specific goals to the year 2015. Goal 7c consisted of halving the proportion of people without sustainable access to improved drinking water sources and to improved sanitation between 1990 and 2015. The results were apparently outstanding: access to "improved water supplies" increased worldwide from 76% in 1990 to 91% in 2015; 58% of the world's population already has an outlet of water running to his house; 2100 million people also gained access to improved sanitation (UN 2015). Even so, many countries could not meet these goals, especially in sanitation.

Just 151 countries, out of 225 met the goal on access to water and sanitation (CONAGUA 2016). Mexico reported having fulfilled both goals in water – from 82% to 96% in the period from 1990 to 2015, with an increase of 14%, well above the agreed 9% – as well as sanitation – which rose from 66% to 85% coverage (Ferro 2017, p. 10). These indicators do not reveal, however, deficiencies in the quality of the provision of services.

By 2015, the resolution called "Agenda 2030 for sustainable development" was adopted, in which 17 goals were included, called sustainable development goals (SDGs), expressed in 169 targets and 230 indicators.

In 2011, the CONAGUA estimated that achieving universal coverage of water and sewerage would require investments over $ 215 billion, over 114 billion to increase the percentage of wastewater treatment, letting aside other investments for flood control or water resource management (CONAGUA 2011). According to other calculations, the cost to close the coverage gap has been estimated in $32,200 million per year (ANEAS 2016). If we added the cost of replacement of obsolete infrastructure, these needs would grow considerably.

Today, water sector financing in Latin America goes through a complex circumstance. Countries are investing at a very low rate in infrastructure, in relation to the size of their economies. Despite having achieved remarkable progress, accumulated shortfalls exacerbate within the SDG's framework, since water and sanitation services will be measured not only in terms of their coverage (access) but their quality and affordability also, where countries will have to assure social and gender equity, within an integrated water resource management model, and preserve water quality and aquatic ecosystems, with greater involvement of local communities.

In the words of Gustavo Ferro, the attainment of the SDGs "will require a significantly greater effort than that carried out to achieve the MDGs, and most likely within less favorable economic conditions." Only the goals for urban areas of the Latin American region would require tripling the historical allocations, with investments close to 250 billion dollars between 2010 and 2030, which would be unsustainable if not complemented with enhanced institutions, with sufficient technical capacity, mandates, powers, and budgets (Ferro 2017, p. 12).

In synthesis, achieving the SDGs and the commitments on climate change, and doing so under a model of sustainable development, will only be possible if innovative financial mechanisms are implemented, through a transformation of the institutional framework of water management and services.

This works in both directions: to make sustainable investments, it is not enough to gather the financial resources required, but it is also necessary to improve the policies and institutional frameworks, which today are not conducive to the efficient operation of the systems.

But also, inversely, to attract the financial resources that are needed, it is essential to have institutions that correct the distortions in the current framework, achieve an efficient and equitable allocation of investment resources, induce a sufficient collection of operating revenue, and generate more stable and sustained financial flows. This institutional reform has to include better governance systems, where the informed participation of the society gradually leads to achieve better results and an effective fulfillment of the rights and obligations of both citizens and public servants.

6.3 Overall Assessment and Policy Questions

6.3.1 The Challenges for a Sustainable Operation of Urban Water and Sanitation Systems

As it has been shown, sustainability of services is being threatened because of the lack of an adequate supply and management of the different assets or resources involved. According to the Economic Commission for Latin America and the Caribbean (ECLAC 1991), sustainability requires the balanced use of six kinds of assets or forms of capital: natural, physical, financial, human, institutional, and social, taking into account their relationships of complementarity and substitution. Within this framework, bigger cities face also bigger challenges, not only because of the lack of regular and enough access to the different forms of capital (in some cases, big cities have clear advantages to get resources due to their greater economic and political influence) but because of their very diverse capabilities to administer and use those assets. In general:

- Natural capital is being compromised because of the deterioration of water sources, watersheds, and aquifers, worsened now by climatic variability and growing competition among sectors.
- Physical capital, that is, infrastructure, equipment, and systems, are also subject to two ways of pressure: the need to extend their coverage due to the cities ongoing growth while replacing the assets which have largely surpassed their lifespan.
- Financial capital should be analyzed in two parts. First, investment capital depends heavily in Mexico of the capacity of cities and state governments to get federal subsidies; several programs exist and it becomes a matter of

negotiation. Second, operating capital relates to the capacity of each system to measure, invoice, collect, and administer operating revenues, but it also depends on each utility tariff setting and approval framework, where only a few big cities have been able to reach a virtuous cycle of good services with adequate tariffs. In recent years, it has been more frequent that state governments and even water operators themselves are asked to commit to repay part of the capital financing, as well as setting up efficiency improvement programs, which is shifting the rules toward a more incentive-based framework. Nevertheless, in Mexico a lot remains to be done to effectively link the performance of water operators to their creditworthiness.

- Human capital is usually easier to access by bigger utilities, because they operate in cities where a richer and wider job market exists and they have larger revenues to pay for better salaries; in this direction, the most important threat is the lack of stability and professionalization, due to the excessive intervention of political instances and the lack of effective accountability mechanisms.
- Institutional assets would comprise the set of rules determining whether the relationship between the quality of services in a broad sense and their price is adequate for each city characteristics, establishing clear boundaries between the tasks, capacities, rights, and obligations of the political authority (which beholds the system property in the name of the general population), the operator, and the public, with a proper balance between goals and means. In Mexico there are several institutional settings, but in general there is a lack of a well-structured regulation framework, with the exception (to some extent) of the very few private participation schemes; political authorities are in charge of the process of revision, approval, and implementation of tariff structures and investment programs, but tariffs tend to be kept below inflation rates with no further considerations, while it is usual that utilities provide bad quality services trying to "finance" their operation by delaying maintenance and distributing water intermittently to cope with growing demands with the same (or less) water and financial resources.
- Last but not least, social or civic capital would be the capacity of every system to obtain from their users or customers a level of commitment beyond their formal or contractual responsibilities toward the service, by means of a proper communication and influence on the people's knowledge, behavior, and habits. While most of the utilities have implemented communication programs, only a few have effective mechanisms for getting feedback and being transparent to their users. Citizens outside the utilities governing bodies usually find it difficult to access its information and to get their demands and expectations answered in a structured manner. Citizen observatories, implemented in the cities of Xalapa, Tuxtla Gutierrez, Saltillo, Ecatepec, and San Miguel de Allende, did not succeed in having concrete incidence on the practices and results of water utilities.

Clearly, many of our urban water and sanitation systems are operating under vulnerable circumstances, and it is hard to identify the key elements, which would bring them to a sustainable path. Even if the federal authorities achieved to balance water budgets and to assure every water system, as a national waters user, a reliable access to good quality water in their sources, other challenges would prevent our systems to be sustainable. The most difficult one seems to be the institutional instability, along with the pressure coming from the lack of a planned and effectively enforced land use development. Since these three issues seem to be beyond the scope of utilities – assuring an effective water resource regulation and preservation, getting a more balanced institutional arrangement, and controlling land use development – a double effort needs to be done: first, promote a sectorial reform to change the external rules which define the way water and sanitation systems access natural and financial resources and are forced to show results and, second, keep on working on the internal strengthening of water utilities.

6.3.2 Some Guidelines to Improve the Sustainability of Urban Water and Sanitation Systems

In the same terms presented in the former section, the next guidelines are proposed to improve the sustainability of our urban water and sanitation systems:

- Mexico needs an authority fully committed and devoted to the custody, restoration, and preservation of its national waters and related public goods. Water rights should serve primarily to set up a strong capacity to measure, register, and administer water usage and discharges, to enforce law and to supervise the state of the nation's watersheds and water bodies. It is true that public and productive systems often need financial and technical support to be efficient and thus to reduce extraction, but having this task assigned to the same authority could be leading to a conflict of interest. The recourse to coordination mechanisms such as basin or aquifer councils has been thought as a means to promote consensual solutions to specific issues, such as temporary limitations to water extractions during droughts, or water uses prioritization for administrative purposes; but users participation does not exempt federal authority of the fundamental responsibility for preserving and maintaining the quality and availability of the national waters.
- Water supply and sanitation are capital-intensive activities. So, infrastructure and equipment are instrumental to have sustainable and reliable services. Water utilities and municipalities should be forced to have an updated registry of all the water and sanitation assets they operate, so they can get accountable for their conservation, improvement, and expansion. Specific responsibilities over the state of the assets should be assigned to municipal authorities and utilities. Setting up proper registry, accounting, and asset management procedures is urgently needed in order to be capable to design a national strategy for infrastructure renewal and expansion.
- Financial system for water and sanitation development needs to be revised and restructured. Mexico needs to leave the current discretionary subsidy-based system, where federal and state programs operate as "relief funds" for rescuing

eternally developing utilities, with a financial support that is often variable, unpredictable, and insufficient. Concrete steps must be taken to set up a financial system that promotes performance improvement, creates real responsibility on the side of the municipality, and operates under clear, equitable rules. Those utilities with better institutional arrangements and managerial practices should get incentives; those showing clear trends of performance improvement should also get advantages. Funding should act as a catalyst for financial self-sufficiency and not as a life saver. This restructuring of the financial system goes along with a deep institutional reform.

- Human capital is perhaps a key element in this puzzle. In spite of the existence of several regulations trying to set up a civil service system in the sector, there is a lot to be done to achieve this goal. Funding programs could also set more stringent conditions to the utilities and municipalities in terms of the profiles and certification credentials of their staff, as a condition to get the funds. In the other hand, federal, state, and municipal authorities need to commit to give water and sanitation services the professional level that is already given to health or security services. It is true that unions sometimes block any effort to set up more efficient staffing practices, but this is also a matter of negotiation and, in the long term, of designing a financial and political solution. Water and sanitation utilities are fundamentally operating enterprises which rely on their ability to manage capable, well-trained, and committed personnel. Capacity building programs are necessary, but not sufficient, to strengthen the human capital of this sector. An effective accountability framework should drive the sector toward its professionalization, where political interference would find it hard to support improvised or incompetent managers.
- Institutional reform could be the key to develop every other improvement. The National Association of Water and Sanitation Utilities (ANEAS) has proposed to enact a national reference law in order to set up the minimal conditions to have, at the state and municipal levels, a well-structured arrangement, where goals match the resources available, the quality of service corresponds with its price, and the relationships between political authorities, utilities, and the public reach an adequate balance, that is, to set up proper economic regulation structures. For some bigger and well-performing systems, new financing mechanisms should be explored, as well as the implementation of better corporate governance mechanisms. Urban water utilities should operate within a proper regulatory framework, with a clear contractual agreement between political authorities, service providers, and users, under an appropriate law enforcement environment and, most importantly, responding and promoting a more informed and active citizen participation. Information must be the key to restore the capacity of the system to acknowledge its current problems and to design the path toward a more sustainable operation. The enactment of a constitutional reform, recognizing the constitutional right to water, should spur all the actors on to find effective ways to ensure an equitable supply for all.

- Finally, sustainable urban water and sanitation services call for a more informed and participative user. It is true that people cannot devote to take part of the decisions regarding every aspect of their lives as citizens, since they have already delegated the responsibilities of taking charge of public services on their municipal authorities, through their political representatives. But there is a general rule in management science: authority can be delegated, but responsibility remains. Citizenship implies sharing responsibility with authorities, which means to be informed, understand, and have incidence in the decisions. Utility managers must also realize that transparency and openness to dialogue can act in favor of the system stability, protecting it from political and influence group intervention. Citizen observatories should be supported and encouraged, as well as public communications that go beyond the messages for promoting water savings; civic culture is a fundamental asset for every modern utility.

On the other hand, utilities can do a lot to promote improvements in the three areas mentioned before, as framing conditions normally out of their scope:

- They can help federal authority to better measure, supervise, and control extractions of other users within the same watershed or aquifer. They can set up agreements with other users to help them use the water more efficiently, if possible to promote exchange agreements of water for treated wastewater, but mainly as a stewardship function for which they are especially well positioned. Water is the main input of every utility: anything they can do to support its conservation works in favor of their sustainable operation.
- Water utilities should strive for having proper regulation mechanisms and social participation schemes, whether it could seem contradictory. A better set of rules would favor the foundations of the utility to work with more stability, gain creditworthiness, and create a space for professional development for its members. Water utilities need to have a leading role in the discussion of the mechanisms that need to be set in place to ensure human right to water, instead of regarding these efforts as a threat.
- Finally, water utilities should also participate in every initiative in order to set up better rules for a better control of land use and development. In the context of climate change, the presence of more frequent droughts and intense precipitation events poses additional challenges to our utilities. New urban design patterns and the use of sustainable urban drainage systems should be a part of our land use regulations.

These proposals call for the effective adoption of the integrated urban water cycle management model, where every part of the cycle is designed considering the full cycle, and land use and design become part of the water and sanitation management system within cities.

6.4 Conclusion

Cities represent a fundamental link in the economic chain of a country. In Mexico, their importance will grow in terms of their economic contribution to national wealth but also as development centers where an equitable access to water and sanitation can help fulfilling every citizen's right to health and work.

Federal, state, and local authorities have implemented huge financial and administrative efforts to support the expansion of urban water and sanitation systems and to improve their performance. Infrastructure has grown at an unprecedented pace, and many programs have been put in place to give water utilities several ways of getting financial and technical supports. But the challenge is also huge, which calls for a profound reform in the way water resources and services are being managed in Mexico.

First, there is an urgent need to set up an effective resource regulation system which leads to the restoration and preservation of water balances in many of our watersheds and aquifers. No infrastructure or money will suffice if we run out of clean water sources.

Second, a deep institutional reform must take place to set up the proper incentive framework to ensure an appropriate relationship between political authorities, utility operators, and citizens. Utilities must be empowered and provided with enough resources to achieve their goals but also must be obliged to be accountable and transparent. Water users must be encouraged to be informed, understand, and participate in the decisions affecting them. Good corporate practices must be implemented in the boards, so they effectively defend the people's interests. Local political authorities should be fully accountable for the state of the assets and processes associated to water and sanitation services. A more mature and informed dialogue should take place between authorities, operators, and citizens.

Third, the financial system should be revised and restructured to become a catalyst for efficiency and accountability. Performance improvement should be rewarded, and funds should be equitably and transparently allocated. Funding should become more predictable, stable, sufficient, equitable, and productive for water utilities. Budgetary support programs should promote self-sufficiency and responsibility. Better management models and capacities are needed in order to achieve a more efficient use of financial funds within the utilities and their regulatory counterparts. To close the financial gap arising from the commitment to achieve the SDGs, attracting alternative sources of capital financing will be mandatory, thus pushing the sector to perform a profound institutional reform.

And finally, water utilities must take part of the political efforts seeking for a more effective land use planning and development. New urban design practices must be implemented into the utilities' processes for the approval of new developments, taking into account the need to better manage precipitation events, to favor rainwater infiltration and detention, and to contribute with the proper management of urban rivers and water bodies.

Integrated urban water management needs to become the model for urban water management in Mexico. Demand management needs to be taken seriously, since public policies in this sector remain clearly in the supply management side. A lot of well-intended negotiation and communication between federal, state, and municipal authorities must occur in order to achieve a new institutional framework. The size of the threat calls for an unprecedented coordination effort.

References

Aboites, L., Cifuentes, E., Jiménez, B., & Torregrosa, M. (2008). *Agenda del Agua*. México, D.F.: Academia Mexicana de Ciencias – Red del Agua.

ANEAS. (2016). *Página de la LXIII Legislatura*. Obtenido de Cámara de Diputados: http://www5.diputados.gob.mx/

Banco Mundial. (2016). *México – Revisión del gasto público*. Washington, DC: World Bank Group. Recuperado el marzo de 2017, de http://documents.worldbank.org/curated/en/284151472615491033/Mexico-Public-expenditure-review

Campanaro, A., & Rodriguez, D. (2014). *Strengthening the financial system for water in Mexico. From a conceptual framework to the formulation of pilot initiatives*. Washington, DC: World Bank.

CCA. (2011). *Gestión del Agua en las Ciudades de México*. México: Consejo Consultivo del Agua.

CONAGUA. (2011). *Agenda del agua 2030*. México: SEMARNAT.

CONAGUA. (2012). Situación del Subsector Agua Potable, Alcantarillado y Saneamiento Edición 2012. México, D.F.: SEMARNAT-CONAGUA.

CONAGUA. (2015). Estadísticas del agua en México. Edición 2016. México: SEMARNAT.

CONAGUA. (2016). Situación del Subsector Agua Potable, Drenaje y Saneamiento. Edición 2016. Mexico: SEMARNAT.

ECLAC. (1991). *El desarrollo sustentable: transformación productiva, equidad y medio ambiente*. Santiago de Chile: UNO-ECLAC.

Ferro, G. (2017). *América Latina y el Caribe hacia los Objetivos de Desarrollo Sostenible en agua y saneamiento – Reformas recientes de las políticas sectoriales*. Santiago: CEPAL.

INEGI. (2010). *Censo de Población y Vivienda 2010*. Aguascalientes, Ags: INEGI.

INEGI. (2015). *Encuesta nacional de Calidad e Impacto Gubernamental 2015*. Obtenido de http://internet.contenidos.inegi.org.mx/contenidos/productos//prod_serv/contenidos/espanol/bvinegi/productos/nueva_estruc/promo/encig15_principales_resultados.pdf

INEGI. (2016). *Encuesta nacional de los hogares 2016*. Obtenido de http://www.beta.inegi.org.mx/proyectos/enchogares/regulares/enh/2016/default.html

Lentini, E., & Ferro, G. (coords.) (2014). *Políticas tarifarias y regulatorias en el marco de los Objetivos de Desarrollo del Milenio y el derecho humano al agua y saneamiento*. Santiago de Chile: ONU-CEPAL.

Montoya, M. Á. (20 de septiembre de 2017). V. La CONAGUA en el Proyecto de Presupuesto de Egresos 2017. *Ágora – Boletín del Grupo Parlamentario del Partido de la Revolución Democrática*, 15–17.

Morgan, J. M. (Enero de 2016). Instituto de Ingeniería. Gaceta Electrónica. Obtenido de Plantas de Tratamiento de Aguas Residuales en la república mexicana: http://www.iingen.unam.mx/es-mx/Publicaciones/GacetaElectronica/GacetaEnero2016/Paginas/PlantasdeTratamientodeAguasResiduales.aspx

ONU HABITAT. (2011). *Estado de las Ciudades de México 2011*. México: SEDESOL.

Saltiel, G. (2008). In R. Sandoval & R. Olivares (Eds.), *Problemática del sector agua potable mexicano – análisis, ejemplos y propuestas* (pp. 229–237). México: ANEAS.
UN. (2015). Millennial Development Goals. Report 2015. New York, USA: United Nations Organization.
UN-Water. (2013). *Mexico country brief.* Nueva York: FAO-UN.

Ricardo Sandoval-Minero is a Consultant on water resources management, as well as water and sanitation technical and institutional development. He is a civil engineer with a master in science from the University of Paris-Dauphine, and a master in science in Operations Research from the National Autonomous University of Mexico (UNAM). He was Former General Director at the Guanajuato State Water Commission and at the Guanajuato City Water and Sanitation System, and has worked also at the federal level in Conagua and for private consulting firms, international organizations, and development banks as a consultant. He is presently a contractual for the Inter-American Development Bank as Water and Sanitation Consultant.

Part III
Institutional Issues

Chapter 7
Water Price Policy and Its Institutional Role as an Economic Instrument for Water Management

Hilda R. Guerrero García Rojas, Diego Garcia-Vega, and Hugo Amador Herrera-Torres

Abstract An analysis on the reform of the Federal Law of Rights that establishes four availability zones in Mexico for water use, catchment, or exploitation and its role in the tariff policy as an economic instrument for water resource management is presented. Before the reform, until 2013, Mexico was classified in nine availability zones, where Zone 1 presented lesser availability levels of water and hence a higher tariff or price and vice versa in Zone 9, besides the fact that zone classification complies with a municipal geographic division criteria with no distinction between hydrological basins and aquifers. As of 2014, Mexico is classified in four availability zones, but current rates or prices respond to a relative availability criterion for hydrological basins and to an availability index for aquifers, even if located in the same territory, which promotes heterogeneity of rates for the collection of fees. It has been considered that although the new classification of availability zones will yield a higher fee revenue, that does not necessarily mean that water management, in terms of efficiency and sustainability, is the most appropriate, since the new classification of zones may force users to migrate to areas where fees are lower and generate pressure on the water resource.

Keywords Water price · ZD · RHA · LFD reform

7.1 Introduction

In Mexico, the human right to water and sanitation is part of the constitution since February 8, 2012, when Article 4 was amended. This right is stated in the sixth paragraph of Article 4 as follows:

> Every person has the right of water access, disposal and sanitation for personal and domestic consumption in a sufficient, safe, healthy and affordable manner. The State shall guarantee this right and the law will define the bases, support and modalities for having access and an

H. R. Guerrero García Rojas (✉) · D. Garcia-Vega · H. A. Herrera-Torres
Facultad de Economía, Universidad Michoacana de San Nicolas de Hidalgo (UMSNH), Morelia, Michoacan, Mexico

H. R. Guerrero García Rojas (ed.), *Water Policy in Mexico*, Global Issues in Water Policy 20, https://doi.org/10.1007/978-3-319-76115-2_7

equitable and sustainable use of water resources. Establishing the participation of the Federation, States and Municipalities as well as the participation of the citizens for achieving those goals. (DOF 2017a: 8)

The way that national waters in Mexico are managed and preserved is through the demarcation of hydrological administrative regions (RHAs). Starting from 1997, Mexico was divided into 13 RHA; basin groups in turn form these regions. Basins are considered as the basic management units for water resources, and their geographic margins correspond to the municipal margins to facilitate the integration of the socioeconomic data. The National Water Commission (CONAGUA) performs its water managing duties through basin organizations, whose field of competence is the RHA (CONAGUA 2016).

The 13 RHAs are (I) Baja California Peninsula, (II) Northeast Pacific, (III) Northern Pacific, (IV) Balsas, (V) Southern Pacific, (VI) Rio Bravo, (VII) central basins of the North, (VIII) Lerma-Santiago-Pacific, (IX) Northern Gulf, (X) Central Gulf, (XI) Southern Border, (XII) Yucatan Peninsula, and (XIII) waters of the Valley of Mexico, and they are presented in Fig. 7.1. In the same way, Table 7.1 shows, among others, some relevant data of the 13 RHAs, as surface, amount of renewable water, population, and percentage contribution to the gross national product (GNP).

Fig. 7.1 Hydrological administrative regions (Source: CONAGUA 2016, p. 215)

Table 7.1 Characteristics of the hydrological administrative regions

RHA	Mainland surface (km^2)	Renewable water resources 2015 (hm^3/year)	Population as of mid-2015 (millions of inhabitants)	Population density (inhabitant/km^2)	Per capita renewable water resources 2015 (m^3/inhab/year)	Contribution to the national GDP 2014 (%)	Municipalities or delegations of Mexico City (number)
I	154,279	4958	4.45	28.8	1115	3.61	11
II	196,326	8273	2.84	14.5	2912	2.86	78
III	152,007	25,596	4.51	29.7	5676	2.88	51
IV	116,439	21,678	11.81	101.4	1836	6.14	420
V	82,775	30,565	5.06	61.1	6041	2.29	378
VI	390,440	12,352	12.3	31.5	1004	14.29	144
VII	187,621	7905	4.56	24.3	1733	4.19	78
VIII	192,722	35,080	24.17	125.4	1451	19.08	332
IX	127,064	28,124	5.28	41.6	5326	2.24	148
X	102,354	95,022	10.57	103.2	8993	5.62	432
XI	99,094	144,459	7.66	77.3	18,852	4.93	137
XII	139,897	29,324	4.6	32.9	6373	7.38	127
XIII	18,229	3442	23.19	1,272.2	148	24.49	121
Total	**1,959,248**	**446,777**	**121.01**	**61.8**	**3 692**	**100**	**2 457**

Source: CONAGUA (2016, p. 20)

7.2 Availability Zones for the Collection of Water Rights Until 2013

Since the National Waters Law (LAN) establishes that for the exploitation, use, or catchment of national waters, the corresponding authority should issue a concession or assignment title; it is natural to establish fees for the water use. These fees are known as charges for water exploitation, use, or catchment. Mexico is currently divided into four availability zones (ZDs) for charging for those fees. This classification is included in the Federal Law of Rights (LFD). The charge per cubic meter of water is higher in zones of lesser availability, and it gradually decreases in higher availability zones.

Until 2013, and according to the provisions of the LFD in that year, the Mexican Republic was divided into nine ZD for the collection of rights over water exploitation, use, and catchment. The list of municipalities within each availability zone is included in Article 231 of the abovementioned law, which is yearly updated. Apart from agricultural use or hydroelectric generation (DOF 2013), the way in which uses and fees for ZD were classified until 2013 is presented in Table 7.2. This table describes eight uses of water, of which the general regime refers to as any different use to those previously mentioned. Fees are expressed in cents of a peso per cubic meter of water.

It is important to emphasize that the charge per cubic meter of water in the case of the agricultural and hydroelectric generation sector is not applied according to the aforementioned, as in the agricultural sector the charge is null as long as it does not exceed the limits of the concession; while in the case of hydroelectric generation, the charge is constant in the nine zones.

The distribution of the nine ZD in 2013 can be seen in Fig. 7.2. In this figure it is possible to visualize that Region XIII waters of the Valley of Mexico is the region with less water availability. On the other hand, Region XI Southern Border is the region with the highest availability. In general, the northern zone of the country is where the zones with lesser availability are located, while the south presents the highest water availability (CONAGUA 2014).

7.3 Situation Before the Reform

The LFD, in its Article 222, establishes that both natural and legal persons are obligated to pay for the right to national waters they use, exploit, or catch, whether in fact or under the protection of allocation, concession, or authorization titles or deeds granted by the federal government according to the ZD where extractions are conducted, in accordance with the territorial division included in Article 231 of this Law (CONAGUA 2014).

In the same sense, those persons that permanently, periodically, or accidentally discharge wastewaters into rivers, basins, seawaters and other water deposits or

Table 7.2 Duties for the use of the nation's water, according to water availability zones, 2013 (Mexican cents per cubic meter)

	Availability water zones								
Use	1	2	3	4	5	6	7	8	9
General regime	2,050.42	1,640.28	1,366.89	1,127.70	888.45	802.97	604.37	214.72	160.92
Drinking water, consumption more than 300 l/inhab/day	81.24	81.24	81.24	81.24	81.24	81.24	37.83	18.89	9.41
Drinking water, consumption equal to or less than 300 l/inhab/day	40.62	40.62	40.62	40.62	40.62	40.62	18.92	9.45	4.7
Agricultural, without exceeding the assigned volume	0	0	0	0	0	0	0	0	0
Agricultural, for every m^3 that it exceeds the assigned volume	14.52	14.52	14.52	14.52	14.52	14.52	14.52	14.52	14.52
Spas and recreational centers	1.17	1.17	1.17	1.17	1.17	1.17	0.57	0.27	0.13
Hydropower generation	0.43	0.43	0.43	0.43	0.43	0.43	0.43	0.43	0.43
Aquaculture	0.33	0.33	0.33	0.33	0.33	0.33	0.16	0.08	0.04

Source: CONAGUA (2014, p. 121)

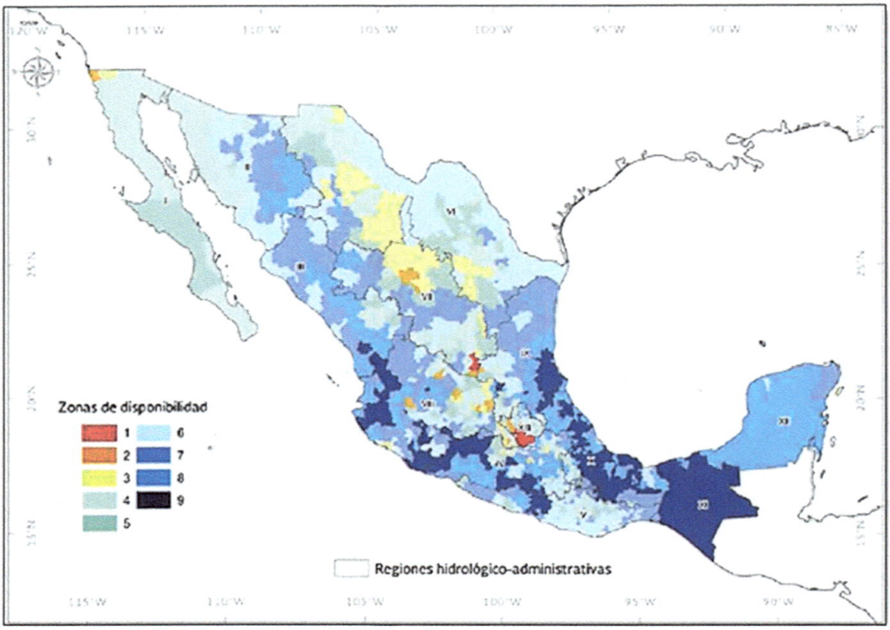

Fig. 7.2 Distribution of water availability zones by hydrological administrative regions, 2013 (Source: CONAGUA 2014, p. 120)

streams, as well as soils or infiltrations into lands considered as national assets or that may pollute subsoil or aquifers. In the same context are those that use, enjoy, or take advantage of public domain assets of the federation in ports, terminals and port facilities, federal maritime zone, dikes, streams, reservoirs, current zones, and national property deposits. It is worth mentioning that there is no payment for seawater extraction or brackish waters with more than 2500 mg/l of total dissolved solids, certified by CONAGUA.

For the collection of fees due to wastewater discharges, receiving bodies (rivers, lakes, lagoons, among others) are classified in three types, A, B, or C, according to the contamination effect inflicted. Type C-receiving bodies are those that have the largest contamination effect. The list of receiving bodies belonging to each type is included each year in the LFD. Fees for the discharge of wastewaters are related to the discharged volume and the pollutant content, which is available in Article 278 C of the LFD. Article 223 of LFD in 2013 establishes that for the exploitation, use, or catchment of national waters, a water right shall be paid according to the ZD where the extraction is conducted and in accordance to fees stated in the DOF (2013).

The logic for charging for water rights is still the same, that is, in the ZD where the water resource is less, the fee charged is higher, and vice versa, in those ZD where the water resource is higher, the fee charge is lower. Guerrero et al. (2015) present an analysis of the structure of water fees in the nine ZD.

7.4 Situation After the Reform

Currently, until 2017, ZDs are basin and aquifer determined (surface water and groundwater). Regarding the management of superficial waters in Mexico, 731 basins have been defined. Basins cover around 65% of the national territory. Among these superficial waters, the following rivers stand out, where most of the water flows every year: Grijalva-Usumacinta, Nazas-Aguanaval, Rio Bravo, and Rio Balsas. Among these, Rio Grijalva-Usumacinta is the one where the largest amount of water flows, which is equivalent to 26% of runoffs in the year. In 2013, 627 basins had availability and 104 deficits. For the management of underground water, 653 aquifers have been defined. In Mexico 37% of water used (except that for hydroelectric power) comes from the subsoil. According to 2013 figures, 55.2% of the underground water used comes from 106 overexploited aquifers. In that same year, 458 aquifers had availability, while 195 had deficit (CONAGUA 2016).

On December 11, 2013, a decree amending, supplementing, and repealing different LFD dispositions was published in the DOF, becoming effective on January 1, 2014. On February 26, 2014, the agreement announcing the value of each variable which integrates the formulas to determine, during the 2014 fiscal year, the ZDs referred to in Article 231 Sections I and II of the LFD, was published, effective as of January 1, 2014. That agreement presented the agreements published in the DOF, where the geographical boundaries of 653 aquifers and 731 basins in the United Mexican States were disclosed.

Article 231, Section I of the LFD, states that the ZD of basins in the country needed to calculate the amount of the right for the use, exploitation, or catchment of national waters will be determined by placing within the following ranks the result obtained from the formula provided in the abovementioned section (DOF 2014). Table 7.3 shows these ranks.

Article 231, Section II of the LFD, states that the ZD of aquifers in the country needed to calculate the amount of the right for the use, exploitation, or catchment of national waters will be determined by placing within the following ranks the result obtained from the formula provided in the abovementioned section (DOF 2014). Table 7.4 shows these ranks.

The last paragraph of Article 231 of the LFD states that regardless of whether taxpayers can determine the corresponding ZD of the hydrological basin or aquifer where the extraction is conducted, CONAGUA, as managing entity, will publish no later than the third month of the fiscal year in question the corresponding ZD for each hydrological basin and aquifer in the country.

Therefore, the relative availability and corresponding ZD to each basin in the country during the 2014 fiscal year are established, based on the formula provided in Article 231, Section I of the LFD, and the values of variables contained in the agreement that discloses the value for each variable of the formulas to determine ZDs during 2014 fiscal year, stated in Article 231, Sections I and II of the LFD, effective as of January 1, 2014, published in the DOF on February 26, 2014. On March 27, 2014, the Ministry of the Environment and Natural Resources (SEMARNAT)

Table 7.3 Range to establish the amount for the use or exploitation of basins

Availability Zone 1	Less than or equal to a 1.4
Availability Zone 2	Greater than 1.4 and less than or equal to 3.0
Availability Zone 3	Greater than 3.0 and less or equal to 9.0
Availability Zone 4	Greater than 9.0

Source: Own elaboration with data of DOF (2014, p. 1)

Table 7.4 Range to establish the amount for the use or exploitation of aquifers

Availability Zone 1	Less than or equal to − 0.1
Availability Zone 2	Greater than − 0.1 and less than or equal to 0.1
Availability Zone 3	Greater than 0.1 and less than or equal to 0.8
Availability Zone 4	Greater than 0.8

Source: Own elaboration with data of DOF (2014, p. 1)

published through the DOF an agreement disclosing the ZDs corresponding to basins and aquifers in the country for the 2014 fiscal year, in terms of the last paragraph of Article 231 of the current LFD.

Regarding payment of rights for the exploitation, use, and catchment of national waters, Article 231 of the LFD for 2017 establishes fees according to the ZD and basin and aquifer where the extraction is conducted. In general, the cost per cubic meter is higher in lesser availability zones, as Table 7.5 shows for wastewaters and Table 7.6 for underground waters. In both tables, the concept of "general regime" refers to any use other than those previously mentioned (DOF 2017b).

Regarding drinking water use, considerations are (a) those assigned to federal entities, municipalities and parastatal and paramunicipal organisms, (b) those granted to companies providing drinking water or sewage services and those that through an authorization or concession provide that service replacing legal entities referred in subsection (a), and (c) those granted to neighborhoods constituted as legal entities that, according to the concession of legal entities referred in subsection (a), provide drinking water supply for households. Fees referred in this section shall be applicable to subjects mentioned therein when the water consumption in the period is less or equal to a volume equivalent to 300 liters per person per day, according to the population described in the final results of the previous fiscal year, referred solely to population from the last General Population and Housing Census published by the National Institute of Statistics and Geography (INEGI).

Tables 7.5 and 7.6 show the rights for exploitation, use, and catchment of superficial water and groundwater, per ZD for 2017. Values of both tables are taken from the DOF issued on December 7, 2016, Chapter VIII and Article 223, and refer to updated amounts established in the LFD in 2017.

As shown in Tables 7.5 and 7.6, the agricultural sector benefits from economic incentives by having a zero tariff, as long as the water consumption does not exceed the granted amount. It is worth noting that the agricultural activity consumes about 76% of the water used in Mexico. Forty-nine percent of the consumption comes from

Table 7.5 Duties for the use of the nation's surface water resources according to availability zone, 2017 (Mexican pesos per cubic meter)

	Availability water zones			
Use	1	2	3	4
General regime	15.19440	6.99510	2.29360	1.75380
Drinking water, consumption of more than 300 l/inhabitant/day (on the excess)	0.90315	0.43318	0.21632	0.10768
Drinking water, consumption equal to or less than 300 l/inhab/day	0.45158	0.21658	0.10816	0.05385
Agriculture and livestock, without exceeding the concession	–	–	–	–
Agriculture and livestock, for every m^3 that exceeds the concession	0.000172	0.000172	0.000172	0.000172
Spas and recreation centers	0.011189	0.006245	0.002913	0.001201
Hydropower generation	0.005221	0.005221	0.005221	0.005221
Aquaculture	0.003754	0.001873	0.00086	0.000399

Source: Own elaboration with data of LFD, 2017, p. 161. Prices are from Ley Federal de Derechos, Capítulo VIII Agua. M$/m3 = (Mexican pesos per cubic meter in current terms)
Note 1: For reference 20.62 Mexican $ = 1 USD in December 23, 2016. Date of the LFD 2017 publication (BANXICO, 2016). Note 2: All the amounts reported in the 2017 LFD established for the year 2017 have been updated based on the "quota without adjustment" of Annex 19 of the "Miscellaneous Fiscal Resolution for 2017 and its annex 19," published in the DOF on December 23, 2016. Note 3: It is important to notice in Tables 7.5 and 7.6 that the fees charged to the general regime are, by far, higher than those for other users. That is the reason for leaving so many decimals. The same for Table 7.2

Table 7.6 Duties for the use of the nation's groundwater resources according to availability zone, 2017 (Mexican pesos per cubic meter)

	Availability water zones			
Use	1	2	3	4
General regime	20.47400	7.92510	2.75950	2.00590
Drinking water, consumption of more than 300 l/inhabitant/day (on the excess)	0.94277	0.43471	0.24507	0.11424
Drinking water, consumption equal to or less than 300 l/inhabitant/day	0.47139	0.21735	0.12253	0.05712
Agriculture and livestock, without exceeding the concession	–	–	–	–
Agriculture and livestock, for every m^3 that exceeds the concession	0.00017	0.00017	0.00017	0.00017
Spas and recreation centers	0.01326	0.00653	0.00320	0.00143
Hydropower generation	0.00522	0.00522	0.00522	0.00522
Aquaculture	0.00412	0.00191	0.00095	0.00044

Source: Own elaboration with data of LFD, 2017, p. 161

superficial sources and the remaining from underground sources (CONAGUA 2016). According to CONAGUA, from the existing 653 aquifers, 448 have availability and 105 are overexploited; meanwhile from the 731 existing basins, 627 have availability. Obviously, both overexploited aquifers and basins are located in zone 1. Geographically, ZDs for *superficial* waters are shown in Fig. 7.3, and the ZDs for *groundwater* are shown in Fig. 7.4.

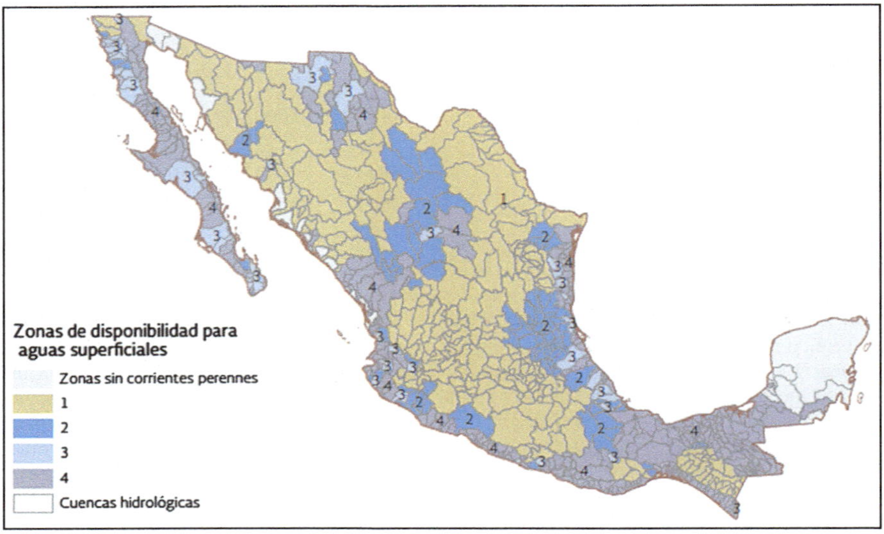

Fig. 7.3 Availability water zones for surface water, 2015 (Source: CONAGUA 2016, p. 146)

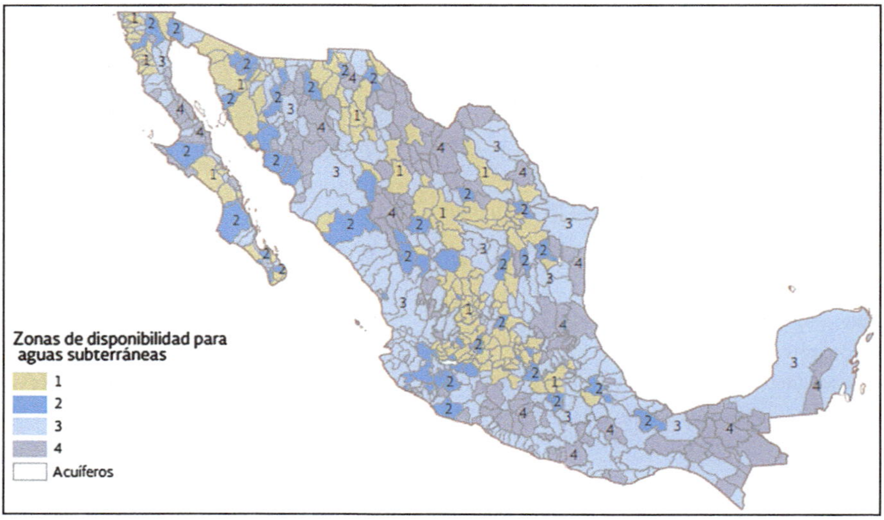

Fig. 7.4 Availability water zones for groundwater, 2015 (Source: CONAGUA 2016, p. 146)

Classification per ZD, basin, and aquifer is based on formulas whose variables define relative availability, regarding basins, or that define the availability index regarding aquifers. One of the variables in the denominator of the formula for calculating the relative availability of water in basins is the annual volume of superficial water extraction (VAEAS); in the case of aquifers, calculating the availability index considers the variable mean water availability (DMA) in the numerator of the formula. For calculating the variables, both VAEAS and DMA consider the water volume granted as an important element according to data registered in the Public Registry of Water Rights (REPDA), where the maximum extraction limits are clearly defined, but it does not mean that users with extraction titles have used that granted volume, which could be more or less. For example, how do we effectively ensure that agriculture production units or industries are really using the permitted water volumes? This leads us to question whether the objectively declared water volumes reflect the real water consumption or if they just reflect the quotient between the water consumption invoice in the economic units and the existing fee for such a sector or economic unit. This analogy would make us rethink if the management of water demand through ZDs and the tariff structure turns out to be really effective.

7.5 Analysis of the Reform

Since January 1, 2014, a LFD reform on national water use and wastewater discharge to national assets became effective. Some of the important aspects of this reform are the ZD reclassification and the methodology for the exemption of payment for wastewater discharge, which constitute elements that will undoubtedly present a greater impact on revenues. The LDF reform brings relevant changes that, depending on the extent, will have an impact on the operations of national water users and/or generators of wastewater discharges of the federal type. What is important in this moment is to determine the impact and effects of the reform, especially in the no less important subject of competitiveness.

The peculiar way to manage water in Mexico demands the consolidation of the reforms regarding water use that generate coherence among judicial, administrative, and fiscal levels and changes as those of LFD, with no solid judicial foundation and sometimes inconsistent with principles of legality and taxation, which may generate excessive and discretionary acts of the authorities. The foregoing reveals the lack of legal certainty for users of the water resource.

Table 7.7 compares the main changes in the 2014 LDF related to 2013 LFD in terms of the classification of new ZDs, as well as the payable amounts for the exploitation, use, or catchment of national waters. The reasons for the reform changes in 2014 LFD related to 2013 LFD are exposed in the decree initiative that reforms, adds, and repeals various provisions of the LFD sent by the President of the

Table 7.7 Comparative between the Ley Federal de Derechos 2013 and 2014

Topic	LFD 2013	LFD 2014
Availability zones	Nine zones	Four zones
Definition of availability zones	By municipality without distinction between surface water and groundwater	For surface water: through the formula in which the relative availability of the hydrological basin is calculated. For groundwater: through the formula in which the aquifer availability index is calculated
Distinction between surface water and groundwater	None	Different quotas are established for surface water and groundwater
Highest quota	$ 18.29, in zone 1 without distinction between surface water and groundwater	$ 18.62, in zone 1 for groundwater
Lowest quota	$ 1.43, in zone 9 without distinction between surface water and groundwater	$ 1.60, in zone 4 for surface water

Source: Own elaboration with data of LFD 2013 and 2014

Republic to the Congress to be discussed and subsequently approved in 2013 and to become effective as of 2014. Roughly, the main reasons are as follows (Initiative 2013). Water is acknowledged as an asset of federal public domain, which is vital, vulnerable, and finite, with social, economic, and environmental values whose preservation in terms of quality and quantity is a fundamental task of the state and society. That is, the Mexican state acknowledges its responsibility as a guiding entity in the design and implementation of the water policy in the country.

As established in the National Development Plan 2013–2018, there are several courses of action that should be implemented, as ensuring sufficient water of suitable quality to guarantee human consumption and food security, ensuring water use and catchment in basins and aquifers affected by deficit and overexploitation, promoting sustainability with no limitation on the development, and strengthening technical and financial capabilities of water operators for the provision of better services.

Current water conditions imply that its management should be conducted through classification, that is, per basin (superficial) or aquifer (groundwater) given that water that precipitates is concentrated in a shallow way in basins, while other infiltrates aquifers, which represent a variability in vulnerability and disposition of the resource, since while superficial water in a basin is susceptible to evaporation and direct contamination, groundwater in aquifers is stored with no risk of evaporation and more protected from contamination.

Until 2013 in the LFD, the calculation for use, exploitation, and catchment rights of national waters did not necessarily address the real water availability that each basin and aquifer had; instead it was determined based on the territorial extension of the municipality, that is, while determining ZDs, the real abundance or scarcity of the resource was not reflected; so in the same municipality, a variety of aquifers and basins could concur with its own conditions and characteristics directly affecting the availability of the resource. Therefore, as of 2014, the establishment of fees for use, exploitation, or catchment rights for aquifers and basins ceases to be uniform for the same territory, since currently in the same territory the charged fee could be heterogeneous for a basin or aquifer depending on a water availability level.

Based on the proposed methodology, the existence of four ZDs is justified, for that reason a reduction of the ZDs from nine to four is suggested. Therefore, it is proposed to establish fees for four ZDs depending on the use of the water resource and differentiation fees according to the extraction conducted in a basin or an aquifer. In this context a reform of Articles 222, 223, and 231 of the LFD is proposed, with the purpose of provision in such a methodology through which ZD classification will be determined in relation to the extraction source, differentiating between aquifer and basin. In this sense, availability of superficial and underground water could be recognized and assigned to the corresponding ZD for the effects of the calculation of the right for use, catchment, and exploitation of national waters.

The intention to express the methodology in the LFD will permit taxpayers to determine by themselves the corresponding ZD for payment of the rights, both for superficial waters and groundwater; thus CONAGUA will publish in the DOF, as managing entity, the values of variables that integrate each of the formulas at the beginning of every fiscal year. In that same sense, CONAGUA will publish, as managing entity, in the same issuing organism the corresponding ZD to each hydrological basin or aquifer in the country.

In order to grant legal certainty to taxpayers, it is proposed as an alternative that they can determine the ZD on their own and corroborate through the list of CONAGUA that the result is the same.

Some of the most relevant considerations of LFD 2014, according to IDEA (2014), regarding the right to use, catch, or exploit national waters are that the new methodology to determine the payment for the use of water is established according to a formula whose factors, to say the least, uncertain, will be published by CONAGUA and other entities; other factors could be obtained applying norm NOM-011-CNA-2000, being necessary to refer to the base of the use of water to determine the certainty of payment. The gap between the new ZDs, independently of the calculation conducted, will show the reality for most of national water users that are out of the metropolitan area of the Valley of Mexico; it is enough to see in Tables 7.8 and 7.9 the relation of amounts until 2013 and since 2014.

Since 2014, in this tax code, the figure of "transfer" (*trasvase*) is created; it is considered as the use of national waters moving from one basin to another to be used in different locations from the extraction site, so it is necessary to evaluate the impact in the cost of drinking water which is also part of such resource supply in diverse sectors, as an example Mexico City, where the estimation of the cost of crude water may suffer a 10% increase in drinking water.

Table 7.8 Availability water zones until December 2013

Availability Zone	ZD1	ZD2	ZD3	ZD4	ZD5	ZD6	ZD7	ZD8	ZD9
Tariff $	$20.50	$16.40	$13.67	$11.28	$8.38	$8.03	$6.04	$2.15	$1.61

Source: Own elaboration with data of IDEAS (2014, p. 2)

Table 7.9 Availability water zones as of January 2014

Availability zone	Surface water	Groundwater
ZD1	$13.82	$18.62
ZD2	$6.36	$7.21
ZD3	$2.09	$2.51
ZD4	$1.59	$1.82

Source: Own elaboration with data of IDEAS (2014, p. 2)

7.6 Conclusions

Establishing a rate or price for the use, catchment, or exploitation of water in Mexico is highly complex and difficult because the price of the rights should not just include financial and administrative expenses incurred for its distribution, sanitation, and supply, but it should also include the real costs generated as a result of environmental degradation, contamination, scarcity, and level of availability of water resources, as well as the degree of pressure and demand exerted by the society over this vital resource. While it is true that until 2013 the classification of nine ZDs did not respond to quantitative criteria or estimations that would classify water availability per hydrological basin and aquifer, even in the same territory, to geographic criteria, i.e., according to a municipal division, it does not necessarily mean that since 2014, with the new calculation methodology, that is, through the proposed algorithm, a more efficient water management is obtained, since, as an example, large users whether commercial or industrial can investigate which geographical areas under the new four ZDs have a larger water availability and therefore are cheaper, generating a greater water demand in those basins or aquifers classified in ZD 3 or 4.

Available evidence in diverse analysis shows that water demand is sensitive to the evolution of the income, population growth, price and other secondary prices, demographic and socioeconomic characteristics of households and industries and also weather, temperature, and rainfall. It is important to mention the inelasticity of water demand as a function of income and price, especially price. This suggests that water consumption will continue increasing. The use of economic instruments is important to control consumption, but it also has its limitations and should include other social and environmental considerations. It is also worth noting that the temperature increase and modifications in rainfall patterns will affect the trajectory of water consumption. Especially, a temperature increase will translate into an increase in the demand for water, which will intensify the pressure on the resource and, as a result, on rates and prices.

As Guerrero et al. (2008: 109–110) state,

> even though since 1999 in Mexico a series of decrees and programs have been produced to exert pressure on operating organisms to apply better actions in the collection, and that these reforms have generated immediate effects, the global consequence has not been greater; and despite an increase in the collection, this does not mean a generation of a direct recovery of water supply costs. Decrees and programs have been mainly modified and created considering that the operating organisms direct their actions towards improving efficiency and infrastructure of drinking water, sewage and waste water treatment.

In general, the current situation of water resources in Mexico is a consequence of an inefficient use and catchment, of a merely engineering direction of the government, a lenient legal application, a decreasing productivity of the resource, and increasing financial and environmental costs that are not reflected on fees, as well as the existence of large subsidies which make impossible a more real quantification of the economic, social, and environmental costs.

References

BANXICO. (2016). Comunicado de prensa del tipo de cambio peso/dólar estadounidense para el día 23 de diciembre de 2016. *Banco de México (BANXICO).*

CONAGUA. (2014). Estadísticas del agua en México. *Comisión Nacional del Agua (CONAGUA). Secretaría del Medio Ambiente y Recursos Naturales (SEMARNAT),* 1–239.

CONAGUA. (2016). Estadísticas del agua en México. *Comisión Nacional del Agua (CONAGUA). Secretaría del Medio Ambiente y Recursos Naturales (SEMARNAT),* 1–242.

DOF. (2013). Ley Federal de Derechos. *Diario Oficial de la Federación (DOF),* 1–476.

DOF. (2014). Ley Federal de Derechos. *Diario Oficial de la Federación (DOF),* 1–76.

DOF. (2017a). Constitución Política de los Estados Unidos Mexicanos. *Diario Oficial de la Federación (DOF),* 1–296.

DOF. (2017b). Ley Federal de Derechos. *Diario Oficial de la Federación (DOF),* 1–476.

Guerrero, H., Yúnez-Naude, A., & Medellin J. (2008). El Agua en México. Consecuencias de las políticas de intervención en el sector. *Fondo de Cultura Económica. No 100 de Lecturas El Trimestre Económico.*

Guerrero, H., Gómez, F., & Rodríguez, J. R. (2015, October). Water pricing in Mexico: Pricing structures and implications. In A. Dinar, V. Pochat, & J. A. Murillo (Eds.), *Water pricing experiences and innovations, Global Issues in Water Policy.* Cham: Springer Publishing.

IDEAS. (2014). Boletín de la Reforma a la Ley Federal de Derechos (LFD) para el año 2014 en materia de uso de aguas nacionales y descargas de aguas residuales a bienes nacionales. *Iniciativa para el Desarrollo Ambiental y Sustentable, S.C. (IDEAS),* 1–3.

Iniciativa. (2013). Iniciativa de Decreto por el que se reforman, adicionan y derogan diversas disposiciones de la Ley Federal de Derechos en materia de Zonas de Disponibilidad del Agua para el año 2014. *Presidencia de la República (PR),* 1–121.

Hilda R. Guerrero García Rojas is a Professor of Economics and Environment in the Facultad de Economia, Universidad Michoacana de San Nicolas de Hidalgo (UMSNH), Morelia, Michoacán, Mexico. She holds a PhD in Economics from the Toulouse School of Economics. She is among the leading water policy experts in Mexico and contributed to policy work in Mexico. Her PhD thesis focused on "economic assessment of policy interventions in the water sector; industrial water demand in Mexico: econometric analysis and implications for water management policy." She worked for Instituto Mexicano de Tecnología del Agua (IMTA), and as a consultant, she has been collaborating with the World Bank, Colegio de México (COLMEX), OECD, and with the Inter-American Development Bank (IDB). Publication topics involving Guerrero include "Water pricing reforms in Mexico: The case of manufacturing sector"; "Applying water demand: Case for industrial use on Mexico"; "Water in Mexico: Impact of policy interventions in the sector"; and "Water pricing reforms in Mexico: The case of the manufacturing sector", among others. Guerrero's areas of interest are natural resources and environmental economics, integrated water management (IWM), economic instruments applied to natural resources, water price and its reforms, applied economics, environmental economic valuation, micro-econometrics, climate change in water sector, and economics of climate change.

Diego Garcia-Vega has a Master in Public Management of Sustainability at the Facultad de Economía, Universidad Michoacana de San Nicolas de Hidalgo (UMSNH). He has professional experience in the area of financial services and has participated in several projects and academic studies, as well as research assistant. He has attended various forums, congresses, and some diplomats, national and international, as well as some publications in indexed journals. His research areas include sustainable development, environmental economic valuation, climate change, sustainable tourism, water economics, and socioeconomic studies.

Hugo Amador Herrera-Torres earned a PhD in Political Science from the Government of the Republic of Cuba (University of Havana), a PhD in Regional Development Sciences, and a Master in Strategic Management of Development, both from Universidad Michoacana de San Nicolas de Hidalgo (UMSNH). He is a Professor at the Faculty of Economics, UMSNH, and a Member of the National System of Researchers (Conacyt - Mexico). His research lines are government management models, as well as public sector economics. He has published books, book chapters, and articles in refereed and indexed journals, highlighting municipal performance evaluation, between them a methodological proposal for the semi-urban municipalities of the State of Michoacán, Mexico, for INAP, which won National Award of Public Administration, 2011. Others are for government policies and public policies, where the cycle of development policies of the government of the State of Michoacán, 2003–2010, Mexico, for INAP won Latin American Public Administration Award, 2012.

Chapter 8
River Basin Organization, the Best Path Towards Integrated Water Resources Management?

Jorge Arturo Hidalgo-Toledo, Cipriana Hernández-Arce, and Sergio Vargas-Velázquez

Abstract Since 1992, Mexico has been formally implementing several principles of the policy approach now known as Integrated Water Resources Management (IWRM), with the passing of the National Water Law. Several years before, innovations in water management were already being made. More decisively, the institutional arrangement and the 2004 amendment to the Law were reorganized by promoting the creation of executive basin organizations with the confrontation of different views. In spite of the importance of those changes, there are still serious problems surrounding water and there is no simple way to reorganize the role of the federal government as the leading entity in water management. What do we need? Some parties insist on an increased decentralization and involvement of various levels of government; social organizations contemplate a joint management and even delegating functions. On the other hand, a different sector of officers and researchers are aware of the lack of a real water authority capable of motivating and penalizing, as well as encouraging basin management based on a new territorial water management. This shows that the IWRM program is still in an institutional transition regarding its application, since there is no sufficient results so far which justify the institutional roles proposed for its operation. This chapter presents a reflection on what has occurred regarding integrated water resources management, and on whether those basin organizations have responded in a proper way to tasks they were entrusted with, as well as on which are the most convenient institutional roles required in Mexico to be successful in water resources management.

Keywords Water resources management · Water institutions · Decentralization · River basin organizations

---

J. A. Hidalgo-Toledo (✉) · C. Hernández-Arce
Instituto Mexicano de Tecnología del Agua, Jiutepec, Mexico
e-mail: jhidalgo@tlaloc.imta.mx

S. Vargas-Velázquez
Universidad Autónoma del Estado de Morelos, Cuernavaca, Mexico
e-mail: sergio.vargasvme@uaem.edu.mx

H. R. Guerrero García Rojas (ed.), *Water Policy in Mexico*, Global Issues in Water Policy 20, https://doi.org/10.1007/978-3-319-76115-2_8

8.1 Historical Evolution of Water Resources Management in Mexico

One of the main points of GWP has been to reconstitute water resources management based on a management unit which takes into consideration environmental characteristics, such as geomorphology, climatic regimes, and biotic systems by correlating hydrology and ecosystem aspects with the need for regulating water uses and economic, political, or social groups that move and fight for water. Globally, this has led to a shift from environmental and water resources management units, based on administrative-political divisions to some other within compact physical spatial units which can be defined as hydrographic/hydrological basins, watersheds or "planning" regions for environmental management.

A hydrological basin is a designated area that is framed by the use of the same water source and by social relations internally generated around its resources. The basin, on the side of natural resources management, enables us to study the ecosystems function, to establish the behavior of runoffs and its anthropogenic alterations, as well as to analyze the effects of its disturbances and to evaluate water management alternatives. From a social perspective, we are able to visualize the social, economic, productive, and cultural network that surrounds the use of natural resources – in this particular case, water. An important aspect to consider is the incorporation of social factors in the study, planning, and management of the environment by establishing regionally defined study units whose structures and social relations are attached to certain ecosystems. A basin is defined by a perimeter associated to the watershed or divide, that is, by the highest topographic points on the ground, but it should be even more important to define it as a designated area for water planning and management and regulation of interests, as they are, in this case, management units. This redefinition is based on the recognition of natural land and water processes in a basin – supported by international experience and research – as the most appropriate foundation for defining "management units" and for "intervention" of policies for sustainable development, as well as the establishment of instances of negotiation and consensus among social and institutional stakeholders (Newson 1992).

This new proposal is intended to regulate natural resources accessibility to curb environmental degradation, which is quite different in programming, organizing, technical, and conceptual terms to the proposals for the generation of development poles from executive commissions per river, which existed in our country between 1940 and 1970. Along with these principles, and with no contradiction in the sense of economic policies of trade liberalization and state decentralization and deregulation, identified as neoliberal policies, several radical transformations have been proposed in natural resources management that reshapes regional development. In our country, the National Water Commission (Conagua) has considered hydrological basins as the best natural units for integrated water resources management (units between hydrographic basins and aquifers from a large-scale drainage area) for which it has transformed the previous management of political and administrative division into a hydrological region management.

As the federal government transferred functions and responsibilities to other levels of government and water user associations (WUAs), it became fundamental to promote an active participation of organized water users in water management. In this sense, environment and water management aims to engage all the social sectors, groups of interest and stakeholders of the conflicts in the solution of problems, but this mainly depends on the political forms of each regime. Nevertheless, a great number of reforms have been promoted which radically transform the management of resources. In this way, we can explain how diverse users occupying a defined territorial space regarding water use in the same basin are able to implement joint actions to stop the severe deterioration regarding available amounts and quality of water resources and, if applicable, to propose solutions to environmental problems.

It is therefore possible to define an integrated management plan for the management of natural resources – renewable and non-renewable – in hydrological basins, to rationalize interventions. Table 8.1 represents the way in which experts view this change in the centralized management model that dominated throughout the twentieth century and what is expected as a transition of institutional arrangements with IWRM. The future perspective implies changes aiming to achieve a higher level of complexity in water resources management, since the decentralization towards a local scope involves huge consequences in the organizational, budgetary, and political levels to reach an agreement among different levels of social and government players.

For just over a 100 years, Mexico was characterized for its water resources management through a centralized water model, which contemplated the

Table 8.1 Trends in basin management

Aspect	Dominant modes	Current trend	Future perspective
Institutional coordination	Centralized	Decentralized	Integrated, based on local actions
Aims of the approach	Aimed at natural resources management	Aimed at natural resources management based on human needs	Aimed at preserving quality of life and environmental balance
Responsible institutions	National responsibility for basin management	National, governmental and non-governmental and producers facilitating the implementation of basin management	Local organized community and producers managing, implementing, and monitoring basin management
Problem analysis	Timely vision and corrective solutions	Integrated vision, corrective and preventive actions	Integrated vision and preventive actions
Type of basin management plans	According to regulatory systems with top-down processes	Designed based on biophysical and socioeconomic characteristics and community consultations	Based on a situational strategic planning with an active and responsible community participation

Source: Own development based on Aguilera Klink (1998)

intervention of the federal government as the unique authority on water resources, as well as for major public investments in hydro-agriculture infrastructure, mainly towards the construction of major dams and for the drinking water supply and sanitation of cities. Emphasis was placed on ensuring water supply for economic development, with no further social or environmental considerations. In Latin America, Mexico turned into the most extreme case of centralization: "many other countries have created similar institutions, but the Secretariat of Water Resources (Secretaría de Recursos Hidráulicos, SRH) has remained as the most powerful of all similar entities in the region. The lack of similar institutions in other countries serves to emphasize the special nature of this institutional evolution in Mexico. Only Cuba, another very special case, has created a centralized entity of this kind, but in a lesser competence extent than the Secretariat of Water Resources" (CEPAL 1989, p. 26). This kind of management became highly questioned since the 1970s and got into great difficulties during the 1980s, when in several hydrological regions of the country, as well as of the world, an overexploitation of water resources – already then – was recognized, detecting also several deteriorated and contaminated bodies of water, which made the expansion of irrigation areas and the fulfilling of the urban systems demand impossible as a result of the low economic and technical efficiency that they had (Comisión de Plan Nacional Hidráulico 1981).

The centralized model is characterized by budgetary and action control by the federal government as the sole representative of the nation in terms of water. The implementation of actions was subjected to corporative negotiations among interest groups, formed around development policies, which were typical of the policy regime in Mexico after the revolution in 1917. The governing party was able to make successful interventions for the development of hydraulic infrastructure because the political regime so allowed; besides, for many social groups, the promise of development was effective, in spite of its costs. This can be illustrated by the expansion in the irrigation agriculture frontier, based on large infrastructure projects and the creation of Irrigation District Steering Committees, as well as by the expansion of drinking water and sewage infrastructure in large urban concentrations, such as Mexico City (Vargas and Sánchez 1996). Especially during the Miguel Aleman's presidential administration, a number of large hydraulic structures were built between 1940 and 1960 (Martínez 2004) and the systematic knowledge of water resources was promoted over large hydrological regions, which provided an extensive knowledge regarding various problems. One of the most important concepts that emerged during this time, and which is related to water management, was that of the rational and multiple uses of dams, which allows for a much more feasible cost-benefit ratio for major investments in the construction of these works. But it wasn't until the 1970s that efforts in the organization of available information began to emerge in a systematic way within a planning scheme made by experts. For such a purpose, the first planning area in SRH was created in 1970, which was in charge of preparing and publishing the first National Hydraulic Plan in 1975, aimed at rationalizing the use and management of water in an integrated manner. In the devising of it, the problem of water resources was analyzed from a broad perspective, considering its geographical distribution, incidence; uses; current and potential harvesting; conditions for its preservation, purification, reuse, and economic importance; and as a factor of social wellbeing. Its development enabled the quantification of water resources, the definition of programs, and the establishment of the measures and policies needed to achieve predefined objectives and goals; all this with no consultation to representatives of interest groups nor any sector of society. Its creation was surrounded by a group of experts, mainly civil and hydraulic engineers, which defined the direction of water policies in our country. Ever since, they have identified scenarios of conflict and restriction, which could be avoided with technical solutions, that is, "as long as there was an efficient use of water" (Aldama et al. 2006). In the case of agriculture, this water management model laid the foundation upon which a "dual" agricultural and livestock sector was settled; and in which peasant agriculture for the domestic market and commercial and exportation agriculture assumed functions at successive stages of development. Hydro-agricultural policy became a key factor of economic development. On the one hand, the large irrigation organizations, called Irrigation Districts (Distritos de Riego), received all support and government subsidies, and on the other hand, small-scale irrigation, made up of several small organizations self-managed by their users, had less governmental support. The participation of social groups was tailored to this Project, according to negotiation and conclusion schemes, where the government sector was predominant. In many instances, as it happened with the executive commissions per river basin, these assumed supra regional powers (Barkin and Timothy 1979).

Despite the strong process of centralization, a great multiplicity of social organizational forms subsisted regarding water. In most cases they kept their decision-making capacity in a restricted territorial scope, like irrigation units and water committees in rural areas, among others, which did not take on the organizational forms promoted by the public entity regulating water resources. Planning according to hydrographical basins intended to promote Mexican development, for which, essentially, the proposal was to make investments in hydro-electrical and basic infrastructure works, usually complemented with agricultural and livestock development programs. Morphological and economic characteristics of the defined regions based on the criteria of hydrographical basins were not adequate for shaping the total structure of the national coverage for regional planning.

The planning agency, where the 1975 plan was configured, later became – as a result of the change of political administration in the country – the National Water Plan Commission (Comisión del Plan Nacional Hidráulico), in 1976. This new agency was in charge of following up the programs and projects set forth in the Plan. In 1981, as a result of a change in public administration, a new National Hydraulic Plan was issued (Comisión del Plan Nacional Hidráulico 1981), aimed at formulating and updating the national regional plans for the use and preservation of water and to monitor their observance. In its formulation, studies included the quantification of hydrological cycle elements. Unlike the 1975 plan, its development enabled a water policy that viewed water from different perspectives, i.e., water as an economic, environmental, and social asset. The National Water Plan of 1981 is characterized by taking into account the quantification of the elements of the water cycle, the definition of the water management framework and investment programs for the 1983–1990 period in agricultural infrastructure, flooding protection, water quality control, drinking water and sewage, industry, hydroelectric generation, and aquaculture. The National Water Plan Commission would become thereupon in what today is known as the Mexican Institute of Water Technology. However, it wasn't until 1990 when major transformations began in the management of water resources in Mexico, which were stated in the National Water Law of 1992, and which were later enhanced in the amendment of that Law in 2004. This amendment included proposals created by technical panels of experts, which were discussed at the time in several forums and through international organizations, leading to the so-called Integrated Water Resources Management (GWP 2000). Thus, the National Water Commission, (Conagua[1]) was created as the national water authority, assuming, as per the last amendment, an administrative organization by hydrological region, based on basin organizations, with basins and aquifers as main management units. In the 1990s, state water and sanitation commissions were created in the federal entities, establishing their own state laws as well – in several cases confronting Conagua in the definition of their own jurisdictions. Also, a constitutional reform was implemented, which, since 1983, gave municipal governments the responsibility of drinking water and sanitation at the local scope (Pineda 2002). Regarding agriculture, 3.4 million hectares out of 6 million with irrigation infrastructure were transferred to WUAs, which were previously operated by the federal government. Councils, commissions and committees for basins and aquifers were created to promote social participation in water resources management, albeit in an advisory capacity, despite attempts to broaden their function. During the first 10 years of Conagua there were several planning efforts, but it wasn't until the 2001–2006 National Water Program that a large national scheme was developed once again. In 1999, diagnoses of the administrative regions were conducted with the intention of giving solutions to problems regarding shortages of supply sources; setbacks in coverage and quality of drinking water, sewerage and sanitation services; inefficiencies in water use, contamination of rivers and water bodies; overexploitation of aquifers; and an increasing competition for water among users.

At this point, some sort of public consultation began, which in fact was intended solely for experts and Conagua officials. However, it is important to emphasize this inception of social or public participation in planning processes, following the principles mildly established in the Water Law of 1992, but which were explicitly stated in the amendment of the National Water Law of 2004. This process ran parallel to the creation of participative entities, which are now officially the foundation for participating in water policy, such as basin and aquifer councils, commissions, and committees. With transformations such as these, we have tried to move from a centralized water management in the federal government towards a management model by hydrological basin, based on the

[1]Nowadays the Comisión Nacional del Agua acronym changed to "Conagua", but before its acronym was "CNA".

decentralization of management, the involvement of multiple stakeholders and the incorporation of a mechanism for the regulation of water demand.

8.2 Decentralization of Government Decisions in Water Resources Management

In the 1980s, the water sector in Mexico lost the authority that has not been recovered until now: the Secretariat of Water Resources ceased to exist, which seemed to be the ruin for some people, because the Secretariat became the Undersecretariat of Hydraulic Infrastructure of the Secretariat of Agriculture and Water Resources (SARH). Then, the entity of "Irrigation District" disappeared and the leading role of the Secretariat faded away. Water districts were added to rural development districts, where rain-fed agriculture is practiced, and a gap was created. From this decision, irrigation districts began to decline. The level of self-sufficiency of irrigation districts droped considerably and, in 1986, the tariff that users payed for water services was about 14% of its actual cost, which was extreamely low (Palacios Vélez 1994) This generated an internal discussion in the Undersecretariat of Hydraulic Infrastructure regarding the specificity of water resources. Thus, there were several issues with which the water sector was striving at that time to increase subsidy or change the line of action. At the beginning of President Carlos Salinas de Gortari's campaign began also the formulation of the National Program for the Modernization of the Countryside (*Programa Nacional para la Modernización del Campo* [Pronamoca]) which identified several issues regarding the transfer of irrigation districts and the idea of defining water management were. This meant water management in a territorial context, as a cross-cutting axis for the management of other resources, that is, the concept of "River Basin". In addition, there was an amendment in 1984 to Article 115 of the Mexican Constitution, where municipalities were given the responsibility of providing drinking water and sanitation services. Throughout this period, and until the mid-1990s, many local water management units fell: some systems were managed by the Secretariat of Human Settlements and Public Works, others were managed by the Health Sector, others were self-managed systems on a lesser scale, and yet others were managed by federal councils for drinking water and sewerage created by the SRH and operated now by the SARH. This chaotic state of affairs posed the Federal Government the challenge to provide a general scheme for providing guidance and order.

In another context, in 1982, Mexico had over its head the crisis of the external debt and a loan from the International Monetary Fund, which was heavily, discussed issues. This arose within the water sector another factor for changing the financial scheme in effect at that time. Therefore, water reforms in Mexico were part of a set of neoliberal government policies enacted during Carlos Salinas de Gortari's presidential term (1989–1994), with the financial support of international agencies (Rap et al. 2004, p. 58).

At the same time, at the international level, beginning with the United Nations Conference on Water, which took place in Mar del Plata, Argentina in March 1977, Mexico participated in every international water forum regarding water policy aspects considered by diverse international entities, and in specialized meetings, many of which were conducted in academic institutions, while others took place in government conferences where legal and institutional aspects were considered.

One of the aspects that arose from the discussions and consensuses reached in these forums, and summarized and exposed by the Global Water Partnership (GWP), is that for solving the problems regarding water resources, an IWRM essential, which is defined as “a coordinating process promoting management and use of water, land and related resources in order to maximize the resulting social and economic well-being in an equitable manner without compromising the sustainability of vital ecosystems”. At this point, a continuous organizational process began by gradually establishing itself in the world and which promoted the application of new water management policies.

Furthermore, in 1989, the first major conflict in Mexico regarding water distribution emerged in the hydrologic region of the Lerma-Chapala basin. This is a semiarid region with a strong pressure on water resources, due to an increasing water demand resulting from major industrial and economic activities, which have prompted disputes over water resources among the comprising states and user sectors, the most important one being the agricultural sector. Environmentally speaking, another issue is the preservation of Lake Chapala, the largest and most important lake in the Country. In the 1980s, after several years of low rainfall, a progressive decline in Lake Chapala began, exacerbating the disputes over the surface distribution of water in the region. To confront this conflicting scenario, the Regional Management of the Lerma-Chapala Basin was created, one of the six regional official entities of the Secretariat of Agriculture and Water Resources, established as a part of the National Water Plan of 1975. This action was considered the first step in the country, on the side of the Federal Government, towards the recognition that the centralized management pioneer in several initiatives that contributed to the transformation of the institutional roles of the water sector in Mexico, since the participation dynamics at a regional management level, proposed at that time for the resolution of conflicts, solved the problems of the basin with a new vision. This initiative enabled the International Network of Basin Organizations to hold the first General Assembly in Mexico, in 1994, and its chairman was precisely the Manager of the Regional Management of Lerma-Chapala Basin at that time (Hidalgo and Peña 2009).

During the 1980s the Regional Management of the Lerma-Chapala Basin was in charge of providing information, improving the plan for water management in the area as well as defining institutional roles in order to involving players or groups of interest of the basin in the decision making. All this through a political control from the federal government. This effort is reflected in the first Coordination Agreement between the federal government and the governments of the five states that share the basin, which was signed on April 13, 1989, before President Carlos Salinas de Gortari and the governors of the states of Guanajuato, Jalisco, Michoacán, Querétaro and Estado de México.

The main purpose of the agreement was to implement a Program of water exploitation and sanitation of Lerma-Chapala basin. To follow up on this program, on September 1st, 1989, the Advisory Council of Lerma-Chapala Basin is created composed of governors of the five states as well as representatives of diverse secretariats and industrial users. Finally, for putting into practice and evaluating the follow up of actions emerged from the program, as a part of the Advisory Council, the Technical Working Group of the Basin Council is created, composed of the state members; its objective was to establish those technical basis of an agreement for water distribution and uses in the basin. During its exercise, the Regulation of the Basin Council Operation was establishes in the Agreement on Special Coordination of the Program of Availability, Distribution and Uses of Surface Waters of National Property of the Lerma-Chapala River Basin (Comisión Nacional del Agua 1991), and the Basin Management and Sanitation Program was agreed upon; having four ruling objectives: (1) Clean up the Basin; (2) Regulate water use among federal entities; (3) Achieve an efficient water use; and (4) Manage and preserve basins and streams, (Oseguera Green 2006).

The experience acquired with the process of negotiation in the conflict of Lerma-Chapala basin was sufficient to shape the amendments of the National Water Law in December 1992, which was reformed including major modifications. One of those was to include, on an institutional level, the creation of advisory and coordination instances that could formulate and execute those actions and programs to improve water management, the development of hydraulic infrastructure and the preservation of water resources, that is; the Basin Councils. So, through these amendments in 1992, the Advisory Council turns into the Council of Lerma-Chapala Basin on January 28th, 1993. By creating this, the first basin council, the water sector policies in Mexico began to consider the creation of these mechanisms and forums to let the parties of interest perform their respective duties in the development and management of water resources, thereby facilitating its participation in the decision making. Therefore, as part of the decentralization process promoted by the federal government in December, 1989, the National Water Commission was established as the sole authority for water, and during the 1990s irrigation districts were transferred into agricultural users; every State establishes a State Water Commission as a facilitator for water authority and, water users are organized in civil associations in order to participate as representative of water uses in Basin Councils (Hidalgo and Peña 2009).

8.3 The Mexican Bimodal Model of Implementation of the Integrated Water Resources Management

On April 30th, 2004, the new reforms to the National Water Law are published, stipulating the IWRM as the best way to face water conflicts and its problems, water resources and its management are considered as a strategic and national security resource and, the basin as the planning and management unit. The strengthening of basin councils is set forth with more attributions and an organizational structure, definition of basin commissions, technical basin committees and technical aquifer committees as participation instances in the different levels and, transforms regional management into Basin Organizations, recognizing them as regional water authorities (Hidalgo and Peña 2009).

In this way, the National Water Commission acts in a technical, ruling and consulting character, in two different scopes described in the Law, one is national and the other regional through Basin Organizations. While Basin Councils are defined as collegiate bodies of mixed integration, for the coordination and agreement, support, consulting and advisory between Conagua, including the corresponding Basin Organization, and those units and entities of federal, state or municipal instances and representatives of water users and social organizations from the respective hydrologic basin or region (Comisión Nacional del Agua 2004). This rationality and restructure of the national system, that emerges with the amendment of the 2004 Law, for planning and water managing in the basin level in the country represents an institutional breakthrough seeking the integration of user participation in the process and widen policies and actions range towards nonstructural initiatives, with the ultimate goal for the preservation and sustainability of the water resource. Nevertheless, we must emphasize that a confusion remains between two scopes of water management that should be clearly differentiated. For water planning it is important to distinguish between the scope of management of water access for different sectors (or large users), related to the determination of usable volumes in a basin and the management of titles permitting an orderly access to those volumes with mechanisms to manage abundance and extreme scarcity; from the scope of management of users, public or productive systems, mainly the subsector of drinking water and sewage and that of irrigation systems. They are scopes and planning objects completely different, since, the first is necessarily linked to the geographic scope of the basin or aquifer; while in the second, the management of WUAs systems, is determined by the limits of each one (once such a system has an assignment in water block), so it is more related to administrative and budgetary structures oriented according to the political and administrative organization of the countries. In our country they are supposed to depend on a sole basin authority (Sandoval 2004). In Mexico, the above mentioned first scope is developed under the coordination of the Federal Government, through Conagua, entitled by the Constitution and the National Water Law to conduct the programming of national waters with the tendering processes of users, citizens and state and municipality governments, through mechanisms established by Conagua, mainly related to the structure of basin councils, but with no effective attachment to the scope of planning of users systems, more than in the domain of regional restrictions to water access and quality conditions of discharges. The second scope, regarding urban and rural public services, is still centered on infrastructure investment, covered under a complex web of budgetary programs and mechanisms which condition the resource concurrence among the three government orders; the evolution of water supply and sewage services belong to the portfolio of actions in the Municipality Development Plan, responds to priorities emerged from rural communities and urban settlements but in its enforcement it faces the application of policies and operative rules defined in the state and federal governments with its corresponding mechanisms of information, "validation" and verification (Sandoval 2004).

In Mexico, this management system is composed of entities which interact from different governments, public and private areas of action, based on decision guidelines mainly defined by the federal government; this rules define the exchange of natural resources (water access), financing (budgetary and those derived from collection of various fees) and information that lead to the response of each entity before its counterparts' behavior. This is not producing effective results, in an efficient manner, elements that characterize ungovernability of the resource in the country. Nowadays, Conagua is working through 13 Basin Organizations, one for each hydrological-administrative region in which the country is geographically divided. There are 26 basin councils. All of them with their respective auxiliary as: Basin commissions (30), dealing with sub-basins, Basin Committees for micro basins (50), Groundwater technical committees (COTAS) for aquifers (88) and Clean Beach Committees in coastal areas of the country (41), reaching a total of 215 auxiliary organizations (Comisión Nacional del Agua 2016).

In spite of this, a structural deficiency remains in the municipal systems development, where there are no real mechanisms to guide public spending towards national o regional development policies. Which makes evident the inconsistency, in the field of public supply and sanitation systems, of a targeted planning approach from the center of the country, either by implementing methods of participation which are finally sanctioned and organized in the Federal District offices against the national planning system that transfers towards investment programs, the priorities obtained from local structures interwoven in a set of programs and operation rules that have an impact upon the municipal budgetary expenditure.

Yet, those planning exercises that have resulted in regional water programs, have had the great virtue of representing a more integrated framework, directed in a lesser manner towards large infrastructure works; have also been short in recommendations at the level of users system management for public or productive use, since its recommendations or strategies emerge from very wide and thorough processes to focus in specific solutions, which show this lack of targeting between a basin scope and a user system scope (Sandoval 2004).

There is a shortage of material and human resources which increasingly limits the most needed surveillance and legal sanctioning function that conditions the effectiveness of water management, as well as a diagnose in terms of such areas involved in water control and use, distribution of capacities among them and the efficiency in information and resources exchanging (Sandoval 2004). Basin organizations are considered more as administrative instances and less as surveillance or management instances. Budget is still controlled and centralized at the federal level. Thou managed through federalized programs, those in several occasions pose so strict operation rules that the resource is not able to reach those municipalities really in need, or the resources are oriented towards constructed works that were no able to be maintained as has occurred with countless built and abandoned treatment plants. The lack of participation of residents in the area is limited.

There is no diagnosis, much less a proposal, towards the identification of inefficiencies in communication, resources transferring, regulation of budgetary mechanisms, and technical and normative supervision, all of these elements that are frequently the main problem for the development of hydraulic services. Something similar occurs in the relation among organized irrigation users and the different areas of Conagua and other federal agencies continuously bound to (Sandoval 2004).

There is a lack of mechanisms to promote a harmonious participation of municipal and state areas in the functions of water management and inherent public goods of national propriety and now under federal jurisdiction (Sandoval 2004). The water planning system nowadays reaches the municipal level through participatory planning exercises but with only consultative intervention. The definition of the necessary actions for investing in drinking water supply and sanitation in rural areas is based on the structure of the National System for Democratic Planning, around the mechanisms for compiling needs translated into municipal programs as well as other development issues; this instrumentation hinders the articulation of processes which are oriented towards assuring the population well-being in a sustainable way, promoting projects, mainly in infrastructure planning, and avoiding or skipping municipal intervention in the promotion of sustainable processes of water services development (Sandoval 2004). States and municipalities have a limited technical and operational capacity for solving problems, and Basin Organizations are also limited to be present everywhere in a technical and operational manner, although it would not be desirable either.

Even thou the Law establish that basin councils shall manage as resources at least water and forests, in practice all issues related to environmental problems are addressed in a separate manner from those solely related to water. The entity of combined collegiate organizations in the councils has limited local participation for having in these instances representatives emerged from the three orders of government that lead and exert pressure on local level problem discussions.

The implementation of councils with no specific goal or problem to be solved promotes the establishment of large hydrological regions, as Rio Balsas basin, which covers an 116,024 sq. km area, comprising the States of Mexico, Morelos, Tlaxcala, Puebla, Oaxaca, Guerrero, Michoacán and Jalisco. This leads to a variety of problems, which can hardly converge in common points, reaching no agreement as a basin or hydrological region. Especially in the Lerma-Chapala basin, although a constant negotiating process with stakeholders has generated a series of treaties and agreements amongst involved states, scarcity of the resource, pollution and degradation problems continue. There are certain groups or NGOs that have locally worked with small groups in the basin, but their results are not known or not discussed in these instances. Usually, basin councils discuss problems and reach agreements which most of the time, as a result of budgetary reasons, are not materialized. As consulting instances, their proposals are not mandatory for the parties, and frequently their actions are not reflected in state or regional programs. There is a lack of awareness regarding their concrete works, except those of Lerma-Chapala basin and just up to a certain extent.

8.4 Results and Impacts of Integrated Water Resources Management

In Mexico, integrated water resources management, in practice is far from crystallize in reality. However, the process which started with the amendments of the Law in 2004, in terms of a decentralization process is not considered a proper route, but as the beginning of a process that needs to be consolidated and settled.

It is important to emphasize that in the international scope, concerns regarding water management and particularly because of the water governance crisis, are being increasingly evident in the international water forums, which have stated for a number of years, that political and economic aspects are the limiting factors for the effectiveness of solutions regarding handling and managing the resource. Mainly because the relative scarcity of some sub-regions and countries exceed the problems of natural availability (geographical factors) over problems regarding: (1) Poor planning, not just water but other sectors as well; (2) Misconception of water resources; (3) Indiscriminate uses and water appropriation; (4) Unsuitable regulations or far from social and cultural context; (5) Coordination amongst policies. Hence, water governance has become increasingly important, standing as a crucial element for global players and in politicies associated to water issues; turning into an external condition prescribed by international institutions for granting international aids and loans, an essential requirement for a successful development and implementation of IWRM programs (Water Forum of The Americas 2011). But what does Water Governance mean? The United Nations Development Program defines Water Governance as the interaction of the political, social, economic and administrative systems to regulate water resources development and management and the provision of water services at different levels in society. It exists when a state policy is effective, bound to an appropriate legal framework to regulate and manage water, so that it responds to environmental, economic, social and political needs of the Nation with the participation of all stakeholders. According to IWRM, a proper governance is the one where the authority sets the rules and offers opportunities for participation to other players and sectors in water management, where the decision making regarding policies established by consensus with all those players involved in water management. For this, it is necessary to have an effective participation in the decision making, acknowledging social and cultural settings to generate new institutional frames and institutional operation (Water Forum of The Americas 2011).

The above mentioned is related to what Kooiman stated about a shift from traditional models where governing was considered basically as "unidirectional", from governors to the governed; towards a "bi-directional" model where aspects, problems and opportunities of the government system and of the system to be governed are considered (Kooiman 2005, pp. 60–61).

This is mainly related to the interdependence among social players, i.e., no stakeholder by him/herself has the knowledge or information enough required for solving complex, dynamic and diversified problems. No stakeholder is experienced enough to efficiently use the required instruments. No actor possesses sufficient action potential to have the domain unilaterally (Kooiman 2005, p. 60).

According to this, to talk about Governance we need to include other levels of government, apart from the central or federal government; and it is even imperative to note that decisions by consensus are more effective in the preservation of water resources; incorporating the points of view of relevant social actors will ensure more sustainable and more socially accepted solutions in the long term. This implies a new challenge to water resources management, since it is not enough to meet sustainably goals, but to achieve a new form of participation which allows on the one hand, co-responsibility of society in water are, along with authorities. This poses to our country a challenge to generate a stronger authority, credible in actions, as well as a more conscious, responsible and mature society in the exercise of their right, but at the same time, of their obligations. To that end, as a principle, it should be clearly stated that federal and state authorities, that is authorities external to municipalities, would only participate in aspects to the extent of problems the immediate order could not be able to solve. In fact, nowadays, some municipalities are more capable of ordering the development of their water and sanitation system than their own state, regional and national authorities.

Besides, it is not deemed for the federal authority to extensively intervene in planning municipal services improvement, except for those related to planning and managing the assignment of volumes in the basin scope, as well as defining treatment, reuse, and discharge conditions under hydrological balance and regional quality criteria that ensure availability and quality of water before productive uses.

Acknowledging and stimulating, as well as encouraging the participation of local players in the execution of planning exercises, contributes to inform and increase the knowledge of more sectors in the population towards water related activities in different scopes and to modify courses of action of participating entities. Decentralized operation, as certified networks of climate and hydrometric stations, can contribute with important assets in the national water planning (Sandoval 2004).

It is essential to conduct a thorough review of current information exchange, financial resources and execution actions; and supervision of programs for the development of drinking water supply and rural and urban sanitation in Mexico. Nowadays, they depend entirely on the uncoordinated definition of operation rules and execution guidelines for each of the parties, with no clear frameworks for the hierarchical organization of actions at state and municipality level. Overlapping consecutive regulatory and supervisory structures, by foreign banking which is the financial support to rural communities, as an example, going through three different areas of supervision in the federal government, plus state area and the intervention of municipal structures of programming and planning, is not a contributor to an efficient solution of problems in coverage and sustainability of rural systems (Sandoval 2004).

The only way that planning acts as a "working system to organize water management" (as stated in the exposition of causes for the amendment of the National Waters Law), as well as achieving water governance and governability, will be by establishing thorough mechanisms that accurately guide and limit the participation of multiple social participation organizations provided by the Law. Likewise, accepting that democratic planning implies the redistribution of faculties to sponsor and implement solutions, for which it is essential to have specific fiscal decentralization policies to simultaneously assist in preventing overlapping or duplicating structures or regulation or supervision (Sandoval 2004).

8.5 Lessons Learned

Mexico is currently facing a different economic and social situation compared to more than two decades ago. Government legitimacy has been transformed from its roots as the transition of the political regime progresses. Now it is rather difficult for the federal water authority to implement diverse actions with no participatory component, information to society, negotiating with local stakeholders or even in coordination and consent of other government levels. Shifting from the centralized management of water model to IWRM implied the redistribution of the decision making, financial resources, knowledge related to water systems functioning, considering no matter how it was done. But the remaining question is precisely if this decentralization and involvement of so many social actors and all government levels is not yet opposed to the Basin Organization design, along with its centralized way of working. There is also the question whether the current institutional agreement is really capable of regulating the dispute over water, which moves diverse interest groups with dissimilar levels of power and action capabilities. Every water conflict that reaches media coverage reveals the enormous prevailing asymmetries among those organized around water, the evident absence of basin and aquifer councils, commissions and committees as instances of agreement, the existence of opportunistic strategies for water use in a context of weak regulation at a local level. This need for change is not solely regarding conservation and sustainability of water resources, but also as a developing institutional and organizational process what implies a change. A change requiring major managing, planning and acting ways where authorities, along with society; define new courses of action.

While mechanisms developed up to date have not responded as expected, they have enabled a major leap. However, consolidation of this process requires Mexico to transform and strengthen the water sector. A coordinating instance at the federal level is needed to lead the design and implementation of policies to guide the works of those stakeholders involved in managing the resource, supported in their execution by regional and local instances and by society represented basin organizations which validate and legitimize the works of those involved for maintaining the sustainable development of water basins in the country.

References

Aguilera Klink, F. (1998). Hacia una nueva economía del agua: Cuestiones fundamentales. *Ciudades para un futuro más sostenido, Boletín CF+S*(8).

Aldama, Á., Gómez Ugarte, L. A., & Hernández, J. (2006). *Instituto Mexicano de Tecnología del Agua: 20 años de innovación con sentido social*. México: IMTA.

Barkin, D., & Timothy, K. (1979). *Desarrollo económico regional (enfoque por cuencas hidrológicas)*. México: Siglo XXI.

CEPAL. (1989). La gestión de los recursos hídricos en América Latina y el Caribe. Naciones Unidas. *Estudios e Informes de la CEPAL, 71* (Nota 30), 26–251.

Comisión del Plan Nacional Hidráulico. (1981). *Plan Nacional Hidráulico 1981*. México: Secretaría de Agricultura y Recursos Hidráulicos.

Comisión Nacional del Agua. (1991). *Consejo de Cuenca Lerma-Chapala, Acuerdo de distribución de aguas superficiales para la cuenca Lerma-Chapala*. México: Colección Lerma-Chapala, Vol. 1, Núm. 5.

Comisión Nacional del Agua. (2004). *Ley de Aguas Nacionales y su Reglamento*. México: Comisión Nacional del Agua.

Comisión Nacional del Agua. (2016). *Estadísticas del Agua en México* (2016th edn.). México: Comisión Nacional del Agua.

GWP. (2000). *Manejo integrado de los recursos hídricos*. Estocolmo (TAC Background Papers, 4): Global Water Partnership.

Hidalgo, J., & Peña, H. (2009). Turning water stress into water management success: Experiences in the Lerma-Chapala River Basin. In R. Lenton & M. Muller (Eds.), *Chapter 8 in integrated water resources management in practice. Better water management for development*. London/Sterling: Earthscan.

Kooiman, J. (2005). Gobernar en Gobernanza. In A. Cerrillo & I. Martínez (Eds.), *La Gobernanza Hoy: 10 textos de referencia* (pp. 57–81). Madrid: Instituto Nacional de Administración Pública.

Martínez, M. A. (2004). *El despegue constructivo de la revolución, sociedad y política en el alemanismo*. México: CIESAS-Porrúa.

Newson. (1992). *Land, water and development: River basin systems and their sustainable management*. London: Routledge.

Oseguera Green, V. (2006). Evolución histórica de los Acuerdos de Coordinación en el Consejo Cuenca Lerma-Chapala. In H. Cotler Ávalos, M. Mazari Hiriart, & J. de Anda Sánchez (Eds.), *Atlas de la Cuenca Lerma-Chapala. Construyendo una visión conjunta* (pp. 90–91). México: Secretaría de Medio Ambiente y Recursos Naturales, Instituto Nacional de Ecología, Universidad Nacional Autónoma de México.

Palacios Vélez, E. (1994). *Elementos para el marco de referencia de la política hidroagrícola de mediano plazo*. México: PROYECTO UTF/MEX/030/MEX CNA/FAO EL.01-94.

Pineda, N. (2002). La política urbana del agua potable en México: del centralismo y los subsidios a la municipalización, la autosuficiencia y la privatización. *Revista región y sociedad, XIV* (24).

Rap, E., Wester, P., & Pérez-Prado, L. N. (2004). The politics of creating commitment: Irrigation reforms and the reconstitution of the hydraulic bureaucracy in Mexico. In *The politics of*

irrigation reform. Contested policy formulation and implementación in Asia, Africa and Latin America. (pp. 57–94). Farnham: ASHGATE.
Sandoval, M. R., (2004). No más planes al agua: hacia una gestión hídrica planificada y eficaz en México. In C. Tortajada, V. Guerrero, & R. Sandoval (Eds.), *Hacia una gestión integrada del agua en México: retos y alternativas* (pp. 47–91). México: Centro del Tercer Milenio para el Manejo del Agua A.C.
Vargas, S., & Sánchez, M. A. (1996). Las grandes tendencias de la agricultura de riego. In R. Melville, & F. Peña (Eds.), *Apropiación y usos del agua: nuevas líneas de investigación.* México: Universidad Autónoma de Chapingo.
Water Forum of The Americas. (2011). *Hacia una buena gobernanza para la gestión integrada de los recursos hídricos*. México: Documento Temático de la Américas.

Jorge Arturo Hidalgo-Toledo is a researcher at the Mexican Institute of Water Technology since 1995. He has earned his Master of Science degree in Planning and Management Infrastructure in 1983 from the Leland Stanford Junior University. He was professor at the National University of Mexico for 12 years. He has published more than 100 papers and reports on hydrology and water resources management. He received the 1980 National Public Administration Award by the Agricultural and Water Resources Secretary of Mexico. His recent research interests are dynamic river basin models and integrated water resources management and planning.

Cipriana Hernández-Arce earned a PhD in Social Sciences at the Universidad Autónoma del Estado de Morelos, and Industrial Engineer and Master in Planning Engineering at Universidad Nacional Autónoma de Mexico. She has been working at the Mexican Institute of Water Technology since 2000. Her main lines of research are simulation of water distribution models, water policies transversally; consensus building base on hydrological models, regional and state planning for the creation of water programs, and more recently, the study of the social impact policies on water basins.

Sergio Vargas-Velázquez is an Economist. He earned a Master in Social Sciences and a PhD in Social Anthropology. He shared 12th in the Leadership for Environment and Development Program in Mexico (LEAD-Mexico). He is a specialist on social issues in water management. His main areas of research are organizational and political processes in irrigation systems, productive change in irrigated areas, water user organization for watershed and aquifer management, and in the last years on social issues in water conflicts. Between 1990 and 2009, he held various positions in the Mexican Institute of Water Technology, where he participated in projects on transfer of irrigation districts, the organization of councils, commissions and committees at basin and aquifer level, characterization of water conflicts, and negotiation processes.

Chapter 9
Payments for Environmental Services: Between Forest Resource Management and Institutional Building

Ricardo Hernández-Murillo and Sazcha Marcelo Olivera-Villarroel

Abstract Environmental services from the perspective of the State have as a public policy objective the conservation of natural environments as well as the coverage of plants that provide environmental services to society, while, from the owners' point of view, decisions on the use of forests, whether through forest extraction activities or the change of land use for agricultural or urban activities, depend on the opportunity costs, the conditions of production, and access to economic resources. This chapter conceptualizes the understanding of environmental services from the economic point of view. The contrast and the workings of the payments for ecosystem services and the community forest management are analyzed; whereas in the last section, it examines the current design of ecological flows to appreciate the environmental services provided by rivers and wetlands. This is exemplified in a case study (Capulalpam de Mendez, Oaxaca). From this perspective, this chapter analyzes the current institutional framework that regulates environmental services such as the recognition of the ecological flow and the payment system for environmental services, emphasizing the contrast with the management of forest resources by forest owners and managers.

Keywords Payment for environmental services · Institutional framework · Forest · Ecological flow · Capulalpam de Mendez Oaxaca

9.1 Introduction

One of the most important causes of the loss of forest areas and the drying of wetlands around the world is driven by the growth of farming regions and the change in land use for agriculture and livestock purposes (FAO-CONAFOR 2005;

R. Hernández-Murillo
World Bank (in memory), Mexico, DF, México

S. M. Olivera-Villarroel (✉)
Departamento de Teoría y Procesos del Diseño, Universidad Autónoma Metropolitana – Cuajimalpa, Ciudad de México, Mexico

H. R. Guerrero García Rojas (ed.), *Water Policy in Mexico*, Global Issues in Water Policy 20, https://doi.org/10.1007/978-3-319-76115-2_9

Aukema et al. 2017). This change involves the loss of services generated by these ecosystems along with their social, environmental, and economic influence; the current biodiversity in those systems and their contribution to the climate change process by reducing the amount of already present carbon sinks (Hernández 2008; Farley and Constanza 2010; UNEP 2011).

The purpose of this chapter is to analyze the current institutional framework that regulates the payment for ecosystem services in contrast to the management of forest resources done by private owners and administrators as well as recognize ecological flows in wetlands around the country as suppliers of environmental services. This chapter will conceptualize the understanding of environmental services from the economic point of view. The contrast and the workings of the payment for ecosystem services and the community forest management will be analyzed; whereas in the last part, the article will examine the current design of ecological flows to appreciate the environmental services provided by rivers and wetlands. This will be exemplified in a case study.

The recognition of the contributions made by ecosystems such as drinking water in urban and rural areas, agriculture, and fishing has evidenced the requirement for enforcing laws that allow governments to maintain the previously stated services. In addition, these laws would provide the owners and administrators (of woods, wetlands, pastures, etc.) with economic incentives so they would avoid changing the land use for activities that pay for their efforts in the short term and that would grant them access to commodities such as clothing and food. The most important instruments generated by the public policy to recognize the importance of the environment and its services are the creation of Natural Protected Areas, a compensation for the Payment for Ecosystem Services, and the establishment of Ecological Flow Regimes in stream banks and aquatic ecosystems.

9.2 Environmental Services and Economic Thought

The importance of the functions of forests, wetlands, and natural surroundings in providing the essential ecosystem services for society only highlights the discrepancy between the rationality of social interests and the individual behavior of the agents in that social order.[1]

In other words, the evidence of the continuous loss of forest cover, the desiccation of wetlands, and the land-use change that the country registered (CONABIO 2009) indicate that the sum of all individual behaviors seeking their economic benefit does not move toward a common good.

[1]Mangrove forests provide services at a local level, as a place where several commercial species reproduce; worldwide this carbon fixation rate is three times higher compared to that of other types of forests (UNEP 2011).

This means that an individual gets an individual benefit from a conduct that may impact negatively the common ownership if attached (Hernández 2008; FAO 2009). For example, agriculture provides food and energy, but it is often also associated with considerable negative environmental externalities. Changes in agricultural land-use strategies and production technologies can potentially trigger large (positive or negative) environmental impacts (McNeely and Scherr 2003).

The owner of the resources (forest or wetland) or whoever has the right to enjoy a harmful usufruct (including the change of forest soil or the development of fishing activities toward other uses) internalizes an economic benefit even if this decision has negative consequences for the social context. However, the decision of preserving forests or wetlands in the short term will not make profits for the owner who decides to sacrifice the benefits that would be obtained.

Economic growth in these areas causes environmental degradation and will continue to do so without the creation of effective policies. Many argue that the policies that would change this would be too expensive and that they would slow growth and destroy jobs. However, environmental degradation itself is costly (Hallegatte et al. 2012). Because of this, those who are involved must search for a resolution in which both, economy and environment, can coexist in equilibrium.

If there is no legal and institutional framework responsible for recognizing the ecosystem services provided by the natural surroundings and that warns of the costs of the negative consequences, the owner's incentives will not match the social objectives. In its current configuration, the legal and institutional framework enables the benefits realization at the expense of the common good. "During the deliberations of the United Nations Intergovernmental Forum on Forests, it was agreed that the underlying causes of deforestation and forest degradation are interrelated. The underlying causes include" (FAO 2012):

Poverty	Absence of a supportive economic climate that facilitates sustainable forest management
Lack of secure land tenure patterns	Illegal trade
Inadequate recognition within national laws and jurisdiction of the rights and needs of forest-dependent indigenous and local communities	Lack of capacity
Inadequate cross-sectoral policies	Lack of an enabling environment, at both the national and international levels
Undervaluation of forest products and ecosystem services	National policies that distort markets and encourage the conversion of forest land to other uses
Lack of participation	
Lack of good governance	

The economic incentives offered by the failures in the institutional laws and the unfair exchange conditions are so powerful that the owner or the people with no right over the natural resources continue illegal logging and changing soil use despite having sanctions and the fact that these actions jeopardize the quality of life for this and future generations.

It is worth mentioning that, to understand the payment for ecosystem services (PES), we should avoid certain prejudices such as:

- To compare PES to a strict conversation, when it can be compared to different extractive activities
- To relate PES to the income obtained by a production cycle, when it could be possible to take the net present value (NPV), according to the terms in the PES contracts versus the expected net income in the same period with a looting use
- To compare the PES income to selling products that would not be obtained by means of an alternative use without deducting supply expenses, the workforce, and financing risk of both options
- To relate PES to "doing nothing" since in a country with high deforestation rates, the agreements of the program clearly involve an active forest conservation (Chapela and Lara 2007)

9.3 Environmental Services As Public Assets

Environmental services generate public assets from private actions that work in different regions and scales depending on the service we refer to (Heal 2000). The classic definition of public assets where they are defined as assets whose nature determine their non-rival character and for which it is impossible to establish a selective exclusion of their benefits turns out to be inadequate to deal with environmental services and assets (Kelsey et al. 2008; Kelsey 2013). This definition makes us think which economic instruments are likely to design the administrative structure that brings together public assets and market rules while at the same time it will establish organizational limitations.

From the economic perspective, the conservation of the services provided by a natural surrounding is a public asset due to the positive externalities. These allow the society to keep its current quality of life and to maintain the environment productive condition. Rivalry is described as the degree in which the use of goods by an individual reduces the availability to others. Goods such as music being listened in a public square, highways, or information are not rival goods as long as they do not reach a saturation point where every additional user's benefit decreases both the quality and consumption for him and even for the group as a whole. So the rivalry, unlike saturation demand of a limited good and the economic shortage threat, leads the user to guarantee his rights of future access and, thus, to the demand for property rights over the stocks or environmental services (Chomitz et al. 2006).

Excludability is a key element to recognize externality. To confirm the rights of movable property and real estate, alienation is the capacity par excellence, partly because excludability is taken for granted and it distinguishes public from private. Nevertheless, there seems to be a problem when the information asymmetry and the capacities of alienation (or power) are such that it is not possible to guarantee the excludability because of the lack of physical resources to demarcate and protect the limits of a good or a service (even with a title deed) and the absence of institutional resources to enforce property rights.[2] The problem with excludability has been particularly ascribed to the characteristics of the good or service we are dealing with, if we compare this problem to the physical challenge of excludability due to demarcation and surveillance "To try to solve the distinction between public and private goods." Even though if this classification criterion has been accepted, it would be difficult to find public or isolated goods.

In the best-case scenario, private and public goods are related in such form that optimal functioning of a market of private goods is in itself a public good (Olivera 2007). In this respect, the outcome from those private actions delivers value to society, in a non-rivalrous way, and therefore could be also framed as a public good. The production of the (normally) private goods can also embrace aspects of public goods, for example, where the production of agricultural products or particular agricultural systems and the landscapes they produce are a part of the local identity (Novo et al. 2017).

History and geography demonstrate that any good or service, no matter how intangible it is, can be subject to private appropriation. Unless there are appropriate rules and organizations to guarantee them, any good or service, tangible or personal property, will be difficult to control. The reason is that the organizations responsible for respecting property rights and carrying these laws out are insufficient.

There is a restriction to environmental services (ES) in the markets, and this is the cost of excluding external agents from the benefit flows. Some critics of the payment of ES claim that to demand payment from the users, producers of ES should provide evidence of the relationship between their activities and the provision of these services. Jurists, on the other hand, mention that, in order to talk about the payment for ES. It is necessary that the law recognizes them without taking into consideration the certainties obtained as a result of scientific knowledge, no matter how rich it was (Andaluz-Westreicher 2016).

Jurists argue that formal recognition must precede the contractual agreement. In accordance with this belief based on legal technique, transactions cannot be made on anything that is not considered subject to trade. This means that there must be at least the will to acknowledge the existence of a service coming from the natural resources so we can talk about PES.

The problem of the inexistence of recognizing a good like a subject of rights and exchange (only recognizing it as an externality, intangible, and public good) can now be solved by the explicit recognition of the property rights by one of the parties, enforced by the State and the society. This phenomenon is the result of a simple concurrence of wills, an expression of trust, self-interest, and even certain trust in formal or informal institutions, which guarantee these transactions. The fact that

[2]In Mexico's countryside, the assumed uncertainty in the possession of the land is not caused by the absence of a recognized legal document, but by the inability to guarantee the excludability, establish limits, and mobilize the institutional resources needed to enforce their rights.

these transactions are regularly registered and that they work well implies a bigger social consequence.

The persistence of these transactions and their social sanction has become a great legal phenomenon. The contracting parties and the society as a group believe that a fair act is being performed. From the ethical point of view, remuneration is given in the sense that it should be fair or proportionate. The legal consequence of PES is the confirmation that remuneration is justifiable with the necessary jurisprudence to contribute to casuistry developed, so in this way, we are building the legal framework of PES (Andaluz-Westreicher 2005, 2016).

While environmental goods and services were defined as stocks of infinite flows present in a territory likely to be extended, and whose exploitation did not require rationalization, assignment of property rights, cost internalization, and valuation, it was possible to consider them to be public goods. However, as pressure over these resources was growing, as well as the subsequent perception of a real or potential shortage, and as the nature of production cycle and biological reproduction (in geological cases like oil and fossil aquifers) were known, people became aware of the fact that the constant availability in a permanent way and the inevitable development of the rivalry in their use would be present.

However, the user can still perceive the environmental services as public, non-excludable, and non-rival. Therefore, the excludability or restriction of consumption to future users or beneficiaries depends not only on the physical characteristic of goods but also on the quality of the institutions. Excludability of ES, in some cases, can be carried out through physical barriers (this gets complicated when dealing with services such as "fresh air" or the beauty of a landscape), the efficiency of the excludability will depend on the assignment of property rights and the institutional arrangements for their defense. For these goods and environmental services to be appropriate in a private way, institutions are being built to lay down rules and to guarantee mechanisms that allow making effective the selective excludability or commercialization rights. It is a must to identify that there is indeed something in nature and that the role of goods and environmental services makes us consider them, partially, as public goods.

9.4 Case Study: Capulalpam de Mendez

Capulalpam de Mendez is a municipality located in the state of Oaxaca, Mexico, which has been a part of the PES program, with a population of approximately 1300 inhabitants and an irregular surface of 3850 hectares. It is a community that has been promoted as "magic town" due to the condition of its forests which give this place its picturesque aspect, so common in Oaxaca's northern mountain range. Forest goods and services are conceived as non-excludable within and managed by the community in a process to ask for the recognition of the national society for the positive externalities of the local landscape.

Capulalpam de Mendez has a forest that was almost unused until the 1950s. In 1956, it was first given recognition of its value when the lands for wood exploitation were granted concession by the federal government. Despite that, this community was able to get some benefits of the eminent domain and established timber rights on behalf of the licensee and which was managed by the federal government in benefit of the affected communities by the forest exploitation. Since then, the forests have financed the government investment in public services for the communities of this area of the Oaxaca mountain range. In 1981, the decision to renew the forest concession was revoked and the forest became community usage. By doing this, it provided the community with a series of management options that the community had to consider.

In the case of this community, it is limited to collective action problems, which we analyzed in the previous sections of this paper since a good resource management requires an efficient collective action, common goals, and an understanding of their surroundings (forest), regulatory system, and internal surveillance. These actions should be based on economic and sociocultural decisions that restrict and allow decision-making in changing legal frameworks. This community has the support of organizations like the Consejo Regional de Recursos Naturales (Regional Board of Natural Resources) and the Unión Zapoteca Chinanteca (Zapoteca Chinanteca Union) which provides technical forestry services to four forest communities in the area.

Capulalpam de Mendez bases its social organization in a cargo system, understood as a non-paid job of cultural roots which corresponds to the indigenous social structure of this region of the Oaxaca's northern mountain range. To sum up, it is a Zapotec indigenous community where the Comisariado de Bienes Comunales (Communal Goods Commissary) is a lawful effective authority. They know that work is a community service; therefore, the person in charge will need a second job to have an income. Cargo system is based on customs in which the "cargo" (position) lasts between a year and a year and a half. There are two types of cargo: citizens imposed by the citizens' assembly (their responsibilities include to serve the urban area and the municipal services) and the community cargos imposed by the "comuneros" assembly (according to the National Institute of Statistics and Geography, "comuneros" in Mexico are individuals who belong to a rural community and have property rights over lands and can enjoy common goods. They are granted this "comunero" (indigenous farmer) status for being members of the farmer population) who serve the community as a group. Cargos are somehow important in the social and economic life of the community; the positions can go from security guardianship to presidential positions in community institutions like forestry and mining industry, both the citizens and community assemblies consider the economic and family situation of individuals before assigning them a job.

Capulalpam is a multi-active community; it works as a complex production unit with a diversified strategy in the use of their carefully thought out resources to increase family income of the members and the resources of their community.

Among the mentioned strategies, we can find immigrant consignments, forestry and mining work, carpentry, goldsmith, and in a subsidiary way tourist activities, farming, and stockbreeding activities. Since the community administers the utilities generated by such activities, the latter can be transferred to communal enterprises and are considered to fulfill certain communal services such as: religious activities, maintenance of infrastructure facilities, scholarships, and educational trips.

The community and the municipal territory are identical, and the number of domiciles with no rights is so reduced that in practice they operate with an almost perfect complementarity that equals a shared government, where the city council works as a counterpoise to that of the commissary of communal goods and vice versa. In this community, it is possible to observe that enterprises finally have dual purposes by providing jobs and producing economic rents for common use by exploiting natural resources of common property.

The Environmental Services Program for the Capture of Carbon, Biodiversity and Agroforestry Systems (Programa de Servicios Ambientales por Captura de Carbono, Biodiversidad y Sistemas Agroforestales – PSA-CABSA – which is a governmental program created in 2004 as a strategy to promote mechanisms for the payment of environmental services in Mexico) is the biggest step toward the recognition of the community's territory and the reappropriation of the forest. The contract signed with CONAFOR (National Forestry Commission) was created in 2001; one of its functions is to develop and encourage production activities, and preservation and restoration of forests leads us to a recognition of rights that reinforce the role of the community as landlord not only of the forest but also of the environmental services that it produces. This contract forces and at the same time recognizes the work being done to preserve their environment (CONAFOR 2001, 2003).

The payment for environmental services is a rent perceived by natural capital controlled by the assembly. The decision to preserve the forest by guaranteeing the flow of environmental services (ES) is part of the community decisions on the management of their natural patrimony and their product diversification. If the preservation commitments required by the PES are at odds with the exploitation interests of the Empresa Forestal Comunitaria (Community Forest Enterprise) or the requirements in farming, livestock lands, or of other uses, the cost-benefit analysis will be of greater importance in the owner's decision, since the decision would correspond to a potentially more productive asset and not to the rent of an estate highly valued for its option value and its contribution to risk reduction and economic and politic vulnerability.

Even in this case, it is difficult to claim that PES is in itself an instrument capable of inducing a change in the owner's (community) behavior in favor of the preservation of the forest. Since the PES should at least offer a competitive amount compared to the profitability of the forestry exploitation, if what we want is a preservation instrument that modifies the current tendencies in the degradation and deforestation processes, even though a lot of communities claim to be part of the CONAFOR's PES program or similar compensation schemes for environmental services that provide their forest (CONAFOR 2006). The analysis of the Capulalpam case suggests that the income reported by the PES program may not be the

Table 9.1 Main economic interrelations between institutions, production, and payment of environmental services. 2006. Value added (2006) in Mexican pesos

Source/ concept	Incomes	Operating cost, investment, devaluation	Operation excess	Community transferences	Investment fund
Forestry enterprise	2,766,696	2,291,651	475,045	101,181	373,864
Mining enterprise	2,398,439	1,952,466	445,993	244,463	201,530
Forest rights	344,232	0		344,232	
PES	50,000	0		50,000	
Municipal income	1,708,992	1,708,992	0		
Total	7,268,359	5,953,109	921,038	739,876	575,394

Sources: Hernández (2008)

determining factor in their decision to preserve the forest, because in the documented case, it represents a lower proportion compared to the incomes from endeavors carried out as a community and certainly does not compete positively with the alternative soil uses.

In Capulalpam, the forest, its preservation and health, is mentioned as a responsibility in the citizen perception of the type of town (understood as the territorial area with a clear assignment of the proportions devoted to the urban, agriculture and livestock, mineral extraction, and forest exploitation) in which they want to live and leave to their offspring. This collective perception of what is desirable turns into the decision to act as a collective individual who does what it takes to lead the general development process in a way that includes the forest preservation.

In this case study, we could identify the isolated elements of an integrated program of PES. The institutional arrangements of the community allow to administer the services within and look for recognition of its benefits outside by means of the PES. However, the voluntary incorporation to the program is accomplished with the subsidy of its other successful activities to the community; in other words, the incomes for other productive activities and the desire to keep the opportunity service of its forests are decisive to the preservation nor are the conditions that the program offers itself (see Table 9.1).

9.5 Ecological Flow

Evidence indicates that the dramatic upheaval in the world's ecosystems is due to pressures brought by land-use change, increasing human populations, the pattern of their settlements, and increasing levels of natural resource consumption. These trends also threaten the future human supply of food and water, conditions of quality of life, and survival of other species (Ortega and Ramos 2016). For forests, there is a consensus that standing natural forests are good at providing clean and relatively stable water flows, yet the impact of tree cover on dry season flows and storm flow protection is highly site specific, asymmetric between forest conservation and reforestation, and sometimes disputed among hydrologists (Ingram and Hong 2011).

As we discussed in the first part of this chapter, a formal recognition from the government and the society is necessary to precede the contractual arrangement that permits to make a payment or compensation for the preservation of environmental services, like the ones provided by ecological flows in aquatic ecosystems and their interrelation to forests preservation. To do this, it is necessary to implement a regulation that recognizes the existence of an environmental service provided for the preservation of natural conditions, as well as a second regulation system that permits the conditions to protect the chosen ecosystem.

That is why in 2004 in Mexico, the National Waters Law was modified to include, among others, the use of national waters for "the ecological protection or environmental use" and, in 2012, the procedure to determine ecological flows in drainage basins was established. This regulation seeks to keep the balance in the natural elements that intervene in the water cycle, and at the same time, it will allow the protection of riparian, aquatic, terrestrial, and coastal ecosystems through a regime of ecological flow in the currents.

This is due to the national problem of the reduction of water in riverbeds, caused by the competition between its uses and the lack of regulation depending on the availability of the resource, for example, the current demand for water in catchment systems in watersheds does not consider the benefits generated by the existence of water in rivers like fishing or forest preservation like mangrove swamps and gallery forests. In this case, subterranean water does not consider groundwater discharge to surface water bodies so affecting an aquifer desiccates natural springs, which protect drinking water services and preservations of habitats in areas away from extraction wells of the aquifer (NMX-AA-159_SCFI-2012).

A way to apply the concept of ecological flow to the instruments of public politics is by guaranteeing water supply in the future by doing a new water use, a use necessary to preserve environmental conditions and the balance of nature (Pfister et al. 2011). Likewise, it permits the management of water resources in a more efficient way so it can face the challenges that supply for human consumption mean and to guarantee supply to present and future generations. It should also be capable of facing environmental challenges like climate change, which demands the protection of watersheds.

It is necessary to reorient water policies that join social and economic policies assuring the social appropriation of the resource, not only as a human activity but also as part of the ecosystem as a whole. It is also important to preserve the services provided by these natural surroundings and provide economic options to owners and administrators of these services which are part of the natural water supply system (woods, wetlands, and meadows) avoiding to change the soil use to economic activities that pay for their efforts in the short term (Rosa et al. 2004).

A viable option to assure the supply of services is by implementing the regulatory methods foreseen in the Mexican Constitution like the preservation of national waters. Water reserve assures the appropriation of water resources for the ecosystem; the objective of the National Waters Reserve is to guarantee the minimum flows for ecological protection, including the preservation or restoration of vital ecosystems associated with the watershed basins. A requisite for achieving this goal is not only the administration of national waters but to bind such administration to the biodiversity that depends on the flow of that basin: this is the creation, protection, and strengthening of protected natural areas near the river's basin we are talking about. Having said that, the proposal is essential to preserve and maintain those areas, some of which are forest areas likely to be granted the provision.

The application of this regulation and the social acknowledgment takes us to the opening discussion on the limitation of private activities in benefit of a common good for society as a whole. In this situation, the regulation is in its early stages compared to the PES' ones, but we can learn from previous experiences when there was an attempt to recognize the environmental services provided by the forest and to include a comprehensive vision of the water cycle and to recognize socially, economically, and legally the compensations and limitations that must be done to owners and administrators of forests and wetlands in the country.

9.6 Discussion

Throughout history, deforestation (and other harmful activities) has accompanied economic development. It was primarily in response to deforestation that the concept of sustainable development originated and evolved within the forest (FAO 2012).

In this community (Capulalpam), the CONAFOR'S program for the payment of environmental services is not a determining factor in the owner community's decision to preserve the forest and to continue providing environmental services. The existence of the program and the participation of the community in the announcement did not modify the social or productive conducts since the community strategy of making traditionalist exploitation, improve the forest and keep it for future generations, was a decision made before the PES program.

The acceptance of the payment for environmental services and the decision-making process of the community owner can only be understood by distinguishing the importance of community institution both formal and informal in its economic and social organization, and that grants its restrictions to the economic decision-making process that goes beyond the perception of an immediate economic benefit.

In a context where the loss rate of forest cover is high, in 2004, it was registered a loss of 600,000 ha (Esteva 2004), and it was reduced approximately to 500,000 ha per year (Rosete-Vergés et al. 2014); the existence of nonfragmented and in a good preservation state forests seems to be associated with an indigenous institutionalism and/or solid community (Esteva 2004), characterized by:

(a) Community status that offers the legal framework for the functioning of efficient management institutions (low transaction costs in a credibility framework of the institutions responsible for enforcing the rules) that effectively coordinate the collective action for the good management of common resources
(b) The existence of access rules which control the access to common resources and that establish performance mechanisms and efficient sanctions
(c) The use of planning tools like the community territory legislation and techniques for the good management of common goods

The existence of well-preserved forest areas may be attributed to one of the following conditions or a combination of them:

(a) Well-preserved forests are mainly located in inaccessible areas where forest exploitation is not profitable, which would be consistent with the minute trip component in the proposal of deforestation risk index of the INE
(b) Community property of forests, particularly those communities where there are strong community institutions, possess a very low discount or negative rate, in spite of being poor communities where there is a high rate of discount compared to their natural resources and which grants a high value to the preservation of the biological resource for future generations
(c) The community territory ordering reveals the opportunity costs of the forest since historically it has been granted high value to the future option that forest resources preservation is a financial option with easy monetization. In the case of an emergency, the community can opt for the extraction of wood for fast sale in the market.

The extension of PES may occur if the schemes manage to demonstrate the growing effects in terms of forest preservation in contrast to the predefined deforestation lines (Wunder 2007) and considering organization aspects of the communities in the management of forest areas.

On the other hand, if the purpose of the policy behind the PES is truly to accomplish long-term conditions in relation to the baseline, which represents the high deforestation rates in Mexico, the intervention would have to focus not only to give a compensation to forests owners but also it should examine policies and encouragement programs, in general, to adjust them to this objective.

Nowadays, the huge offer of sectorial programs that emerged as responses to pressure groups with different points of view, which range

from forest conservation to enhanced living conditions, were translated into the creation of the conditions for the active forest preservation through the reconstruction of community institutionalism and the strengthening of encouragement policies for the promotion of economic activities, including a good forest management. This last, to obtain different kind of products (timber and nontimber) and other income alternatives that would allow generating the basis for guaranteeing the community reproduction, reducing the forest pressure and complementing the assistance programs for short term might improve the living conditions of the community.

Besides preserving ecosystems, forest conservation allows us to protect the quality and water supply in the fluvial systems and wetlands in general. Therefore, recognizing that the lack of trees in a system generates an absence of water supply in the long term, in other words, failing to acknowledge that the societies living in the area are part of a far more complex and interrelated natural system will develop in a mismanagement of the resources of that area.

9.7 Conclusions

The creation of limitation rules of uses (like the ecological flows and the PES in its multiple versions) can be seen as a strategy to preserve them and looks for explicit recognition of the associated costs to preserving the ecosystems provided by environmental services. It also looks for incentive mechanisms that solve the interest divergence between the producers and the users.

As non-developed areas and natural habitats reduce, the previously guaranteed environmental services (in a freeway) are being threatened. These rising shortages make them dangerously marketable to the environmental services.

The existence of several contractual arrangement examples in both developed and developing countries to protect water sources, biological diversity, landscapes, carbon sinks or natural barriers which protect us from natural disasters evidences the idea that a strong economic assessment is not always necessary to calculate the consumer surplus or an opportunity cost study which allows us to calculate the producer surplus.

References

Andaluz-Westreicher, A. (2005). *Bases conceptuales para un enfoque de los Servicios Ambientales*. Manual de derecho ambiental. Editorial, Proterra. Lima.

Andaluz-Westreicher, C. (2016). *Para una adecuada gestión ambiental (V)*. Portal Perú. Perú. Diciembre del 2016. http://portalperu.pe/columnistas/31-carlos-andaluz-westreicher/476-para-adecuada-gestion-ambiental-v

Aukema, J. E., Pricope, N. G., Husak, G. J., & Lopez-Carr, D. (2017). Biodiversity areas under threat: Overlap of climate change and population pressures on the world's biodiversity priorities. *PLoS One, 12*(1), e0170615. https://doi.org/10.1371/journal.pone.0170615.

Chapela, F., & Lara, Y. (2007). El Ordenamiento Comunitario del Territorio- un esquema para hacer compatibles los objetivos de conservación y los derechos sociales e indígenas. En: *IUCN Commission on Environmental, Economic & Social Policy: "POLICY MATTERS. Conservation and human rights"* (Issue 15, pp. 288–29).

Chomitz, K. M., da Fonseca, G. A. B., Alger, K., Stoms, D. M., Honzák, M., Charlotte Landau, E., Thomas, T. S., Wayt Thomas, W., & Davis, F. (2006). Viable reserve networks arise from individual landholder responses to conservation incentives. *Ecology and Society, 11*(2), 40.

CONABIO. (2009). *El capital natural de México: Vol. I*. Conocimiento actual de la biodiversidad.
CONAFOR. (2001). *Plan Estratégico Forestal*. Versión 2, 18 de agosto de 2001. consultado en: www.conafor.gob.mx
CONAFOR. (2003). *Reglas de operación del PSA-H*. Diario Oficial de la Federación, México, 2003. CAPITULO III. Beneficiarios. SECCION I. Criterios Técnicos. Artículo 10.
CONAFOR. (2006). *Reglas de Operación*. Diario Oficial de la Federación 16 de febrero, Ciudad de México.
Esteva, G. (2004). *Vacas, Milpas o Bosque: Una Exploración Econométrica De Las Causas de la Deforestación En México* (p. 23). Tesis de Licenciatura. Instituto Tecnológico Autónomo de México, México.
FAO. (2012). *State of the world's forests*. Rome: Food and Agriculture Organization of the United Nation.
FAO – CONAFOR. (2005). Revisión del Programa Estratégico Forestal 2025 (PEF 2025) y del Programa Nacional Forestal 2001–2006 (PNF 2001–2006), en el marco del proyecto UTF/MEX/072.
FAO, OAPN. (2009). *Pago por Servicios Ambientales en Áreas Protegidas en América Latina*. http://www.fao.org/3/a-i0822s.pdf
Farley, J., & Costanza, R. (2010). Payments for ecosystem services: From local to global. *Ecological Economics, 69*(11), 2060–2068.
Hallegate, S., Heal, G., Fay, M., & Treguer, D. (2012). *From growth to green growth – A framework*. USA: National Bureau of Economic Research. http://www.nber.org/papers/w17841.pdf
Heal, G. (2000). Valuing ecosystem services. *Ecosystems, 3*(1) (January–February), 24–30.
Hernández, M. R. (2008). *El Programa de Pago de Servicios Ambientales y su aplicación en un núcleo agrario con manejo comunitario del bosque*. Tesis que para obtener el grado de Doctor en Economía, División de Estudios de Posgrado de la Facultad de Economía de la Universidad Nacional Autónoma de México.
INE. (ca.2006). *Situación general existente en las comunidades agrarias con respecto al manejo de los recursos naturales*. www.ine.gob.mx
Ingram, G. K., & Hong, Y.-H. (2011). *Climate change and land policies* (pp. 277–300). Cambridge, MA: Lincoln Institute of Land Policy.
Kelsey, J. B. (2013). *Constraints on the adoption of agricultural technologies in developing countries*. Agricultural technology adoption initiative, J-PAL (MIT) and CEGA (UC Berkeley). https://www.povertyactionlab.org/sites/default/files/publications/ATAI%20Review%20Paper%20May%202013.pdf
Kelsey, J. B., Kousky, C., & Sims, K. R. E. (2008). Designing payments for ecosystem services: Lessons from previous experiences with incentive-based mechanisms. *PNAS, 105*(28), 9465–9470.
McNeely, J. A., & Scherr, S. J. (2003). *Ecoagriculture: Strategies to feed the world and save wild diversity*. Washington, DC: Island Press.
Norma mexicana de caudal ecológico NMX-AA-159-SCFI-2012. (2012). *Diario Oficial de la Federación*.
Novo, P., Slee, B., Byg, A., Creany, R., & Faccioli, M. (2017). *Providing smart delivery of public goods by EU agriculture and forestry*. USA: The James Hutton Institute. http://www.provide-project.eu/documents/2016/10/d2-2_conceptual-paper-on-the-unpacked-notion-of-public-goods.pdf
Olivera, V. S. M. (2007). *Explotación de Recursos No Renovables en Áreas protegidas: valoración de las Áreas protegidas en Bolivia*, KIPUS – Bolivia.
Ortega, G., & Ramos, J. M. (2016). *Ecología y conservación: Los árboles al servicio del ambiente*. Revista de divulgación cientifica y tecnológica de la Universidad Veracruzana. México (Vol. XXIX, num. 3, Septiembre–Diciembre).

Pfister, S., Bayer, P., Koehler, A., & Hellweg, S. (2011). Environmental impacts of water use in global agriculture: Hotspots and trade-offs with land use. *Environmental Science and Technology, 45*, 5761–5768.

Rosa, H., Kandel, S., & Dimas, L. (2004). *Compensación por Servicios Ambientales y Comunidades Rurales: Lecciones de la América y Temas Críticos para fortalecer Estrategias Comunales*. Programa Salvadoreño de Investigación sobre Desarrollo y Medio Ambiente (PRISMA).

Rosete-Vergés, F. A., Pérez-Damián, J., Villalobos-Delgado, M., Navarro-Salas, E. N., Salinas-Chávez, E., & Remond-Noa, R. (2014). *El avance de la deforestación en México 1976–2007*. Madera y Bosques, Instituto de Ecología. Xalapa, México (Vol. 20, núm. 1, 2014, pp. 21–35).

UNEP. (2011). *Water footprint and corporate water accounting for resource efficiency*, Paris: United Nations Environment Programme. http://waterfootprint.org/media/downloads/UNEP-2011.pdf (Feburary, 2018).

Wunder, S. (2007, February). *The efficiency of payments for environmental services in tropical conservation. Conservation Biology, 21*(1), 48–58.

Ricardo Hernández-Murillo World Bank (in memory).

Sazcha Marcelo Olivera-Villarroel is a Professor in the Department of Theory and Design Processes at the Universidad Autónoma Metropolitana – Cuajimalpa, México. Originally from Bolivia, he received his PhD in Natural Resources and Environmental Economics in 2007 from the Universidad Nacional Autónoma de México. His Master's degree is in economic arts from the Pontifical Catholic University of Chile. His research interests include resources and environmental economics, theories of human behavior, climate change, and value theory. He is a Member of Latin American and Caribbean Environmental Economics Program, the National System of Researchers (Mexico), and works as an independent consultant for the World Bank and the Economic Commission for Latin America and the Caribbean (ECLAC), among others. He was a visiting professor at the London School of Economics, and is currently spending his sabbatical year at the University of Maryland Center for Environmental Science – Chesapeake Biological Laboratory.

Chapter 10
Water Management Instruments in the National Waters Law

Luis Enrique Ramos Bustillos and Gustavo A. Ortiz Rendón

Abstract This chapter draws on the need to rethink regulation with respect to what the current National Waters Law defines as "gestión del agua." The concept "gestión" can be translated as "management," but, as it will be explained in this chapter, in existing Mexican law, it has yet other meanings that need to be acknowledged. Clarifying this definition might foster better regulation of different instruments that might improve water public policy. Many of these instruments are currently scarcely and dispersedly regulated in the law. The consequence is a lack of implementation of many of them, e.g., ecosystem services or environmental flow. Analysis departs from main categories of instruments given by the Law of Ecological Balance and Environmental Protection (LGEEPA) mainly political, financial (fiscal), and economic instruments. Yet, it focuses on how some of these instruments are recognized in the National Waters Law (LAN). Proposals of legal reform of each one of these categories are taken into consideration to improve regulation of the current legislation, which could be useful for a possible enactment of the General Water Law, currently in discussion at the Mexican Congress. It is because one of the fundamental matters of analysis is the significance of the term "gestión del agua" that it will recurrently, as an exception, be referred in Spanish.

Keywords Water governance · Multiple stakeholder participation · River basin councils · Political · Economic and tax tools · Water financial system · Wastewater discharge fees · Market instruments · Environmental costs

L. E. Ramos Bustillos (✉) · G. A. Ortiz Rendón
Independent Consultant, Jiutepec, Morelos, Mexico

H. R. Guerrero García Rojas (ed.), *Water Policy in Mexico*, Global Issues in Water Policy 20, https://doi.org/10.1007/978-3-319-76115-2_10

10.1 Regulation of "gestión del agua" and Integrated Water Resources Management (IWRM) in the National Waters Law (LAN)

The 2004 extensive reform of the National Waters Law introduced, among others, the concept of *gestión del agua*. This concept could be translated as "water management," but according to the definition as provided by the law, it refers more to the authority to transact, conduct, or carry out certain acts related to water management issues. "Gestión del agua" was defined generically as a "process, sustained by the assembly of principles, policies, acts, resources, instruments, formal and informal norms, goods, resources, rights, powers and responsibilities, through which The State, the water users and the organized society, coordinated with each other, promote and carry out to achieve sustainable development for the benefit of the human beings and their social, economic and environmental milieu."[1]

The 2004 reform of the National Waters Law introduced as well the term integrated water resources management, which was conceived as well as a "process" to foster the coordinated management and development of water, land, and related resources and of these with the environment.[2] The definition of both concepts aims to integrate processes, actors, and actions. Paradoxically, their definition reveals a gap, which is to be seen in several forms. The first one is that it creates a multiplicity of interpretations with regard to the precise content of both definitions. When asking officials and employees of the government agency National Water Commission (CONAGUA), some of them had not even heard about IWRM, while others would define them as a process for the coordination of government agencies, and only 5% would know the definition on IWRM as given by the law.[3]

Another element that reflects the gap of both concepts is the fact that they lacked further regulation for their effective implementation. When analyzing the LAN, it can be seen that the concepts were introduced conceptually but without further regulation to provide a more solid content where water policy and acts could be based on. For example, IWRM is considered of public utility and as a matter of national security,[4] but regulation is scarce when trying to integrate water public policy within the framework of IWRM.

In English, both the term **gestión** and the term **administración** can be translated as "**management**," but the term "gestión" could be also translated as "transact,

---

[1] Art. 3, XXVIII LAN.

[2] Art. 3, XXIX LAN.

[3] Inquiry conducted in 12 courses regarding National Waters Law given between 2007 and 2009 to personnel of River Basin Organisms of CONAGUA, which are the water authority in the different hydrological regions in which Mexico is divided. Mexico has 13 hydrological regions, each one with its own authority.

[4] Art. 7, I LAN.

conduct, carry out" or to "take measures to do something or direct a project."[5] The content and reach of the term "gestión" might produce confusion if not clarified. In our concept, the main difference between these concepts lies under the framework of water governance, which requests a co-participation of different actors, transcending the paradigm that water management is an issue of government authorities. Water governance requires new and improved ways of co-participation, responsibility, and solidarity to include the private and the social sectors.

The *Dictionary of the Spanish Royal Academy* defines both "gestión" and "administración" as the action and effect to manage. However, when defining the verb "to manage," it defines it as the act to govern, exercise authority and command over a territory and the persons that inhabit it, direct an institution, order, dispose, organize, and hold a position, while the verb "gestionar" is defined as dealing or carrying out a business.[6]

National Waters Law does not define the term "water management." However, this law does have a whole chapter referred to water management. Analyzing this chapter, it is to be seen that most of its regulations are referred to issues pertaining to water authorities, pertaining originally to the president of Mexico directly or through the federal agency National Water Commission (CONAGUA). Most of the powers of different government agencies are defined in this chapter. However, CONAGUA has the power as the supreme authority for IWRM, which includes management, regulation, control, and protection of the hydric public domain.[7]

The rules that define the specific authority of the different river basin management authorities of CONAGUA do differentiate the terms *administración* and *gestión* handling them separately when defining that the general director of CONAGUA *can exercise the authority concerning the transaction of water resources and the management (administración), conduction (gestión), and custody of national waters.*[8] This is to be seen when analyzing the authority of the Subdirección General de Administración del Agua (general subdirection of water management of CONAGUA) where both concepts are clearly differentiated.[9]

Prior to the 2004 reform, LAN already contained a chapter for water management which included, among others, the making of plans and programs, control, evaluation, participation, regulation, and surveillance but that included as well acts, e.g., the granting of concessions, authorizations, residual water discharge permits, and water right transfer authorizations. LAN included as well other issues such as the inspection and surveillance that water rights are paid appropriately, initiation of administrative procedures, and imposition of sanctions.[10] Therefore, it is to be seen

---

[5]Diccionario de María Moliner, as quoted in http://www.diclib.com/cgi-bin/d1.cgi?l=es&base=moliner&page=showindex seen April 26, 2013.

[6]Consulted in http://lema.rae.es/drae/ January 16 2012.

[7]Art. 9 LAN.

[8]13, II.

[9]24,I.

[10]Arts. 9 y 12 bis LAN and Arts. 13 y 24 of Reglamento Interior CONAGUA.

that LAN locates "water management" within the framework of acts that imply powers for a certain authority that would include their formal materialization in an administrative act considered for legal purposes as an act of authority, which could be contested before the administrative courts.

On the contrary, "gestión" as contained in LAN is perceived more as a *process* sustained in the assembly of principles, policies, acts, and resources. This concept is circumscribed within the framework of water governance as defined by UNDP in the Water Governance Facility as "The political, social, economic and administrative systems that are in place, and which directly or indirectly affect the use, development and management of water resources and the delivery of water service delivery at different levels of society. Importantly, the water sector is a part of broader social, political and economic developments and is thus also affected by decisions outside of the water sector."[11]

This concept does not deal exclusively with acts within the water sector as many of the water- related conflicts are located outside this context, e.g., agriculture or industry to name two of them, which have an impact on water policies, conflicts that transcend the traditional framework of water management understood as a set of acts of government water authorities and which involved other parties and actors, both from the private and from the social sector. It is within this model that the term of "gestión" and IWRM introduced by the 2004 legal reform should be understood if they are to have any real meaning in water public policy and practices in Mexico.

In this sense, it might be important to distinguish clearly both concepts. It seems that when reforming the National Waters Law (LAN), legislators located a broad definition of "water management" understood in a macro category, where both powers of government authorities would be considered within this framework but understanding that water management goes beyond these powers referring to coordination, co-participation, and solidarity of the government with water users and organizations of the private and social sector and that water must be seen in a larger context integrating land and other resources. It is in this context that the concept "gestión" acquires importance in the definition of the water policy within the framework of water governance, to include regulation on principles and mechanisms that would foster coordination, cooperation, and participation in the definition and implementation of water plans and programs.

In our opinion, the terms "gestión" and "administración" are complementary and interrelated concepts which content is important to differentiate, on the one hand, to clarify and expressly define the part of water management that is circumscribed and limited to specific acts of water authorities in order to strengthen this authority while, on the other hand, to expand the definition and vision of the meaning of water management. In the latter sense, it would be important to have a separate chapter in the National Waters Law referred to "gestión" understood as a process to conduct, transact, and carry out water projects, which involves other actors and not only as a government issue. Within this chapter, topics that are currently disperse or that lack

[11]Consulted in http://www.watergovernance.org/whatiswatergovernance February 9 2013.

regulation could be fortified, e.g., environmental services, aquifer management, environmental flow, and social and private participation.

It would be important to clarify issues regarding water management, i.e., the making of certain administrative acts from the management of water as a resource in matters referred more to plans and programs to be implemented. The first requires a certain act from the water authority, while the second one would be located more within the framework of a process of coordination and cooperation when implementing a certain plan or action.

For instance, pollution control could be an act of "gestión" (to carry out a project) or of "administración" (an exercise of a specific act of authority). Acts referred to the exercise of authority would fall under the umbrella of a specific administrative act, while the formulation, implementation, coordination, and planning would be deemed within the context of "gestión." Evidently, any management act can be contained in a broader context of "gestión del agua" and vice versa, but it is important to differentiate acts that belong exclusively to water authorities of those located in a broader context of participation of other actors and sectors with respect to water management. It is in this broader context that it makes sense to differentiate both concepts toward a more efficient implementation of water public policy.

10.2 Water Governance and Water Management Actors or Platforms

Governance implies the possibility for public participation through social organization. *Participation transforms the democratic system and gives it another dynamism. It produces a channel of permanent relationship between government and governed. Joint action allows for decisions to be more reasonable, being product of consensus, where community problems are acknowledged in a better way and where solutions are found jointly* (Carmona Lara 2011).[12] It must include as well integration across policy sectors (Grigg 2008).

LAN (1992) determined that together, river basins and aquifers were the constitutive elements for the unity of water management.[13] The 2004 reform established other levels of governance by establishing that the river basin is composed by sub-river basins and the latter are integrated by micro-basins.[14] The novelty of this reform is that integrated river basin management, river basin, and aquifers were all considered of public benefit.[15]

An important progress in regulation consisted in the recognition of actors that prior to the 2004 reform lacked legal grounding. LAN (1992) ruled River Basin

[12]pp. 126–127.

[13]3, f. IV LAN (1992). After 2004 it changed to XVI.

[14]3, XVI LAN.

[15]7, I y 7 bis 1 LAN.

Councils scarcely in one article.[16] The rights and obligations of the River Basin Council were better defined with the 2004 LAN reform; their action range was broadened to include, among others, harmonization of duties, building of knowledge, divulging, participation in the definition of goals, fostering coordination, and providing an improved legal framework to develop their tasks.[17] IWRM was taken into consideration as framework toward transferring authority to the River Basin Councils (Vargas 2002). At the same time, the 2004 reform acknowledged different actors at the different levels of water management. Administratively, according to LAN, water can be managed at smaller units within the river basin, which the law denominates as SUB or MICRO. For the so-called sub-river basins the management platform is through commissions, while for the micro-basins it is a committee, and for the aquifers (underground water technical committees, known as COTAS), the range of activities of all of them and how they are to interact with each other and with other actors need yet to be defined to foster not only their constitution but the effectiveness of their work as well.

Regarding the River Basin Councils, it can be seen that LAN established in detail the form for their integration, even going to the extent of defining their structure and the officer positions that each of them needs to have. While focusing in structure regulation, it established a tight system of creation and functioning of these Councils, which might be partially responsible for the rigidity and lack of dynamism of many of the River Basin Councils. Regulartion coud offer some tools to prevent for River Basin Councils to be co-opted by vested interests and to expand water users' interests beyond the negotiation of water quotas without taking into consideration a broader perspective regarding improvement of how water is used, diminishes pollution, and contributes to the sustainability of the basin.

It is important to acknowledge that there has been an improvement in access to information in the water sector. CONAGUA, as part of the National System for Waters Information (SINA), publishes diverse statistics of the water sector in Mexico. An example of these publishings is the digital atlas of water in Mexico, which is a useful tool when noting that in Mexico, there are 26 river basin councils, 32 commissions at sub-basin level, 39 committees in the micro-basins, and 82 underground water technical committees (COTAS).[18] However, it is important to notice that there is a contrasting disparity on how these are all distributed. For instance, in the south of Mexico, the micro-basin is the level that has been privileged, while in the center and north of the country, the sub-river basin and aquifer organizations have been promoted. For example, in 2010, of the 25 micro-basin committees existing to that date, 20 were concentrated in four states: Guerrero, Oaxaca, Chiapas, and Tabasco. Regarding the sub-basin level, 15 of the then 21 commissions were located in three basin organizations: Northwest, Lerma-Santiago-Pacífico, and Aguas del Valle de México.

[16]Art. 13.

[17]13 bis 3 LAN.

[18]Estadísticas del Agua en México, edición 2013, CONAGUA, p. 114.

While having these different levels of possibility for participation, the number does not necessarily mean improving water management. Many of these organizations lack sufficient resources, e.g., financial, technical, or human, often functioning through strong government support but without a real process of grassroots participation where a wider myriad of actors are considered. In many occasions, these platforms are utilized by vested interests to obtain objectives that focus more on a private (collective or individual) interest but without the vision of sustainability. It is urgent to expand the forms of social and private participation in these platforms while at the same time improving the current ways of organization, taking into account broader and more democratic forms of participation, with objectives and goals that could be measured. Watershed or regional development initiatives require the establishment of an operation of different bases: political/legal, economic/financial, social, and organizational (Dourojeanni 2001).

Although IWRM has been a concept widely accepted and promoted as a framework for water management (Camargo 2005; Amaya Navas 2005) and as stated before it has been recognized by the Mexican National Waters Law, it has been questioned by others for being vague, abstract, and of difficult achievement, only accepted in label with no change in how water is managed factually, raising concerns about the need of dealing with water management through this paradigm (Biswas 2004). Integration of concepts in the water legislation in Mexico must acknowledge the criticisms to IWRM specially when dealing with its effective implementation in order to promote a sound water management.

10.3 Dispersion of Articles of Law Referred to Water Management in the National Waters Law (LAN)

There is a tangible confusion regarding the meaning and scope of the term "water management" since the provisions related to this subject are not concentrated in one section or chapter of the LAN. Concerning the sphere of authority, the CONAGUA is given the capacity to act as the authority in the field of "water management."[19] It is established that the Basin Councils are responsible of participating in the definition of the objectives and criteria for the formulation of water management programs and of contributing to IWRM.[20] The right of citizen organizations to participate in the water management is also stipulated in LAN.[21]

The management of national waters is considered as a basic instrument for the national hydric policy[22]; nevertheless, ambiguity persists regarding the scope of this

[19] 9, I LAN.

[20] 13 BIS 3, I, IV LAN.

[21] 14 BIS LAN.

[22] 14 BIS 6, III LAN.

concept. This can be noticed in the 2014–2018 National Hydric Program and its predecessor of 2007–2012 National Hydric Program, both of which scarcely mention IWRM and only in a conceptual manner within one of the strategies to consolidate the quality of water as part of the objective to promote an integrated and sustainable water governance, but yet need to be enhanced and clarified properly in regulation. Yet, the 2014–2018 does include a provision for a needed reform of the water legislation.

On the other hand, the 2004 reform introduced concepts and subjects that were regulated in a fragmented and disperse manner. For example, the need to preserve wetlands was established, which requires their delimitation and inventory among other actions.[23] However, the regulation of wetlands was included in the chapter related to the prevention and control of water contamination, being this one of the aspects of wetland management but not the only one.

A similar situation occurs with the introduction of the environmental flow, which was regulated using different terms: environmental use, minimum ecological flow, minimum flows for ecological protection, or ecological reserve. This already demonstrates a lack of clarity, but in addition its regulation is dispersed within the law, without any sort of link with key subjects that affect it such as the granting of water rights (Bustillos & Enrique 2007). The normative integration of this matter in a water management chapter could contribute to a better implementation of the ecological flow within the hydric policy.

Similar situation occurs with other subjects like ecosystem services or water desalination. In the first case, there is no regulation that allows to maintain a water policy for this matter, despite the fact that the LAN considers the full incorporation of the environmental variable of public interest.[24] In turn, the desalination is only marginally mentioned when it stipulates that this process will require a concession[25] but without further regulation beyond the mere reference of the need of obtaining such concession.

Dispersion is to be discussed in this chapter concerning fostering water awareness, which was placed under the title regarding the uses of water, adding it as chapter BIS, even when the promotion of water awareness goes beyond the subject of uses and users.

Similar situation occurs with the "Water Basins" as entities in charge of conducting the transfer of rights-regulated operations. These were included in the 2004 reform[26] in the chapter regarding the transfer of titles, but these entities are only a part of the operation concerning the supply and demand of water rights.

[23] 86 BIS1 LAN.

[24] 7 bis, VIII.

[25] 17 LAN.

[26] 37 bis LAN.

10.4 The Need of a Specific Title in the National Waters Law to Regulate "gestión del agua"

Scarce regulation of many of the abovementioned topics and the fact that they are scattered in LAN reflect the lack of clarity in the meaning and possibilities of water management. Legal reform is necessary to go beyond the concept definition toward defining a clear set of rules in fundamental topics, e.g., wetland management, environmental flow, or when fostering water awareness, just to mention some examples. It is important to integrate these concepts within the framework of water management, as all these issues require the coordination and co-participation of authorities, users, and society for an effective implementation. It would be advisable to include in the National Waters Law a title regarding **gestión del agua**. This would allow for the concentration of the different matters that are currently disperse in the law as well as to regulate the subject with greater depth. Other subjects such as environmental flows or the water financial management can be regulated in chapters within the framework of **gestión del agua**.

The management of underground waters, a vital issue in water management, could be included in this title, which could include provisions regarding the technical committees of underground waters (COTAS) as currently these are only mentioned in the law as participation entities that need to be promoted.[27] This regulation could serve as a supporting base for the public policy concerning the management of overexploited aquifers. Key issues identified by Carabias and Landa (2005) could be included here, such as the preparation of comprehensive studies on aquifers, water restriction orders, as well as a potential governing legislation for periodic handling and measurement.

Water management and climate change are an issue that would be important to include within water policy and law. International agreements ratified by Mexico need to be taken into account for this purpose. For instance, the chapter regarding protection against flooding is currently considered within the title of uses of water when it could be placed in a broader context of water management and climate change.

10.5 Political and Economic Instruments

Beyond the previous considerations and taking into account the basic differences between *gestión* and *administración* here above explained in detail, it is advisable to consider the different political instruments mentioned in the Law of Ecological Balance and Environmental Protection (LGEEPA) and its relation with LAN, the Government Fees Federal Law, and other regulations. When doing so, it is relevant

[27] 14 bis LAN.

to have a comprehensive/integrated water management, understood as *an ensemble of processes that involve coordination, negotiation and exercise of policies, acts, resources, instruments, formal and informal norms, goods, assets, properties, rights, powers, and responsibilities* of the different authorities (federal, state, municipal), of users and society which has as goal of promoting the coordinated management and development of water, land, and resources related to these and to the environment.[28]

The LGEEPA provides that economic instruments will be used as an incentive to encourage the objectives of the environmental policy to promote behavioral changes in people involved in activities in industry, commerce, or provision of services, so that their activities are compatible with environmental protection and sustainable development. Pricing system includes information of environmental costs and benefits so that those responsible for environmental harm accept the responsibility of such costs and in this way promote greater social equity in the distribution of the associated costs and benefits and, furthermore, grant incentives for those who work in specific actions to protect, preserve, or restore ecological balance. LGEEPA classifies economic instruments subdividing them in normative administrative mechanisms of different nature: fiscal, financial, or market instruments. Those of fiscal nature must encourage the compliance of the environmental policy objectives and should never have as an exclusive purpose money collection.

LGEEPA also defines financial instruments as loans, surety bonds, and insurance for civil liability, funds, and trusts, when their objectives are directed to the conservation, protection, restoration, or better use of natural resources and the environment. It also considers as financial instruments those directed toward financing programs, projects, studies, scientific research, technological development, and innovation that promote ecological balance and environmental protection. Lastly, it defines the market instruments as concessions, authorizations, licenses, and permits that correspond to preestablished volumes of pollutant emissions in air, water, or land.

A fundamental element where a linkage between LGEEPA and LAN and more precisely between environmental authority SEMARNAT and water authority CONAGUA is hardly needed is with regard to the making and implementation of ecological zoning. Zoning is an instrument of environmental policy regulated in LGEEPA. There are four types of ecological zoning[29] (OET). Although the National Waters Law (LAN) recognizes the need to link water to its related resources when defining integrated water resources management (IWRM) in practice, this becomes difficult when assessing not only the current structure and number of officers and employees CONAGUA has, but it also raises a practical issue: If the water authorities face great challenges for water management, it seems difficult to request from them a broader vision to link it with other resources. It is important to acknowledge that IWRM transcends water authorities. Including a title in the water legislation dedicated to "gestión del agua"

---

[28]Art. 3, XXIX LAN.

[29]General, regional, local, and coastal, 19 and 19 bis LGEEPA.

could be helpful to produce a legislation that may serve better to link policies and regulations toward the implementation of ecological zoning.

For these instruments to work, it is relevant to acknowledge the tension between social participation and expert scientific and rational decision making for resource optimization should be taken into account in order to transcend the technocratic planification (Conca 2006) responding mainly to technical criteria and not to political commitments in key aspects, e.g., regulation and grant concession (Rey Bengoa and Hurtado Mora 2013).

10.6 Fiscal Instruments Considered in the National Waters Law (LAN)

10.6.1 Government Fees

The LAN prescribes that CONAGUA, in coordination with the Basin Councils, has the power to study the government fees and recommended charges for the extraction of national waters, discharge of wastewaters into national receiving water bodies, environmental services, and basin tariffs.[30] CONAGUA has authority to determine, investigate and collect taxes and government charges in the form of public revenues allocated to the Basin Councils.[31] The basin organisms have the power to conduct studies on the economic and financial assessment of water according to the supplying source, location, and type of use. This is to support the definition of basin tariffs and government fees and to spread the results in the region to improve the awareness of prices and costs as well as to strengthen the paying principle among users due to management, water services, and protection or related vital ecosystems.[32]

LAN establishes that water users must pay for its use and exploitation under the principle of "user-payer," pursuant to the Government Fees Federal Law.[33] It clarifies paying for water as a basic instrument of the national water/hydric policy.[34] Concessionaires have the obligation to pay the corresponding government fiscal fees derived from the extraction, consumption, and volumetric discharges. Failure to pay for more than 1 fiscal year will be reason enough for suspension and, if recurrence, the revocation of the concession and the corresponding assignment.[35]

The aforementioned is related to the Eighth bis title "Water Financial System" (SFA) in the LAN, which is in charge of CONAGUA, and is considered as a supporting base of the actions regarding the integrated water resources management

[30]Art. 9 XXVIII LAN.

[31]Art. 9, XXIX LAN.

[32]Art. 23 BIS 6 XVIII LAN.

[33]Art. 14 BIS 5 XVI LAN.

[34]Art. 14 BIS 6 IV LAN.

[35]Art. 29 BIS 5 IV y 92 III LAN.

in Mexico. The SFA should determine the diverse financial sources, ways of obtaining resources, criteria for the implementation of expenses and recovery, which includes accountability and management indicators, as well as implementation goals. LAN includes the need to pay government fees for the provision of different administrative services by the water authority and the rights for the national property that it administers, e.g., rocks and sand of waterbeds.[36] It also provides the possibility of establishing coordination with the state governments, with the federal district or municipalities, under the terms of the Fiscal Coordination Law and the Government Fees Federal Law.[37]

An important aspect set by the law is that the fees and other taxes must be formulated in a way in which they favor the management demand while encouraging an efficient use of water, the rationalization of consumption patterns, and, if the case, the inhibition of activities that require an excessive demand. Currently, these fees are established for the following concepts: fees for administrative services related to water and other inherent resources, for the exploitation or use of water for the use and enjoyment of national real property riverbeds and water bodies and currents, and for transfers between basins, national property deposits, and material extraction of these bodies, as well as wastewater discharges. The amounts charged for the use or exploitation of national waters are established pursuant to differentiated fees according to the availability zones, under the terms defined by the Government Fees Federal Law (LFD[38]) as well as according to the use for which it is intended and the volume of water used. If it is not possible to measure the volume of water due to a change or a broken meter, the law provides several mechanisms to determine the amount to be charged. The law also provides the hypothesis for the presumptive determination of the volume of water consumed. Government fees are updated taking into consideration the Mexico Consumer Price Index.[39] Until 2013, the fees were defined according to the availability zone as defined by LFD for each of the municipalities, situation which did not always correspond to the reality of water availability in the river basin, and were applied notwithstanding if it was an underground or a superficial source. From 2014, the source is differentiated to establish the fees, according to four zones. The fees for 2017 for superficial waters vary from \$15.1944/m^3 for zone 1 to \$1.7538/m^3 in zone 4. For underground water, the fees for the same year are \$20.4740/m3 in zone 1 to \$2.0059/m3 in zone 4.[40]

This fee has been very useful in this area as a basic financing instrument, as well as serving as an incentive for an efficient use of water, due to the fact that the consumer in a rational behavior tends to value this resource when having to pay for it. The fact of using a resource for a personal and economic benefit implies that a

[36]Art. 112 LAN.

[37]Ley de Coordinación Fiscal y de la propia Ley Federal de Derechos.

[38]Arts. 223 and 231 LFD.

[39]Arts. 227, 228 and 229 LFD.

[40]Ley Federal de Derechos publicada en el Diario Oficial de la Federación el 31 de diciembre de 1981 actualizada al 2017, Diario Oficial de la Federación del 7 de diciembre del 2016.

third party will be excluded from this resource, especially in scarcity zones. When paying for water, which is a property of *all*, and having the nature of national property, the said payment compensates society. Payment allows financing of works and actions to enlarge the coverage zones to give access to this resource for other uses and users in the required quantity and quality. This is why, as an exercise of social and distributive justice, all users with no exceptions must contribute with the corresponding payment.[41]

These payments, in accordance with Article 31 of the Mexican Constitution, must be established proportionally and equitably. This means among others that those with higher consumptions must pay more.

10.6.2 Wastewater Discharge Fees

In 1991, the Government Fees Federal Law laid down the amounts to be paid for discharges of concentrations of chemical oxygen demand (DQO) and total suspended solids (TSS), when these were above the permissible limits. Payments are to be made quarterly according to differential fees depending on defined availability zones and the quantity in kilograms of pollutants poured into the receiving water body. The calculation of the discharge fee was simple and it remained with no substantial modifications until 1995. This simplicity caused relatively important results in the collection and in the increase in coverage of the treatment plants, mainly in consumption industries with financial stability. It was until 1997 that the payment was based on the official standard norm *NOM-001-ECOL-1996* which established the government fee amount according to the surplus to the permitted limit of the parameter groups: fecal coliforms, potential hydrogen, basic pollutants, heavy metals, and cyanides. The law is simplified in 2008, again considering the two initial charging parameters of the legal instrument – TSS and DQO – with the resulting risks for environmental water policy. Subsequently, from 2014, when there are municipal discharges, differential rates are established for activities with predominantly biodegradable and non-biodegradable discharges.

Generally, the initial objectives of the discharge fees have not been fulfilled: for the average fees per cubic meter of discharge to be higher than the treatment costs and to encourage the observance of the environmental norms. This has been surpassed by unfortunate changes in the law, by some inaccuracy that have become windows of opportunity to circumvent the law, as well as by temporary and definitive exceptions and exemptions that have hampered the collection of necessary

[41]Even when the Government Fees Federal Law provides an amount to be charged for agricultural use, when there is a higher consumption in the concession title, the fee is not applied in practice due mainly to a lack of measurement and control. Agriculture sector is exempted from payment and it is the use of higher water consumption in Mexico (70% of the total).- Programa Nacional Hídrico 2007–2012. Comisión Nacional del Agua, febrero 2008.

income to invest in sanitation actions.[42] Yet, *the purpose of the environmental economic instruments is not the peremptory prohibition of the unwanted environmental activities, but to give people cues to orient their conduct towards the goals, which are environmentally preferable, and to achieve the purposes planned by the environmental policy* (Mendezcarlo Silva 2006). In other words, although these fees have a clear collection purpose, they should also be instruments of public policy that inhibit pollution and that appropriately internalize the environmental costs.

10.6.3 Collection Through Government Fees

Recovery of money via government fees has been important, although insufficient to govern this activity. Up to 2004, collection of government represented more than 50% of CONAGUA's total budget, while in 2015 this percentage represented 38% of the total budget altering the proportion.[43] Regarding the total CONAGUA budget, the federal agency for auditing (Auditoría Superior de la Federación) revealed that in 2015 CONAGUA spent 42,221 million pesos out of which 36% was generated by CONAGUA itself via fee collection. In the period 2008–2015, tax collection was increased to 11%, similar to CONAGUA budget, since it was 12%.[44]

In terms of the Income Law of the Federal Government (LIF) for 2015, SEMARNAT would receive for this year as fees for the use and exploitation of properties of public domain an estimated amount of $20,241,000 pesos, from which approximately 81% comes from federal taxation from the government, basically for the water use, exploitation, and other duties, which is administered by CONAGUA although it is entirely given to the Federal Treasury.[45]

The trend of expected fee collection compared to CONAGUA's budget if current conditions prevail is shown in Fig. 10.1.

The results of the spatial distribution in the collection of government fees have the following characteristics: the hydrological administrative regions that commonly present greater registries in the collection of fees for the use or exploitation of water are the Valley of Mexico, Lerma-Santiago-Pacífico, Rio Grande, Balsas, Center

[42]The best year in the collection of this government fee was 2011, in which 267 million pesos were collected, according the CONAGUA authorities. "2013 meeting on the reforms to the Government Fees Federal Law," March 2013, CONAGUA, Mexico, D.F.

[43]CONAGUA (2013) and Ortiz and Donath (2013).

[44]Ley Federal de Derechos para los ejercicios fiscales 2008 y 2015 y Estadísticas del Agua en México, CONAGUA, edición 2016, p. 152.

[45]Ley de Ingresos de la Federación para el Ejercicio Fiscal 2015 and Estadísticas del Agua en México, Edición 2016, México, Conagua.

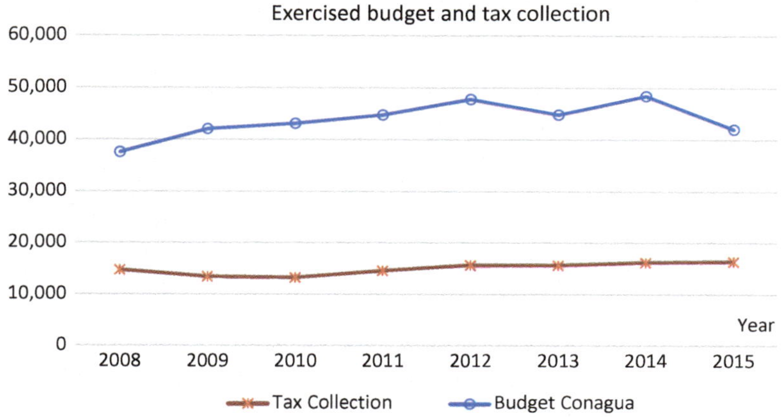

Fig. 10.1 Budget CONAGUA and tax collection 2008–2015. (Source: Own making from Statistics of Water in Mexico 2014-constant prices of 2015)

Gulf, and the central basins on the north (in that order).[46] Regarding wastewater, the hydrological administrative regions that commonly present greater collection registries for government fees of wastewater discharges are North Gulf, Center Gulf, Lerma-Santiago-Pacífico, and Balsas.[47]

Considering only the current budget levels (slightly over forty-two thousand million pesos in a year), the questions that can be drawn from the previous are the following: How much increase in fees is needed? Or, what the collecting and auditing systems would need to improve in order to achieve a financial self-sufficiency of the water public sector? Or better yet: What would be needed to improve on these matters to achieve self-sufficiency before a possible and necessary budgetary growth?

Public policy needs to address the need to take into consideration instruments of control to inhibit the wrong use of water: to consider them as instruments of control to inhibit the wrong use of water, waste, and pollution or, on the contrary, to just look for important collecting registries. If the latter interpretation prevails, the environmental cost could be high. Fiscal-environmental policy must be strengthened with mechanisms that foster extended and equitable payment for water and water bodies, improving enforcement. CONAGUA has statistics on the number of inspections and resolutions. The outcome of both, the inspections and authority resolutions, needs to be measured as well when evaluating proper enforcement. Yet, there is a huge deficit in the field of inspection and surveillance for law enforcement. ASF reports that in

[46]Workshop to actualize CONAGUA's personnel on the water legal framework,. CONAGUA, edición tecnológica, IMTA 2007.

[47]Information provided by personnel of the Coordinación de Revisión y Liquidación Fiscal, CONAGUA (2008).

average, each river basin organism has approximately five inspectors to attend approximately 26,000 annual grants, which makes it hard for appropriate control.

10.6.4 Characteristics of the Improvement Fees

The Federal Public Works of Hydraulic Infrastructure Law (LCMOPFIH) has as main as main purpose to serve as an instrument of the federal government to partially recover investments made in the construction, restoration, or enlargement of such infrastructure. Once deductions have been made concerning subsidies, contributions, and discounts on the total construction cost, the law defines an amount called "recoverable value," to which an additional 90% is applied (i.e., an additional 10% support). The said contribution is to be paid in partial payments by the infrastructure's beneficiaries, for up to 40 years for hydroagricultural works and 25 years for aqueducts and multipurpose projects.

Even though the law was initially enacted in 1986 and brought up to date in 1991, since then until now, there has been very little collection in relation to its potential.[48] The reasons why such contribution has failed are not clear, but undoubtedly, this regulation has generally not been very attractive for users and public servants. One of the shortcomings of this law is that it duplicates the provisions of Article 31 of the Mexican Constitution, which defines the obligation to contribute to public expenses in a proportional and equitable manner. Criticism to this law has been made as well on the grounds that the investment cost is a component already considered in the calculation of fees and tariffs paid for water services and that the improvement fees as a means of contribution for public works generates *double taxation*.

This law needs to be reviewed in order to amend it to make it feasible or abrogate it if deemed obsolete. When reviewing it, it is important to consider that this law could represent an interesting financing source that currently has not been used to its best. It has a high social content that would be important to be acknowledged as a means to recover marginal federal investments of works. This has to be done carefully in order to avoid double taxation related to the costs of works already being recovered via tariffs and fees of the supplying systems.

[48] An amount of 26.8 million is the estimate to be collected under the Federal Revenue Law for fiscal year 2013.

10.7 Financial Instruments

10.7.1 Private and Social Involvement

LAN contains a whole chapter referred to private and social involvement in the making and management of water projects and services. The intention is to foster the possibility for water users, i.e., private or social sector to be involved in the making and management of water projects or services, and to provide rules for access to credit and other financial tools for this purpose. Modalities of participation can go from a contract for the provision of services for the construction, equipment, and operation of infrastructure, but these works could also be granted through a concession.[49]

According to LAN when allowing private or social participation, it should be taken into consideration an efficient water use, the reasonableness of the consumption patterns that inhibit an excessive demand. Law also rules causes for the termination of grants and refers to the bonds that need to be given to guarantee works or services. Currently, there is a federal trust called National Fund for Investment in Infrastructure (Fondo Nacional de Inversión en Infraestructura, FINFRA) that through the specific program on modernization of water provision service organisms (PROMAGUA) provides financial support, an important amount of these resources not to be recovered by the government to investment projects that consider private participation under a scheme of a provision of services contract (CPS). Up till 2015, an amount of 11,965 million pesos was given for this purpose of which around 4780 million pesos were resources "a fondo perdido" (sunk cost). Targets of these resources are to increase efficiency and coverage of water and sanitation and to create conditions for sustainability of the organisms providing these services.[50]

10.7.2 Risk Insurances for Water Flooding

Although it has been a marginal topic within water management, water floodings have taken momentum as part of climate change. This is particularly true for states like Tabasco, Veracruz, and Chiapas that have had tremendous problems due to floods. Therefore, when redefining the agenda and legal reform for water management, a renovated vision of water floods needs to occur in order to allocate sufficient resources and personnel in the prevention and control of these natural phenomena.

LAN defines that river basin authorities must support CONAGUA in the coordination of competent institutions and the establishment of insurances for damages in areas of high risk. There is already a fund that was created for these natural events

[49]Eighth Title, Chapter II. See arts. Art. 5, II and 9, XII LAN.

[50]Arts. 9, 46,47 BIS, 104, 105 y 107 LAN.

called "National Natural Disaster Fund" (FONDEN). This fund assigned for the year 2017 an approximate amount of 6036 million pesos.[51] More than 4000 million pesos of the total were for the coverage of damages and losses owed to hurricanes and flooding. Article 18 of the Civil Protection Law defines that it falls under the responsibility of the governments of the states and/or the government of Mexico City to acquire insurance to attend coverage of damages and losses owed to natural phenomena for which the National Coordination of Civil Protection and the federal fiscal authority propose a model on risk insurances.[52]

It would be important to improve public policy in the making of plans and programs to water flooding and other water-related natural phenomena. Finding methods and amounts for insurance coverage would be vital for many producers that would otherwise be ruined after such an event while on the other hand finding a method for reasonable premiums.

10.8 Market Instruments

10.8.1 Grants, Authorizations, and Permits

Water rights can be given to both private and government agencies. CONAGUA has the authority to decide, upon petition, to give these rights. Water usage is given through grants. The system is based upon the principle of prior request; first to come is first served. Yet, in a new water management paradigm, this system needs to be reviewed to include a broader form of water usage that takes into account sustainability of the resource. In certain hydrological regions, e.g., Lerma-Chapala, there is an overallocation of water rights that paradoxically has produced, and in order to acquire water rights many would have to buy those rights in the black market as onerous water rights transfer is forbidden by the Mexican Constitution. According to the National Waters Law, water right granting should take into consideration average annual water availability. The issue remains on how average water availability can be determined if there is lack or deficiency in measurement, a problem that is coupled with the fact that there has been overallocation of water rights.[53]

Exploitation of riverbanks and extraction of rocks are permitted via authorizations, and residual water discharges to national water bodies like rivers or to the sea are given via permits.

The typology of the different status of water users is described as follows:

[51] http://www.diputados.gob.mx/LeyesBiblio/pdf/PEF_2013.pdf

[52] Art. 19, XXIX Ley de Protección Civil.

[53] In the agreement where average water availability for underground water was established of the 653 aquifers in Mexico (D.O.F. April 20, 2015) where it is stated that the volume of underground water that has been granted is superior to the average annual availability.

In good standing. – Water rights holders that fulfill legal requirements, that have been granted water rights, permits or authorizations and who are current with their payments of government fees or duties.

Users that exceed use and exploitation volumes or are not current with their payments.- Water rights holders that do have proper water rights but that misuse or abuse the authorized volumes or that are not current in their payment of government fees or duties.

Illegal users – Water users that use and/or exploit water or their resources without any grant, authorization or permit.[54]

According to the performance audit made by the Federal Agency for Auditing (ASF) in 2009, the existence of clandestine water users is owed to lack of surveillance from government agencies. Results from this audit show, for example, that in the metropolitan area of Mexico City, 50% of the wells that were taken a census of are illegal. This reveals a lack of control of water authorities and a rampant nonfulfillment of the law on the side of water users. This problem is interrelated with the lack of reliable data and proper measurement on water availability which hampers proper water management.[55]

Water rights have to be recorded in the Public Recording Office for Water Rights (REPDA). The idea behind REPDA is to provide certainty and security with respect to the time and conditions for water users in order to foster investments in the water sector. The existence of grants and of an agency where they are recorded provide a good basis upon which public policy can be improved. CONAGUA reports that accumulative registered acts till December 2015 are of 486,896 titles, of which 144, 832 correspond to grants given to governmental agencies in charge of water provision and the remainder of approximately 342,064 for private users.[56] Together, this stands for a total granted volume of water rights of 266,559 annual million cubic meters, which represents 59% of the annual average availability of renewable water in the country.[57] Around 76.3% of off-stream water is used for agriculture, 14.6% for public supply, 4.3% for self-supplying industry, and 4.8% for energy generation excluding hydropower.[58]

CONAGUA reports that only 43% of water users fulfill all legal requirements, while 57% does not. Most recurrent practices are illegal drilling of wells, exploiting water without any grant, and discharging wastewater without permit or lack of measurement devices. On the other hand, CONAGUA is falling behind in attending legal request for grants or permits. In 2011, only 40% of grant or permit petitions were addressed in the timelines according to the law, and only 66% of the required extensions of grants and permits were processed.[59]

---

[54]In the "Gaceta de Administración del Agua" 2012 most frequent causes of nonfulfillment of the law that were identified are using more water than authorized and using or exploiting it without paying government fees.

[55]Auditoría Superior de la Federación (ASF) Auditoría de Desempeño: 09-0-16B00-07-0016.

[56]Compendio estadístico de administración del agua, CONAGUA 2015.

[57]Estadísticas del agua en México, CONAGUA 2016 p 20. According to this document, Mexico has an annual amount of 446′777,000 milllion of cubic meters of renewable fresh water.

[58]Statistics on Water in Mexico, CONAGUA, 2016 edition.

[59]Compendio Estadístico de Administración del Agua, CONAGUA edición 2013.

In order to improve water management through the system of water rights granting, it is important to improve efforts to have appropriate water availability studies that according to LAN must be reviewed every 3 years[60] while expanding coverage in water use measurements, inspection, and surveillance and to intensify the campaign to standardize water users and to cancel effectively illegal water usage or exploitation. It is fundamental to fortify the role of water authorities, mainly CONAGUA, in this task. For this purpose, it is fundamental that CONAGUA focuses in programs that could include supporting innovation and development of technologies that would increase efficiently in water use as well as to reorganize the current forms for the control and management of water users, improving customer attention, and the times for response when granting or extending water rights.

10.8.2 Water Banks

Many river basins face a severe water scarcity, while demand for water increases, and many water bodies are overexploited. In many regions, it is not possible to obtain water rights through a grant. This is one of the reasons why the figure of water banks was introduced in LAN[61] in order to provide a tool for water rights reallocation and to curtail the black market of water rights and at the same time to release water from an overexploited aquifer or river basin. However, LAN only made a very marginal reference to the water banks without defining the basic elements for their constitution or operation form. CONAGUA has water banks functioning in the 13 river basin institutions, with support offices in the local directions of CONAGUA. Among other things, it is possible for water users to, via the web, publish a potential offer or demand of water rights while providing users with information about the technical and administrative elements of a particular river basin or aquifer. In 2011, 13 water banks through their support offices attended a total of 5396 users.[62]

According to CONAGUA, requests for information and/or attention of future demands of water were located in order first in the Lerma-Santiago-Pacífico hydrological region, followed by Cuencas Centrales de Norte Region, and in third place the Río Bravo Region.[63]

[60]Art. 22 LAN.

[61]Art. 37 bis LAN.

[62]Compendio Estadístico de Administración del Agua, CONAGUA Edición 2012.

[63]Water banks in Mexico, CONAGUA 2012 http://www.conagua.gob.mx/bancosdelagua/SGAA-4-12-BAM-12.pdf. Future water rights transfer demands are estimated according to the volume of water that has been granted for agricultural purposes and the percentage of the gross internal product in the different hydrological regions.

The idea of the water banks is to enhance planning, management, and regulation of water rights transfers for a more efficient usage. However, the lack of regulation and how water banks have been implemented is the seed of a yet to be solved contradiction. While onerous water transfer is forbidden by the Mexican Constitution, water banks as part of CONAGUA try to foster offer and demand for water rights reallocation, but those transfers usually involve an economic compensation; otherwise, it would be difficult that water users would freely transfer their water rights. These policies must necessarily be linked to a reallocation that takes into consideration economic, social, and environmental concerns to diminish impact on the environment and corrects or prevents overexploitation of aquifers, particularly to prevent water coverage in drought years. All of these characteristics should be taken into consideration when redefining the structure and functions of the water banks.

10.9 Ecosystem Services

The environmental variable is to be taken into consideration for water policy as a matter of public interest.[64] One of the forms to do so is through the implementation of a policy and actions that acknowledge the ecosystem services when planning or implementing water projects. LAN defines ecosystem services as the social interest benefits generated or derived from river basins and their components, such as climate regulation, conservation of hydrological cycles, control of erosion and floodings, aquifer recharge, drainage maintenance, soil formation, carbon capture, purification of water bodies, and biodiversity protection.[65] Ecosystem services must be acknowledged, quantified, and paid.[66]

Ecosystem services were acknowledged in LAN in 2004. Yet, regulation focused in the concept and principle definition but still lacks proper legal integration, perhaps to be contained in one chapter within a title of "gestión del agua." Possibly, this is why there has been a public policy gap and water projects from water authorities at all levels of government to this regard. Even if LAN emphasized for ecosystem service purposes the interrelation of forests with water, it is important to acknowledge that this topic goes beyond forests and must have an own space when planning and implementing water policies by water authorities. Lessons could be learned from existing cases in Latin America (Martínez et al. 2007). LAN definition of ecosystem services provides a guide or list of topics upon which specific ecosystem service programs could be built in the water sector, to acknowledge sensitive topics like erosion or flooding control or climate regulation. For instance, these programs could even be implemented at municipal level to promote protection of urban river watersheds currently under serious threat owed to fast urbanization.

---

[64] Art. 7 bis, VIII.

[65] Art. 3 XLIX LAN.

[66] 14 bis 5 XI.

10.10 Conclusion

There is a need to integrate current disperse regulations regarding **gestión del agua** recognized as a separate process within the water management field in order to provide a better framework for regulation of relevant topics within the water field that goes beyond exclusive government participation in water management. Instrument categories as provided by LGEEPA – environmental, economical, financial (fiscal), and political – could serve as a guideline to build specific regulation for legal reform of water legislation.

Government authorities need to be strengthened with necessary personnel and fortified authority while at the same time differentiating the exercise of authority acts with the possibility of participation of both the private and the social sector. Therefore, it is important to acknowledge this participation on a broader frame that offers improved forms of intervention in processes that transcend government's field of action. The role of key actors at the different levels of management should be strengthened, e.g., micro-basin committees, sub-basin commissions, river basin councils, and the technical underground water committees (COTAS).

Market instruments for water rights reallocation or payment of environmental services might be helpful to foster a more efficient use of water. Government fees, as a fiscal instrument, need to be defined not only for collection purposes but also with a view to control demand and produce a more efficient use of water. There are many aspects that would need to be reformed in the law to improve government fee enactment and collection. The following are some proposals:

- To actualize methodology for the calculation of government fees, considering environmental costs
- To establish river basin tariffs attending the availability and economic value provided by water for the production of goods and services
- To start collecting duties or government fees for water usage of the agricultural sector that currently pays nothing
- To provide support and incentives to increase coverage of macro- and micromeasurement
- To improve water authorities' capacities for inspection, control, and collection of government fees, which might include the signing of agreements for the delegation of certain acts at state and/or municipal levels of government
- To develop strategies and methodologies to increase profitability for the collection of government fees
- To consolidate a culture of payment of government fees and of the economic and environmental value of water
- To allocate the government fees derived from water and its resources to the specific works and actions in the correspondent river basin via a specific fund to be built in each hydrological region
- To support an incentive technology transfer for a more efficient, clean, and water-saving technologies via deduction of other taxes, e.g., income tax

- To allocate at least 1% of the total tax or duty collection in water for research and development of the water sector

References

Amaya Navas, O. D. (2005, Septiembre). *La Gestión Integrada de los Recursos Hídricos como herramienta para lograr el objetivo constitucional del desarrollo sostenible en América Latina y el Caribe.* en: Incorporación de los principios de la gestión integrada de los recursos hídricos en los marcos legales de América Latina (pp. 231–290). Universidad Externado de Colombia, primera edición.

Biswas, A. (2004, June). Integrated water resources management: A reassessment. International water resources association. *Water International, 29*(2), 248–256. Consulted in: http://www.tonydorcey.ca/597/Posts/IWRM%20for%20Water%20International.6.pdf. 26 Aug 2015.

Bustillos, R., & Enrique, L. (2007). Aspectos jurídicos relevantes para la implementación del caudal ecológico en México. Coautoría con Ortiz, Gustavo y Serrano, Alejandra. *En Requerimientos para implementar el caudal ambiental en México* (pp. 95–113). – IMTA- Alianza WWF/FGRA-PHI/UNESCO-SEMARNAT, primera edición, México.

Camargo, E. y Machado de Faria Leme Franco, N. (2005, Septiembre). Incorporacao dos principios da Gestao Integrada de Recursos Hidricos nos marcos legais de America Latina: experiencia brasileria e licoes a serem aprendidas. en: *Incorporación de los principios de la gestión integrada de los recursos hídricos en los marcos legales de América Latina* (pp. 107–121). Universidad Externado de Colombia, primera edición.

Carabias, J., Landa, R. y Colaboradores. (2005). Agua, medio ambiente y sociedad. Hacia la gestión integral de los recursos hídricos en México. Universidad Nacional Nacional Autónoma de México-El Colegio de México-Fundación Gonzalo Río Arronte.

Carmona Lara, M. D. C. (2011). *La Constitución y el Agua.* en: E. O. Rabasa y C. B. Arriaga García (Coordinadores), *Agua aspectos constitucionales* (pp. 83–140). Instituto de Investigaciones Jurídicas-Editorial Porrúa.

CONAGUA – Comisión Nacional del Agua: Atlas digital del Agua 2010 visible en http://www.conagua.gob.mx/atlas/

CONAGUA – Comisión Nacional del Agua: Atlas digital del agua en México 2010, consultado en http://www.conagua.gob.mx/atlas/# 9 de febrero 2013.

CONAGUA – Comisión Nacional del Agua: Bancos del Agua en México, CONAGUA 2012 visible en http://www.conagua.gob.mx/bancosdelagua/SGAA-4-12-BAM-12.pdf

CONAGUA – Comisión Nacional del Agua: Compendio Estadístico de Administración del Agua, edición 2012 visible en http://201.116.60.96:8080/wb/ceeaa/compendio_estadistico_de_administracion_del_agua

CONAGUA – Comisión Nacional del Agua: Estadísticas del agua en México, 2008 visible en http://www.conagua.gob.mx/CONAGUA07/Publicaciones/Publicaciones/EAM_2008.pdf

CONAGUA – Comisión Nacional del Agua: Estadísticas del agua en México, 2011 visible en http://www.conagua.gob.mx/CONAGUA07/Publicaciones/Publicaciones/SGP-1-11-EAM2011.PDF

CONAGUA – Comisión Nacional del Agua: Estadísticas del agua en México, edición 2013, visible en http://www.conagua.gob.mx/CONAGUA07/Noticias/SGP-2-14Web.pdf

CONAGUA – Comisión Nacional del Agua: Estadísticas del Agua en México, 2016 visible en http://www.conagua.gob.mx

CONAGUA – Comisión Nacional del Agua: Estadísticas del agua en México, edición 2016, visible en http://201.116.60.25/publicaciones/EAM_2016.pdf

CONAGUA – Comisión Nacional del Agua: Gaceta de Administración del Agua, Subdirección General de Administración del Agua, 2012.

CONAGUA – Comisión Nacional del Agua: Información sobre montos de Recaudación de Derechos, proporcionada por la Coordinación de Revisión y Liquidación Fiscal 2008.

CONAGUA – Comisión Nacional del Agua: Situación del Subsector Agua Potable, Alcantarillado y Saneamiento, septiembre 2012.

CONAGUA – Comisión Nacional del Agua: Programa Nacional Hídrico 2007–2012 visible en http://www.conagua.gob.mx/CONAGUA07/Contenido/Documentos/PNH_05-08.pdf

CONAGUA – Comisión Nacional del Agua: Programa Nacional Hídrico 2014–2018, visible en http://www.conagua.gob.mx/conagua07/contenido/documentos/PNH2014-2018.pdf

Conca, K. (2006). *Governing water: Contentious transnational politics and global institution building*. Cambridge: MIT Press.

Diccionario de la Real Academia Española. Consultado en http://lema.rae.es/drae/el16deenero2012

Disposiciones aplicables en Materia de Aguas Nacionales, Ley Federal de Derechos, Unidad de Comunicación Social de la Comisión Nacional del Agua, ediciones 1991, 1997, 2008.

Dourojeanni, A. C. (2001). Public Policies for Integrated Watershed Management. In A. Biswas & C. Tortajada (Eds.), *Integrated river basin management –the Latin American experience* (Water Resources Management Series, pp. 52–94). Oxford: Oxford University Press.

Grigg, N. S. (2008). Integrated water resources management: balancing views and improving practice. *Water International*, *33*(3), 279–292. https://doi.org/10.1080/02508060802272820. Consulted at: http://www.tandfonline.com/doi/pdf/10.1080/02508060802272820. 29 Aug 2015.

Martínez-Tuna, M., et al. (2007). Payments for environmental services as an institutional tool for downstream-upstream cooperation in watersheds: Lessons drawn from five case studies in Latin America. en: *Gestión Integrada de los Recursos Hídricos* (Tomo II, primera edición, julio, pp. 161–187). Universidad Externado de Colombia.

Mendezcarlo Silva, V. (2006). *Magazine Nuevo Consultorio Fiscal*, number 406, Mexico.

Norma Oficial Mexicana (NMX). (1997). Nom-001-Semarnat 1996, publicada como Nom-001-Ecol-1996 el 6 de enero de.

Ortiz, R. G. & Donath, E. (2013). *y De la Peña Eduardo D.- Instrumentos Legislativos y económicos de política pública: hacia el logro de la gestión integrada del agua en México*.- Instituto Mexicano de Tecnología, primera edición, México.

Rey Bengoa, C. M., y Hurtado Mora, J. I. (2013, Septiembre). La visión integral de la gestión del recurso hídrico como herramienta indispensable para la lucha contra la corrupción en el sector agua. en *Gestión Integrada de Recursos Hídricos –Transparencia e Integridad* (Tomo IV, pp. 63–93). Universidad Externado de Colombia.

Vargas, S. (2002). Agua y organización social: De la centralización estatal a la gestión integral por cuenca. Patricia Ávila García (editora). En. *Agua, cultura y sociedad en México* (pp. 215–231). El Colegio de Michoacán/Instituto Mexicano de Tecnología del Agua.

Luis Enrique Ramos Bustillos is an attorney at Law from the Instituto Tecnológico de Estudios Superiores de Occidente (ITESO) with a Master degree in Law from Postgraduate Division of the Faculty of Law of the Universidad Nacional Autonoma de México (UNAM) and a MA in Development Studies from the Institute of Social Studies in the Netherlands. He has specialized in water law and public policy and has worked as a consultant with local, national, and international organizations. His range of expertise includes the fields of water rights, water governance, sanitation, river basin management, aquifer regulation, environmental protection, water and energy, coastal areas, and climate change.

Gustavo A. Ortiz Rendón has graduated in Economy from the Instituto Politecnico Nacional (IPN), with a master degree in Planning and Development from Universidad Autonoma del Estado de Morelos and a diploma in Investment Projects (OEA-NAFIN). Mr. Ortiz has worked for more than thirty years in the Water Sector in Mexico, and participated as project manager in more than 40 projects related to economic, financial, administrative, and normative studies regarding national waters and environment. He is the main or unique author of eight books and more than 13 chapters in national and international publications. He has been teaching different courses and participated in several conferences and congresses. For more than 15 years, he has taught at the Master level in the División de Estudios de Posgrado of the Facultad de Ingeniería de la Universidad Nacional Autónoma de México and the Master in Water Sciences at Instituto Mexicano de Tecnología del Agua. He has directed several theses for Bachelor's and Master's degree students. First place, in Serfin's National Environment Award. Honorable Mention in BANOBRAS National Award. He has received several national and international acknowledgments for his participation in projects such as: the first proposal in the water duty establishment in 1981 and its reforms, the first version of Improvement Contribution Law for hydraulic infrastructure works in 1986, the first draft for functioning and operation of Water Public Rights Registry in 1994, reforms to the National Water Law in 2004, the first proposal of Lerma-Chapala Water Bank in 2005, and National Water General Law in Nicaragua, among other projects. Currently, he is a specialized external consultant in topics related to the economic and legal aspects and administrative and financial matters of water and environment sector.

Chapter 11
Indigenous People and the Right to Water

Sergio Vargas-Velázquez

Abstract The number of Indian communities in Mexico is imprecise, but the importance of the way they manage water as a common pool resource is critical to understand in many other rural sites, where water is also a common resource under a regulatory framework of legal pluralism. The Mexican government has slightly recognized the sociocultural diversity only after the Zapatista uprising in the mid-1990s, but there are no references regarding cultural and organizational specificity of water management in Indian villages nor in federal law or even in national planning, although these community forms are heavily involved in numerous conflicts over water. Many mestizo communities are also organized around water as a common good and the perception of having local water rights, when in fact the federal government is in charge of managing water rights. The intense process of redistribution of water in basins and aquifers under stress or by the river basin closure is leading to additional pressure on water in the Indian territories.

Keywords Indigenous · Common pool resource · Water conflicts

11.1 Introduction

Social identity is the result of the confluence of a series of cultural traits, the self-assignment of individuals to a group, and the organization of the same social subjects related to other groups. It is an ingredient of every human group that begins with the differentiation between "us" and "others," regarding a wide range of issues (Giménez 2000). It also provides people with a sense of belonging to a cultural, social, economic, or political group, based upon which they are capable of acting in collectivity to manage all kinds of resources, including water and other natural resources. When referring to groups that are different from our own, we use the socially constructed stereotypes regarding the other; but that is where we have to

S. Vargas-Velázquez (✉)
Universidad Autónoma del Estado de Morelos, Cuernavaca, Mexico
e-mail: sergio.vargasvme@uaem.edu.mx

H. R. Guerrero García Rojas (ed.), *Water Policy in Mexico*, Global Issues in Water Policy 20, https://doi.org/10.1007/978-3-319-76115-2_11

differentiate between elements with which a group identifies itself, from those elements that others associate them with. This is important in the case of indigenous people in Mexico, as there are different ways in which government and society identify them, from the elements they use to define themselves.

Social identity is also embedded in a physical space that subjects identify as their territory, which is configured through symbolically established social borders or with explicit physical markers. Each territory consists of a cultural landscape where natural resources like rivers, lakes, springs, vegetation, mountains, and species are usually included. This is the case for the indigenous groups in Mexico as well as for many mestizo peasant peoples in their relationship with environmental assets, particularly water. In Central and Southern Mexico, many mestizo peoples organized around a peasant economy maintain shared characteristics in their social organization with indigenous peoples, in water management as a common pool resource and as a social good.

Identity in traditional societies is associated with more homogeneous ways of life, collective – less socioeconomic differentiation, group management of an economic resource, shared beliefs, and legitimate forms of collective action. They are organized according to specific needs, rather than rationalist criteria typical in governmental bureaucracy. They are much less individualistic than modern societies, because they maintain strong reciprocity bonds among its members, which sometimes are also extended toward nature: human beings are not seen as a separate category of nature, nor nature is necessarily subordinated to the human being needs (Descola 2001). This is expressed in the multiplicity of forms of social organization in areas as varied as the relations of kinship, religious cycle, political representation of their interests, or in economic activities such as the use of a forest, fisheries, or the preservation of a water supplying spring for their daily needs. There are a variety of combinations of features, which produce a multiplicity of preservation levels and cultural resistance of the native peoples, or the loss and dissolution of their social identity.

The capacity for collective action and identity in indigenous societies is placed primarily at the local level, with a group of people that coexists and shares everyday life, recognizing thereby a territory as their own. This occurs most commonly, although not necessarily, in rural areas. In contrast, national identities are based on a great social heterogeneity and individualism and a certain level of cultural globalization, where loyalty toward the national state, understood as a political order in a national territory, stands out. The relationship between social identity of the indigenous and mestizo peasant groups regarding the national identity is a problem, since it assumes in the former the preservation of its autonomy and ways of life in various aspects.

Identity is also an ingredient of social change in traditional communities embedded in modern societies. In here, modern/traditional duality is merely analytical in contrast with opposing social trends. More likely, in terms of water management and territory, a continuum between the two of them must be considered. In this way, we can analyze indigenous peoples as modern in their own way, whereas they coexist with the national society, while their forms of organization may still be functional, not without deep tensions.

In Mexico, there are many collective identities at the local level, organized in various ways regarding water and other resources, with an endless set of combined elements in a heterogeneous manner. Three troublesome premises can be identified that have been usually simplified in the construction of a public discourse about indigenous peoples and, therefore, regarding water rights.

The first troublesome premise lies in the difficulty of counting, classifying, and locating indigenous groups. The government solution has been to define them based on language, but even this has presented significant variations in just a few decades. In the 1970s, 56 ethnic groups were considered; afterward 78 ethnolinguistic groups were finally recognized. In the present century, the National Commission for the Development of Indigenous Peoples (CDI) defines them from 62 indigenous languages, the National Institute of Statistics and Geography (INEGI) considers according to the last population census 85, and the National Institute of Indigenous Languages (INALI) recognizes 364 varieties grouped into 68 linguistic groups. It should be added that many traditional peoples, although they have lost the language, attire, and other traits, still retain their collective identity in a very wide range of transformations: this is the case of the Opata natives, in the state of Sonora, or the *afromestizos* peoples on the coast of Guerrero and Oaxaca and the *Mazcogo* peoples in Coahuila. Sometimes they call themselves "original peoples," as it is in several locations in the center of the country, emphasizing its specificity in relation to the national society. In other cases, they recover some other elements of the social organization to build their collective identity. This leads toward a substantial rise in the population that can be assigned to the indigenous forms of water management, which includes groups within urban areas.

Defining how much indigenous population exists, and where they are to be found, also involves a problem of characterization, because taking into account only the restrictive ethnolinguistic considerations, figures vary between 6.3 and 12.7 million (Vázquez 2010). This amount would be added to many other people, who are not recognized as indigenous but retain the same social organization of many indigenous peoples around water; a good example of this is Tetela del Volcán (Guzmán Ramírez et al. 2012).

The second troublesome premise lies in characterizing a general pattern of the relationship of these indigenous peoples with water. In first instance, because they are located in a variety of possible hydrological situations, both in regions with a natural excess of water as in those of permanent aridity. In Mexico, the diversity of forms of social management of water has no official recognition; they are not even considered important regarding the population they serve for drinking or irrigation purposes. Social management of water – irrigation or human consumption – is part of the reality around the country, but its importance in the official documentation is hardly recognized. Ávila (2007:15–18) notes that while the Mexican Constitution recognizes the cultural diversity of the country and the existence of social, economic, and political institutions of the indigenous peoples, in the National Water Law, there is no reference in this regard, despite the fact that in the rest of the environmental legislation, this already occurs.

The main aspect in which the contrast between indigenous peoples and the government management of the water focuses is in the possibility of a great diversity of forms of social management, usually based on water as a resource for common use, as opposed to the homogeneity sought in government management, and in its principles of bureaucratic technical management and even in the principles of economic efficiency in its use. This is not an obstacle for the settlement of major concentrations of indigenous population in large-scale irrigation or municipal systems of drinking water areas. Mainly in the center and south part of the country, it is possible to find, in almost every small town, a plurality of organizations that take advantage of water under diverse regulatory frameworks, different to the official ones. As an example, in the municipality of Temoaya, State of Mexico (Estrada and Franco 2004), a water utility linked to the respective state water commission, whose powers and jurisdictions are established among federal, state, and municipal laws, can be identified. However, there are local committees of drinking water, which are not dependent in management from the state or municipal governments. They are recognized as a part of an autonomous traditional structure that bans any intervention or participation of people from outside the community. The most numerous committees are those formed in small rural sites, under their own institutions, thus emerging a combination of situations. Some of them are linked with an organization of land ownership, whether communal or "ejidal" (which is an area of communal land used for agriculture, on which community members individually possess and farm a specific parcel and differentiate from commons in the way in which land was given). Some others are independent from land management, but linked to the system of religious charges, while those who reach the management of local water system have to go through a series of requirements such as having served in a religious charge or in the neighborhood organization, like the "mayordomías." In these systems, there is usually a well, spring, or small infrastructure around which individuals are organized to preserve it. There are also other even smaller systems, depending on the characteristics of the water resource. In the Los Altos de Morelos region, there are numerous systems around hoses and water boxes, "jagüeyes" and levees, as well as those organized based on water supply pipe trucks (Guzmán 2009), which are important variations. The same occurs in states with a greater indigenous presence, such as Guerrero, Oaxaca, and Chiapas or Michoacán (Ávila 1996).

Regarding irrigation, something similar occurs. Irrigation in Mexico is divided into large irrigation, currently composed of 85 irrigation districts, which are under the supervision of the federal government via programs for water allotment, tariffs and financing systems, and production support, among others. These comprise approximately 3.2 million hectares. The indigenous population located there and explicitly recognized are Colonias Yaqui and several irrigation modules of Rio Mayo, Rio Fuerte, Tikul, Valle de Mezquital, Tepetitlán, and Tehuantepec, among others. In contrast, there are more than 40 thousand small irrigation systems, historically out of the direct supervision of government entities, managed by its own users that cover about 2.8 million hectares. A significant number of these systems are classified as controlled as they maintain an important bond with water authorities, since they have infrastructure whose origin was in a government program, or by managing infrastructure that requires some governmental intervention. The non-supervised ones are those of which little is known and there is no control over them. There is a great uncertainty about the total number and organization diversity of these systems. Here, the systems of community management in surface water are dominant. The indigenous population that has access to these irrigation systems is organized in a self-management manner. An example of this is the case of the Tzotzil indigenous group, which established an irrigation system with residual water (Burguete-Cal y Mayor 2000).

The third troublesome premise is the characterization of the relationship between state and indigenous peoples regarding water access and use. Governmental bureaucracy has no official account of how many systems are operated by indigenous peoples nor the local forms of water management, but it identifies the most important out of several confrontation points between social and government management.

As a starting point, there is the designation of nation as the original owner of water in the Mexican Constitution, on behalf of which the federal government was established as its representative since the 1920s. This ushered a strong concentration of functions and budgets at the federal level, excluding state and municipal governments for decades. It is up to the 2004 National Water Law reform, when state jurisdiction waters were incorporated after numerous disputes in late 1990s between water state commissions and the federal authority so state governments could serve as adjutants in the management of the water, especially regarding drinking water and sanitation. An amendment of Constitutional Article 115, in 1983, granted the municipality with operating functions for drinking water and sanitation, but the allocation of water is provided by the federal authority, but in many cases, it only covers the municipal centers, leaving the rural environment to their own forms of self-management.

This contrasts with the vast majority of local regulatory systems, mainly in those considering water as a resource for common use, a widespread situation in indigenous communities and many *mestizo* peasant communities. In there, the local water right prevails, usually based on the definition of its territory, sometimes defined upon ways of access or possession of *ejidal* or communal land; these are the so-called uses and customs, which even include forms of local government. Many indigenous peoples and *mestizos* consider that water that flows through their territories belongs to them, since they are already recognized in their right to the territory by the federal authority through an *ejidal* or communal title. This expresses the existence in the local scope of a self-regulatory framework, usually not legally recognized by formal authorities, producing a situation of "legal pluralism" (Roth et al. 2005), in which the decisions of the federal, state, or municipal government are mediated and reinterpreted in the interior of the jurisdictions of social entities. The following aspect of contrast lies in the organization and orientation asymmetry of policy frameworks for water management. The social water management is usually linked with other forms of social organization, as *mayordomías* (religious charges but also civic in traditional villages), *barrios*, or other forms of local organization. The *barrio* is a territorial organization, in some cases, associated with kinship rules, in others with access and management of natural resources. For example, in Pichátaro, a town in the P'urhépecha region of Michoacán, there exists a *mayordomo per each* barrio – *kenhi* in Porhé language and the institution is known as *kenhekua* – to manage religious activities and to regulate the access to forest resources. The scale at which social organizations exist and move generally does not exceed what they can "walk in a day" and the persons with whom they usually relate.

Under this scenario, it is possible to imagine the number of discrepancies and contradictions in the coordination between organizations that respond to policy frameworks and emphasize the negotiation, tension, or dispute processes between social and governmental actors in different scales. A first contradiction arises from water distributive justice in the water territory. It is a territory because it is a space of domination, ownership, and belonging of individuals or communities. One of the characteristics of indigenous groups is that its continuity is subject to the group control on certain natural resources. The change in the forms of access, the loss of some strategic resource, or the erosion of their local rights to the usufruct has produced in many occasions a serious attack against cultural identity.

11.2 Social Water Management and the Political Action on Indigenous Peoples

Social groups, locally organized to take advantage of their water resources on a continuous basis and long enough to apprehend its characteristics, develop more sustainable management practices than those proposed by analytical scientific knowledge (Toledo 2002, 2003). This implies a local knowledge that is directly focused on the use of resources in a coordinated fashion to specific conditions. Modern science has helped an intense anthropization of the hydrological cycle, based on an always-expanding market economy, in which negative externalities that produce water deterioration are not recognized. In this way, there is a contrast between an analytical science proposal that is helpful for justifying the diagrams of integrated management which are now being proposed by various international agencies and the local management carried out for centuries. This is not implying that indigenous peoples do not deteriorate water resources, but they do make a more sustainable management.

Indigenous knowledge is studied and characterized by the ethno-ecology, which explores, in an interdisciplinary manner, the knowledge or cognitive systems and a set of productive practices, including the different uses and management of natural resources, showing that many of these traditional local groups, based on self-management, safeguard an intricate knowledge of their resources (Toledo 2002). In this way, we can speak of an ethno-hydrology (Gelles 2004).

One of the specificities of the local management of the indigenous and peasant peoples on water is their community organization, which implies the absence of many of the limiting effects between land, water, forest, and biodiversity that exist in government management and analytical science that work by parcels of knowledge. This implies a level of political organization that is manifested in the defense of their resources, which claims its community character, as well as the defense of its territory based on organizations that in the 1980s onward began to include ethnic demands. Although many ethnic traits have been lost, communalism is a part of the political culture in large indigenous areas. Official institutions, dissident groups, and communities themselves consider in their speeches various referents regarding communalism. In political terms, it validates the right of the community to exist, which implies a series of claims: the defense of a common heritage, the exercise of self-governance mechanisms, and respect for culture and indigenous language. But not all claims are handled with the same intensity and on behalf of all communities. In general, communalism strengthens the sense of belonging and the possibility of facing problems in a collective manner, particularly regarding water, land, and forest.

This type of political manifestation is supported by the traditional social organization, where the peoples are divided into *barrios*, which are a series of civil and religious positions of diverse degrees of complexity, as they have evolved since the disappearance of the so-called Republic of Indians at the end of the colonial period as unrecognized forms to the central government. Regarding water, there are still several forms of management, some of them linked to the system of civic-religious charges as the *mayordomías* and *atopiles*, in other water boards or water judge figures. In many *ejidos* (form of collective ownership of the land) with mestizo population, these forms of organization for the management of their resources subsist.

There is a long tradition in the peasant and indigenous communities with the administration and management of their drinking water systems as irrigation systems. This is due to the lack of solutions on the part of governments to meet the demand of the fluid, whereupon communities have tended to take responsibility for their own supply, or simply there was never any government intervention for the purpose, on many occasions based on the traditional community organization, to obtain resources and organize work for the preservation of the system and the distribution of water.

On the social organization, a series of relationships are intertwined to give sustenance to the conscious property of a territory as one of the main elements of the community, and the defense of such a territory and the mechanisms that exclude strangers for profits as well as community rules that govern access to the communal heritage are manifested.

In the studies on the types of local power in Mexico (Varela 1984; Padua and Vanneph 1988), one of the main characteristics is the articulation between national and local levels through institutions, organizations, or individuals that perform the function of mediators or political brokers, those who receive support from the local base as well as the recognition from the formal government. Varela shows how the inadequate bases of the central power have led to a concentration of power assuming an authoritarian character, which is associated with low efficiency of public policies (Varela 1984:44–57; Medellín 2006). This articulation of levels of power based on political brokers was one of the axes of the corporate regime of state party between the decades of 1930 and 2000.

With the political processes initiated with the regime change a little more than two decades ago, but not completed (Rodríguez 2009), its fragmentation and a greater difficulty in action are stressed. The role of political brokers between levels has been transformed, retrieving the local action and – assuming the thesis of Flores that analyzes security policies – fragmenting the action of the federal government (Flores 2009). In this context, conflicts over water turn into quarreling spaces over access to the resource but also into claims for a greater participation in the decision-making processes associated with water allocation (Mazahua women, agricultural users of sewage in San Luis Potosi or Valley of Mezquital, the indigenous people in Xoxocotla, Morelos) but, more commonly, of participation in the decision-making in general (access to safe water and sanitation services, transfer of scarce resource, among others). The issue of access to water is reformulated as a human rights issue, as did the Mazahua demonstrators opposing to water transferring for supplying Toluca Valley and Mexico City. The thesis of the biocultural diversity states that the main areas of concentration of biodiversity and water catchment are usually inhabited by indigenous groups and that, when they are under the traditional management of some group, better conditions of ecosystem conservation are noticed. Boege (2008) states that in indigenous territories 49% of the water in Mexico is collected, while this only represents less than 10% of the population. Others have pointed out the coincidence between protected natural areas, Ramsar sites, and coastal wetlands.

The symbolic aspect of water in the social management of traditional peoples, which continues to be the subject of myths and ceremonial practices, because, in addition of being a vital necessity, it is a fundamental part of their sacred life, shall not be overlooked. In principle, there is no separation between nature and society, typical of Western thought. There are pledges for water, rain prayers for a good harvest, and prayers for obtaining water, it is included in oral tradition, it is displayed in art, and it is used in everyday life, or purification rituals are performed with it (Sandre and Murillo 2008). It even is granted a personality and intentionality; for example, in the language P'urhepecha of western Mexico, the cloud is *Tata janikua* (Mr. cloud) and the river is *nana iorhekua* (madam river), explaining the hydrological cycle based on the relationship between both entities.

11.3 The Indians in the Model of Centralized Water Management

For a little over a hundred years, Mexico was characterized by its model of centralized water management, based on the intervention of the federal government as the water authority, through strong public investment in hydro-agricultural infrastructure and supply of drinking water and sanitation for large cities. The emphasis was to ensure water supply for economic development, without major social or environmental concerns. This was achieved through the resources of a political regime of corporate nature that included Indians within this model. The political regime in Mexico has been characterized in many occasions as authoritarian and highly centralized at the federal level (Rodríguez 2009; Meyer 1995). The authoritarianism appears to be linked with the forms of representation of corporate interests, and both features have been combined to argue regarding the persistence of the meager citizen participation and local powers diminished before the central power, that is, the existence of an incomplete citizenship (Zermeño 1996). But the case of the indigenous population is more complex. Although excluded from the benefits of development, they are maintained as organized peoples since they have been able to resist to that centralism and persist in the local control of their natural resources. This means that there are local political orders that screen the decisions of the governmental centralism (Varela 1984, 1987; Padua and Vanneph 1988). In the literature, the corporate bond of the Mexican political system is identified through "caciques" or political brokers, some of them with a clear disqualifying sense, as a result of an anti-modernity bias and in others as a sign of autonomy and resistance to the Western world.

With the creation of the National Commission for Irrigation in 1926, major projects began on adaptation and transformation of the natural environment to build dams, irrigation districts, and resource management focused on promoting an increase of agricultural production and consolidating a sector of agricultural producers closely related to the policies on financing and commercial distribution of the national agricultural production. Since then, a differential development began, leaving out an important sector of peasant rural population as the one constituted by indigenous peoples. Postrevolutionary state considered itself responsible for the development of this hydraulic infrastructure in order to develop modern agriculture and thus to integrate more remote regions into progress. Small-scale irrigation systems, managed by their own users, would be in a second place in the attention of the federal government. This led to the establishment of clear differences between the systems of large irrigation (irrigation districts) and small irrigation (irrigation units). In the districts, organization units would be consolidated with the promotion of the federal government, where, in most cases, even the distribution of water allotment was in charge of federal government officials not permitting any self-management of their own farmers, whereas the so-called irrigation units would be based on the effort and economic capacity of their own beneficiaries. Some of them have survived to this day, from ancient ways of organization for water distribution, as the atopiles, water judges, and mayordomos in villages of central Mexico. These changes supported a new conception of water as a productive resource, by converting it into a strategic element not only for major businesses but also for the purpose of the state administration.

In the same way as the Mexican State implemented its model of centralized water management, it was until the 1980s that heavily integrationist policies regarding indigenous population were included. In the 1920s, various policies aimed toward castilinization and abandonment of the indigenous language started modernization through their incorporation in nontraditional economic activities, new technology, and productive or economic reorganization where a large development project was implemented. This made many indigenous peoples and mestizo peasant institutions and political traditions to be watered down. Integrationist government policies were implemented from the promise of an improvement in the quality of life for the indigenous population.

In the fulfillment of major infrastructure projects, whether for irrigation, hydroelectric dams, or of a different kind, negotiations, agreements, covenants, and compensations were as usual required. But anyway, the federal government had to prevail to achieve the building of large-scale infrastructure; this is the case of the Cerro de Oro Dam (Bartolomé and Barabas 1990), or it had to intervene in the region resulting in greater inequalities among those who agreed to irrigation and those who remained at the margins of the irrigation districts (Finkler 1974; Villagómez 2006; Vargas 2007): suburban areas supplying megacities or receiving discharges (Cirelli 2002; Peña 1997), through regulation policies of water systems and their water resources achievement that worked against the social dynamics of the indigenous peoples (Vargas 2011). In many of these systems, forms of local management prevailed, which have resisted the federal authority in order to maintain control over water resources according to their "uses and customs"; resisting or accepting government policies, the indigenous peoples had to adapt themselves but without a promise of development and better living conditions fulfilled.

Another significant issue is the orientation of the federal government operations in the hydrological cycle. Most major infrastructures were carried out without considerations of its environmental or social impact in the surrounding areas, inhabited by peasant and indigenous communities. The intentions of economic development by the federal state produce not a few conflicts with indigenous peoples, where the management that is promoted by governments has an impact on the alteration of the resources. This contrasts with water management seen from the perspective of the indigenous peoples, which looks forward a much more environmentally friendly use. This difference has led to introducing a dichotomy between modern ecocidal societies versus indigenous peoples as ecologically correct (Toledo 2003). Projects boosted by the federal government consisted in the intervention of bodies of water, thus altering its hydrological cycle, with the purpose of transferring volumes of water to where the massive demand was located, cities such as Mexico City, or irrigation areas, drastically altering the natural distribution of water, making it economically accessible. Boehm et al. (1999: 30) designates this economic-technological rationality as a desiccant policy since it consisted mainly in altering watercourses through diversions, channeling or transfers, or through wetlands drying. This has always confronted with the idea of preservation, respect, and even identity of indigenous peoples with their natural resources, in particular with the water policy that soon became a development policy, mainly for the modern agricultural production, transforming delayed relations of production, as well as a form of political unification. The new regime would try out forms of linkage between farmers and government institutions that would be quite effective for maintaining order and generating public policies oriented toward a domestic market growth, until the fiscal crisis of the state in the 1980s, and the entry of open trade policies and economic deregulation in the 1990s.

11.4 Persistence of Social Inequality

Social situation of the indigenous peoples has historically had the greatest backwardness and marginalization, as well as a lack of recognition of their cultural specificity. Currently, in the indigenous regions, the scarcity of jobs, low wages, and asymmetrical relations between indigenous and non-indigenous population prevail, a situation that exerts an enormous pressure toward its dissolution as organized peoples; the loss of control of a resource such as water can be very adverse to their political and economic organization ways. There are many indicators that show that the indigenous population is at a clear disadvantage in health, education, and particularly in income-generating opportunities. Regarding access to safe drinking water in households and health problems associated with sanitation, they are also a token of social inequality. The federal public expenditure has been focused on those municipalities with an indigenous population with the lowest human development, but there is no conclusive evidence that these resources are benefiting the indigenous peoples (UNDP 2010).

Comparisons in the report of the UNDP are eloquent. In 2005, 16% of the total population in the country lived in municipalities of high marginalization. In contrast, 60% of indigenous population lived in these municipalities themselves. In the official poverty measurement, six social rights are incorporated: education, health, social security, housing, basic services, and food. In the year 2008, 93.9% of the indigenous population was deprived of at least one of these rights and 64.2% of at least three of them, so 70.9% of the indigenous population was classified in multidimensional poverty, defined as the percentage of people with at least one social deprivation and at income poverty (UNDP 2010).

This economic and social situation is closely linked to the fact that the investment of public resources in indigenous regions continues to present significant lags compared with the national mean. The income earned by the indigenous population is extremely low, since 80% of them earn less than two minimum wages, and 40% of working indigenous peoples are in the primary sector, while the national level indicates 16%–45.9% of indigenous men and 18.2% of indigenous women are working in the agricultural sector (UNDP 2010). Here, the unpaid work is inserted, which is the one that supports self-consumption production. The importance of family work among indigenous people reveals that in this type of economies the labor market is not consolidated, which makes migration necessary in search of monetary income to ensure the survival of the family. Of the population aged 15 years and above, 25.5% is illiterate, and 72.5% lack health services. More than 38% of indigenous children suffer from chronic child malnutrition, a figure three times higher than in non-indigenous children (12.5%). In the period 1988–2006, the prevalence of chronic child malnutrition among indigenous populations diminished from 55% to 38%; in contrast, the non-indigenous was reduced from 24.6 to 12.5%.

Indigenous population in Mexico usually settles in rural localities that are characterized by living in precarious conditions in education, housing, infrastructure, and basic services. In the year 2005, the percentage of private indigenous housing with no piped water was 29.5% and the non-indigenous 12.2%. Regarding drainage, almost half (44.4%) of them had no drainage, while this percentage in the non-indigenous was 13.3%. Houses with dirt floors in the indigenous population is 38% and in the non-indigenous population 10% (UNDP 2010).

11.5 Indigenous Peoples and Integrated Water Management

A little more than three decades ago, the model of centralized management of water entered into huge difficulties, when it was officially recognized that in several hydrological regions water was already being overexploited, deteriorating all bodies of water, not being able to further expand the amount of irrigated land nor to satisfy the demand for urban systems with such a low economic and technical efficiency (CNPH-SARH 1981).

From 1990 major transformations in water management began, which are reflected in the National Water Act of 1992, and subsequently expanding with the amendment of the Act in 2004, where proposals developed by the technical staff of experts from international organizations were incorporated and compiled in the so-called Integrated Water Resources Management, IWRM (Agarwal et al. 2000). In 1989 the National Water Commission, CONAGUA, was created, as the national water authority, which assumes in the last amendment of the Act an administrative organization by hydrological region based on executive agencies of river basins, with watersheds and aquifers as management units. In the 1990s, state commissions of water and sanitation in the federal entities are created, establishing as well their own state laws – in several cases in confrontation with the CONAGUA because of the definition of their own jurisdictions. In addition, a constitutional reform that since 1983 granted municipal governments the responsibility of drinking water and sanitation at the local level was implemented (Pineda and Salazar 2008). Regarding agriculture, between 1990 and 2005, 3.4 million hectares formerly operated by the same federal government, out of 6 million that are equipped with irrigation infrastructure, were transferred to farmer's organizations. At the end of the 1990s, councils, commissions, and committees of basin and aquifer were created, although only in advisory functions, to promote social participation in the management of water resources, in spite of attempts to expand these spaces, the first three dating back to the beginning of the decade; they shift from technical bodies to spaces of participation. However, a large sector of small irrigation and drinking water systems that keep their own ways of water organization apart from the governmental supervision present a broad spectrum of forms for social water management defined as legal pluralism (Boelens et al. 2005).

With these changes and some others, a shift from the centralized management model of water in the federal government toward a model of management by river basin has been sought, based on decentralization and participation of many actors and the incorporation of regulatory mechanisms of demand; this at least as a purpose, but the history that can be documented in these years, is quite different. In this context, cities and metropolitan areas reaffirm in their favor the balance of power, but with a margin of maneuver in relation to water that is reduced increasingly fast.

In 1990, several hydrological regions of Mexico were already taking advantage of their water resources to its highest level, even exceeding the border of the hydrological balance. Since the National Hydraulic Plan in 1981, several serious problems of supply were identified in 18 cities, and it was considered that other 17 would be added before 2000. Since then, the solution established as the most viable was to make a change from agricultural use to urban use, which implied closing 115 thousand hectares, in addition to other 110 thousand that would disappear with the same urbanization (CNPH-SARH 1981; 50–52). Likewise, an ominous scenario for 2000 was identified, in which water extraction would be superior to the average availability of annual renewable water. In fact, that was already happening, but it was a reality publicly emerged in the second half of 1990s. The *Water Agenda 2030*, a planning document produced by CONAGUA, states that a significant number of hydrological regions endure a substantial gap between supply and water demand which will not be solved in no other way than by redistributing water between uses and regions plus other costly actions to make a more efficient use.

The National Water Law of 1992 established the Public Register of Water Rights, based on which water rights would be arranged, extractions of watersheds and aquifers regulated, and new policy instrumented; all these aimed to regulate water demand. Water consumption should be given under an efficiency criterion, for which institutional and economic mechanisms to redistribute the water would be created, although this process has been carried out on an irregular basis, persisting several downloads of residual water and irregular royalties, and concessions that do not correspond to the extractions from users.

CONAGUA began in the late 1990s, the publication of decrees of availability, formalizing the existence or not of volumes that are likely to be granted. In this way, it was officially recognized the hydrological closure (Molle and Wester 2009) of several basins and aquifers such as the Valley of Mexico, Lerma-Chapala, Balsas, Rio Bravo/Rio Grande, and around 101 out of 653 aquifers. This affects a significant number of cities and metropolitan areas in the center and north of the country, precisely where the process of increasing urbanization and economic activity is concentrated. In the absence of a water flow in the hydrological cycle to distribute water to a higher human demand, it turns from a situation in which it was possible to supply cities with more infrastructure and transfers in its periphery or neighboring basins and aquifers to a hydrological limit (even without considering the ecological flow necessary for sustaining biodiversity), in which management measures should focus on making a more efficient use and encourage saving and transfer of volumes in uses and users. That is why the literature on this subject is conceptualized as a shift from supply management to demand management, which in the case of Mexico would be better if we analyze it as the shift to the management of deficit. These changes in the water management model are consistent with the neoliberal policies that are implemented, although with a more pragmatic than orthodox profile.

Also the relations of the indigenous peoples with the Mexican State changed significantly, transfiguring inclusive policies in other of recognition of indigenous peoples in certain aspects, mainly educational through bilingual-bicultural education, linguistic with the reorientation of the policies toward the preservation and enhancement of the linguistic heritage of the country, or recognition of their traditions, but not in substantive aspects concerning their ways of political organization, the recognition of their traditional forms of administration of justice, the right to elect their own representatives, to the development of socioeconomic projects based on their worldview, their natural resources, and their needs and respect for their cultural identity. In the indigenous regions, there are numerous social movements since the 1970s with ethnic claims, which bind in most of the cases with the conservation of their territories and the management of their resources (Bartra and Otero 2008); water is one of them.

The most important attempt to transform the relationship Indian peoples and Mexican State in relation to the forms of political organization and management of their resources consisted in the uprising of the Zapatista Army of National Liberation in 1994 and the process that continued afterward until the beginning of the decade of 2000, when an incorporation of the political regime in the recognition of the autonomy of indigenous peoples and the domain of their territories and resources was intended. The legal recognition of political and social rights in culturally distinct peoples in the context of the national state that encompass them should be translated into constitutional amendments, to give them certain rights as autonomy, self-governance, territories, quota of representation at the congresses, as well as the establishment of policies for their exercise.

However, the political establishment opposed, offering more subordinate forms of local government in "uses and customs," instituted in the Indigenous Law in July of 2001, which substantially differs in these aspects of the agreement of the EZLN and the federal government in 1996.

As of the change in presidential administration in 2000, a political transition started, in which it is considered that the competence of political parties has transformed the regime, although there is much literature that considers that there are more elements of continuity that actually a break with the past (Rodríguez 2009). In any case, a huge gap is now open in the interaction between different levels of government, the House of Representatives, Senate, and political parties with the indigenous organizations, allowing the creation of governments according to "uses and customs" in several states and conducting various legal changes toward the recognition of multiculturalism and the respect of languages and traditional indigenous organizations, after decades of assimilation policies. All of this was made possible, in addition to the change of regime in general. The recognition of a multicultural country, in which there are many indigenous and native peoples, organized around water according to "uses and customs," allows the incorporation of a legal pluralism, dealing differently with the model of water management excessively centralized that characterized the country during the twentieth century. What is now basic to recognize is the existence of a diversity of forms of social water management, all of which are effective at the local level and require an instance that coordinates and regulates them in the basin scope, not one that subordinates them to a governmental action that organizes them in an ineffective homogenous manner.

Currently there are deep imbalances regarding surface water and groundwater. This situation has generated recurrent conflicts of interests between the different users, which have been quickly taken up by political elites in the state and federal bureaucracy, in addition to the process of organization of environmentalists groups and farmers of the large-scale irrigation schemes, turning an environmental conflict into a political conflict between the federal and state governments.

The concern caused among ethnic groups and the changes suffered by their respective ecosystems are causing new restatements, mobilizations, and especially the need to be proposed as managers of their own environment, while witnessing governments and authorities violate the right to self-determination and sovereignty, opposing traditional knowledge and practices for the protection of water.

Rapid urbanization, in a phase in which decentralization policies and implementation of market rules seemed to regulate the use and pricing of water, has led to a confrontation of local institutions with the national regulatory framework, prevailing the asymmetry between cities to its environment, producing water territories dominated by urban areas. All this generating a new territorial scale with respect to water, where cities expand their influence areas, their weight in the decision-making process and transform even the hydraulic bureaucracy and the hydraulic regulatory framework in its favor (basin organizations). Rural areas do not generate forms of social organization able to negotiate and fight for water at a basin scale, although they generate multiple strategies of resistance to this process as we have tried to show with the presented cases. There are numerous examples of disputes over water for irrigation or drinking use in where native populations are involved. Most of these cases are an example of the manner in which the loss of local control over water undermines the integrity of communities, even more when they are located in basins or aquifers immersed in a deep environmental crisis, and where a greater pressure on irrigators and rural localities in favor of the cities is to be expected.

11.6 Water Conflicts and Social Movements

In the last decades, IWRM has been installed as the global intervention paradigm in water management across multiple scales and sectors. However, while this approach raises a number of key guidelines to be followed in terms of organization for management, assessments of its progress at national scales reveal differentiated applications that refer to negotiations, disputes, and redefinitions of what IWRM means in specific contexts (Saravanan et al. 2009; Molle 2008; Allouche 2016; Biswas 2004).

The IWRM policy was not a road map, but rather general precepts conceived from international experts with strong professional biases defined from their role in government agencies, many of them still geared toward building infrastructures, with little recognition of the sociopolitical dimensions. That is why the IWRM project has been implemented top-down by governmental organizations embedded in political regimes in which citizen participation is limited, either by its clientele approach, corporate character, or authoritarian ways of carrying out policies. The complications of the Mexican hydrocracy – technical groups of engineers with an economic and political orientation that form a bureaucratic organization that is not neutral in its valuations and interests (Treffner et al. 2010: 254) – in categorizing community water management, as well as recognizing their role in persistent problems and conflicts, have implications for the intensity of conflicts over water and the formation of coalitions with other group citizens and national and international social movements, without CONAGUA having a clear strategy of how to institutionalize and intervene in conflicts.

A recent research on water social movements (Kloster 2016), with hemerographic information, allows us to locate the type of confronted identities (which are those formed by social movements), the repertoires of action used in the struggle (direct, indirect, and nonconformity expressions), the government's response to actions, as well as the difference of strategies of struggle that the social movements with a citizen base follow with respect to the natives and peasants' movements. Kloster compares the years 1990, 2000, and 2010. Considering the bias of the hemerographic information, since it represents events of a scale and intensity that usually surpasses the local scope, nevertheless it allows her to visualize the defense of the citizen character or common pool resource defense with a territorial slant or a policy change that threatens the loss of benefits already built, while, on the other hand, the struggle for the construction of greater citizen rights (Kloster 2016: 52). This allows us to represent a point that is of our interest to highlight the trajectory or directionality of the struggle and the meetings within the legal order or outside it, since the implementation of IWRM and its participatory approach should be achieving some level of institutionalization of the differences in interests and the convergence toward the general objectives of water management.

Kloster presents the different fighting tactics in the 3 years she compares. Here it is interesting to highlight the use of different repertoires of action, which are "a limited set of routines learned, shared and acted through a process of choice

Table 11.1 Contentious actions, 1990–2010

	1990	2000	2010
	(Percentage)		
Direct actions	21.9	54.9	54.2
Political pressure	74.5	26.5	20.9
Nonconformity	1.4	18.6	25.0
Others	2.2	0.0	0.0

Source: Kloster (2016, p. 60)

relatively deliberate … are cultural creations learned, but do not descend from abstract philosophy or take form as a result of political propaganda, but arise out of the struggle" (Tilly 2002, 31–32). In this it is fundamental to differentiate those repertories where physical facts are used including acts of violence, those that are only verbal claims or written complaints. According to Tilly, direct action is the self-organized realization of an individual or group initiative, focused on giving specific responses to specific situations or creating more favorable conditions, using available means.

Table 11.1 shows the significant change in repertoires of action, more than doubling the use of direct actions in relation to other forms, such as exerting pressure. Kloster breaks down the type of actions involved in each of these areas, and, for example, among the most significant in the category of lobbying is the use of the complaint in the press, from 68% in 1990 to 9% in 2000 and to 14% in 2010. However, the use of confrontation as an action tactic ranged from 14% in 1990 to 59% in 2000 and 44% in 2010 (Kloster 2016: 64–66). In summary, the set of fighting tactics, which have confrontation and pressure, went from 24% in 1990 to 77% in 2000 and 72% in 2010. In the protest cycles of the social movements for water, there are normally stages of escalation or reduction of conflicts, and it can be assumed that the repertoires change regularly, but what appears here is rather a structural change in the management regime, in the sense that there is a significant change in the type of action, as well as in the social profile of those who are more mobilized.

A discussion about the meaning of scarcity as a relative and not absolute notion is essential to demonstrate that water scarcity does not directly generate conflict, but that perception of inequality in distribution and access is more significant, as well as the loss of control over the resource that is considered common. For this, Kloster classifies those who perform the actions according to groupings as follows. Neighbors are those subjects identified as they speak or denounce from the local level, without any organizational identity. These range from 29% in 1990 to 41% in 2000 and 31% in 2010. The other subjects that consistently appear are government authorities, from any level, from 15% in 1990 to 28% in 2000 to 32% in 2010. The other change emphasized by Kloster (2016: 111–112) is that of the category of indigenous people, who are not identified in 1990 but become 2% in 2000 and 6% in 2010. Other identified social groups are political parties, farmers, businesses, NGOs, and neighborhood representatives. If the data is organized considering the social identities that carried out more direct actions, it turns out that two groups can be visualized, which correspond to the two types: citizen movements and community movements. On the one hand, there are the natives and peasants' mobilizations, whom 86.6% of their actions are direct – of nonviolent direct actions as well as violent direct actions – and the rest without pressure or confrontation, taking together the records of the 3 years. For NGOs it is 60% of its actions with pressure and confrontation. As neighbors, neighborhood representatives, and political parties, more than 50% of their actions are with no pressure nor confrontation.

The practice of IWRM would entail the creation of ways of expressing diversity of interest in water, with a watershed approach, and the establishment of appropriate instances for the institutionalization of conflicts over water. However, there is a tendency toward polarization, consolidation of social sectors with repertoires of action based on direct actions, as long as their interests are not solved and their perception of water injustice prevails. This was clearly expressed in the analysis made by academics and organizations regarding the proposal for a new General Water Law in 2015, where the dilemma currently faced by water management was manifested: greater facilities for private investment in the solution of problems of urban supply (mainly) through greater interventions in the hydrological cycle through transfers and infrastructures or the solution via social participation and consensus with the social diversity and interests that characterize the majority of the hydrological units of the country, strictly assuming the aim of limiting economic development to the availability of water in the regions and incorporating in the institutional arrangement the different forms of social water management, characterized by their operation at the local level.

11.7 Conclusions

It is very complicated to summarize in just a few pages a process as intricate as the permanence/dissolution of indigenous societies in Mexico, even more in its relationship with water, their struggles, and collective action to have their right to water and water justice recognized.

Transformations originated from endogenous and exogenous processes that lead to an increasingly intensive exchange between indigenous societies and national society. For the first time, there is an absolute demographic decline of the indigenous population in the 2010 national population census, according to the ethnolinguistic criteria used by the federal government. In contrast, their presence becomes increasingly evident in conflicts over water through the media; either in the case of Canal Independencia in Sonora that reduces irrigation water for the Yaqui tribe favoring the city of Hermosillo, all water transfers from peri-urban areas of medium-sized towns such as Oaxaca (Zapotec and Mixtec peoples and peasants), Toluca (Mazahua and peasant villages), Queretaro, among others, the tensions for urban-industrial waste water use in farming areas, such as Valle del Mezquital (hñahñú) which receives those from the Valley of Mexico, or the many conflicts between peoples for water access and distribution.

This situation leads us to five main theses. First, an important part of the national territory is facing a process of hydrological closure or, as expressed by the federal government, inability to close the gap between water supply and demand. The simplest solution is to have a greater intervention in the hydrological cycle (more water from further afield or urban and industrial consumption), which involves the implementation of mechanisms to transfer volumes and water rights, hardly fair in an unequal society, in addition to setting aside the water demand management and the environmental slant of integrated management of water resources proposal. This process involves the indigenous people, because they live in regions that can provide water to the economy, without implying a benefit or solving their current water shortage.

The second thesis affirms that the process of redistribution of water between hydrological regions is plagued by economic and power asymmetries, which show as a very unequal process, either between large urban areas with respect to their periurban and rural environments and with neighboring basins; large corporate farmers with respect to peasants; and social management organizations in relation to government infrastructure projects. Water markets, many outside the control of CONAGUA, water banks, and wastewater exchange for others of first use, among other measures, have already been implemented without a noticeable existence of an impartial mechanism.

The third thesis asserts the continuity of integration processes of the indigenous peoples through an indigenous reform that does not recognize the central points as their territories and natural resources, water among them.

The fourth thesis refers to the persistence of a weak performance of the system of government management of water at the local level and even the largest abandonment before the draft reforms in the institutional arrangement of water in the current federal government. Basin or aquifer councils, commissions, and committees have been displayed as instances based on a top-down model of participation.

The fifth thesis foresees the intensification of the conflict among self-management forms and urban-industrial and agribusiness groups of interest. Conflicts over water, based on water as a resource for common use and local water, imply a growing conflict for water, to which many mestizo peasants adhere.

What remains under discussion is the role of local water management models that should be developed in the next years. Currently, there is a wide variety of organizational forms, among which is the community water management of indigenous peoples, which are not formally recognized by CONAGUA. However, local management turns out to be the best adapted to the specific circumstances at the local level, but with many problems of coordination and functioning that could be the objective of government policy. The CONAGUA concentrates all the powers and budgets; it is in fact an organization incapable of carrying out its policies in a large part of the national territory as a result of the existence of this legal pluralism in drinking water committees and small scale irrigation systems. Besides, the problem of accessing safe water and quality water is an important issue for indigenous peoples, as they are indicators of poverty and lack of access to public services.

References

Agarwal, A. et al. (2000). *Manejo integrado de recursos hídricos*. Estocolmo, Global Water Partnership, 2000. (TAC Background Papers; 4).

Allouche, J. (2016). The birth and spread of IWRM – A case study of global policy diffusion and translation. *Water Alternatives, 9*(3), 412–433.

Ávila, P. (1996). *Escasez de agua en una región indígena: el caso de la Meseta Purépecha*. Zamora: El Colegio de Michoacán.

Ávila, P. (2007). *El manejo del agua en territorios indígenas en México* (Serie del Agua en México, vol. 4). México: Banco Mundial, Departamento de México y Colombia, Región de Latinoamérica y el Caribe.

Bartolomé, M. A., & Barabas, A. M. (1990). *La presa Cerro de Oro y el ingeniero el gran dios*. México: INI, México.

Bartra, A., & Otero, G. (2008). Movimientos indígenas campesinos en México: la lucha por la tierra, la autonomía y la democracia. en S. Moyo y P. Yeros (Coord.). *Recuperando la tierra. El resurgimiento de movimientos rurales en África, Asia y América Latina*. Buenos Aires: CLACSO.

Biswas, A. K. (2004). Integrated water resources management: A reassessment. *Water International, 29*(2), 248–256.

Boege, E. (2008). *El patrimonio biocultural de los pueblos indígenas de México. Hacia la conservación in situ de la biodiversidad y la agrodiversidad de los territorios indígenas*. INAH, Conaculta, CDI: México.

Boelens, R., Zwarteveen, M., & Roth, D. (2005). Legal complexity in the analysis of water rights and water resources management. In D. Roth, R. Boelens, & M. Zwarteveen (Eds.), *Liquid relations. Contested water rights and legal complexity* (pp. 1–19). New Jersey: Rutdgers University Press.

Boehm de Lameiras, B. y Manzo, M. S. (1999). La sed saciada de la ciudad de México y la nueva cuenca Lerma-Chapala-Santiago, un ensayo metodológico de lectura cartográfica. *Relaciones. Estudios de Historia y Sociedad, XX*(80), 15–67.

Burguete-Cal y Mayor, A. (2000). *Agua que nace y muere: sistemas normativos indígenas y disputas por el agua en Chamula y Zinacantán*. México: UNAM.

Cirelli, C. (2002). Supplying water to the cities: Mexico City's thirst and rural water rights. In S. Whiteford y R. Melville (Comps.). *Protecting a sacred gift: Water and social change in Mexico*. San Diego: University Press.

CNPH-SARH. (1981). *Plan Nacional Hídrico 1981*. México: Comisión Nacional del Plan Nacional Hidráulico y Secretaría de Agricultura y Recursos Hidráulicos.

Descola, P. (2001). Construyendo naturalezas. In P. Descola y G. Pálsson (Eds.), *Naturaleza y sociedad. Perspectivas antropológicas*. México: Siglo XXI.

Estrada, A. y Franco, H. (2004). Entre la ley y la costumbre: El uso y manejo del agua potable en el municipio de Temoaya, Estado de México. *Páramo del campo y la ciudad*. No. 7, diciembre. Centro de estudios sobre marginación y pobreza del Estado de México.

Finkler, K. (1974). *Estudio comparativo de la economía de dos comunidades de México: el papel de la irrigación*. México: INI.

Flores Pérez, C. A. (2009). *El Estado en crisis: crimen organizado y política. Desafíos para la consolidación democrática*. México: CIESAS.

Gelles, P. (2004). Etnohidrología, desarrollo y política cultural en la sierra peruana. In F. Peña (Ed.), *Los pueblos indígenas y el agua: desafíos del siglo XXI*. Bogotá: WALIR-IMTA, El Colegio de San Luis.

Giménez, G. (2000). Identidades étnicas: estado de la cuestión. In L. Reina (Ed.), *Los retos de la etnicidad en los Estados-nación del siglo XXI*. México: Centro de Investigaciones y Estudios Superiores en Antropología Social, Instituto Nacional Indigenista, Miguel Ángel Porrúa.

Guzmán, N. B. (2009). La gestión social del agua potable en los altos de Morelos. en S. Vargas, D. Soares, O. Pérez, & A. I. Ramírez (Eds.), *Gestión de los recursos hídricos: realidades y perspectivas* (Tomo II). México: IMTA/SEMARNAT/Universidad De Guadalajara. México.

Guzmán, N. B., Reyes Quintero M., Pérez Rodríguez, A. I., & González Flores, L. (2012). Agua y territorio comunitario: Tetela del Volcán vs. Hueyapan. In S. Vargas, E. Mollard, & A. Güitrón de los Reyes (Coord.). *Los conflictos por el agua en México: caracterización y prospectiva*. México: Conamexphi-IMTA-UAEM.

Kloster, K. (2016). *Las luchas por el agua (1990–2010)*. México: Universidad Autónoma de la Ciudad de México.

Medellín, P. (2006). La política y las políticas públicas en regímenes de 'obediencias endebles'. Una propuesta para abordar las políticas públicas en América Latina. In R. Franco y J. Lanzaro (Coord.). *Política y Políticas Públicas en los procesos de reforma de América Latina* (pp. 101–144). Buenos Aires: CEPAL-FLACSO.

Meyer, L. (1995). *Liberalismo Autoritario, Las Contradicciones del Sistema Político Mexicano*. México: Ed. Océano.

Molle, F. (2008). Nirvana concepts, narratives and policy models: Insight from the water sector. *Water Alternatives, 1*(1), 131–156.

Molle, F. y Wester, Ph. (Eds.). (2009). *River basin trajectories: Societies, environments and development* (Comprehensive Assessment of Water Management in Agriculture, No. 8). Cambridge: International Water Management Institute – CAB International.

Padua, J., & Vanneph, A. (1988). *Poder local, poder regional*. México: El Colegio de México-CEMCA.

Peña de Paz, F. (1997). *Los límites del riego agrícola con aguas negras en el Valle del Mezquital*. Tesis de Maestría, Universidad Iberoamericana, México.

Pineda, N., & Salazar, A. (2008). De las juntas federales a las empresas de agua: la evolución institucional de los servicios urbanos de agua en México 1948–2008. In R. Olivares y R. Sandoval, (Coord.). *El Agua Potable en México. Historia reciente, actores, procesos y propuestas* (pp. 70–88). México: ANEAS.

Rodríguez Araujo, O. (2009). *México ¿un nuevo régimen político?* México: Siglo XXI.

Roth, D., Boelens, R., & Zwarteveen, M. (2005). *Liquid relations. Contested water rights and legal complexity*. Piscataway: Rutdgers University Press.

Sandre, I., & Murillo, D. (2008). *Agua y diversidad cultural en México*. Montevideo: UNESCO-Programa Hidrológico Internacional para América Latina y el Caribe.

Saravanan, V., McDonald, G., & Mollinga, P. (2009). Critical review of integrated water resources management: Moving beyond polarized discourse. *Natural Resources Forum, 33*(2009), 76–86.

Tilly, C. (2002). Repertorios de acción contestataria en Gran Bretaña: 1758–1834. In E. M. Traugott (Ed.), *Protesta social: repertorios y ciclos de la acción colectiva* (pp. 17–47). Barcelona: Editorial Hacer.

Toledo, V. M. (2002). Ethnoecology: A conceptual framework for the study of indigenous knowledge of nature. en J. R. Stepp, et al. (Eds.), *Ethnobiology and biocultural diversity* (pp. 511–522). International Society of Ethnobiology.

Toledo, V. M. (2003). Ecología, espiritualidad y conocimiento. In *De la sociedad del riesgo a la sociedad sustentable*. México: Programa de las Naciones Unidas Oficina Regional para América Latina y el Caribe para el Medio Ambiente-Universidad Iberoamericana.

Treffner, J., Mioc, V., & Wegerich, K. (2010). A-Z Glossary. In K. Wegerich & J. Warner (Eds.), *The politics of water. A survey*. London: Routledge.

UNDP. (2010). Informe sobre Desarrollo Humano de los Pueblos Indígenas en México. In *El reto de la desigualdad de oportunidades*. México: Programa de las Naciones Unidas para el Desarrollo.

Varela, R. (1984). *Expansión de sistemas y relaciones de poder. Antropología política del Estado de Morelos*. México: UAM-Iztapalapa.

Varela, R. (1987). Problemática sobre el sistema político mexicano. *Relaciones* 31 verano, Vol. VIII: 61–76.

Vargas, S. (2007). Agua y sociedad en los sistemas de riego de la cuenca del Alto Lerma. en *Agricultura, Sociedad y Desarrollo* , *4* (1), enero-junio 2007; 1–17. El Colegio de Posgraduados.

Vargas, S. (2011). Los pescadores del lago; entre el manejo comunitario y el deterioro ambiental. en R. Huerto, S. Vargas, & C. Ortiz (Eds.), *Estudio ecosistémico del lago de Pátzcuaro: aportes en gestión ambiental para el fomento del desarrollo sustentable*. Instituto Mexicano de Tecnología del Agua, Universidad Autónoma del Estado de Morelos, Universidad Michoacana de San Nicolás de Hidalgo.

Vázquez de León, L. (2010). De las identidades numeradas, racionalidades divergentes y la etnicidad entre los jornaleros agrícolas indígenas en el Occidente de México. en Multitud y Distopía. *Ensayos sobre la nueva condición étnica en Michoacán*. México: UNAM.

Villagómez, Y. (2006). *Política hidroagrícola y cambio agrario en Tehuantepec, Oaxaca*. Zamora: El Colegio de Michoacán.

Zermeño, S. (1996). *La sociedad derrotada. El desorden mexicano del fin de siglo*. México: Siglo XXI.

Sergio Vargas-Velázquez is an Economist. He earned a Master in Social Sciences and a PhD in Social Anthropology. He shared 12th in the Leadership for Environment and Development Program in Mexico (LEAD-Mexico). He is a specialist on social issues in water management. His main areas of research are organizational and political processes in irrigation systems, productive change in irrigated areas, water user organization for watershed and aquifer management, and in the last years on social issues in water conflicts. Between 1990 and 2009, he held various positions in the Mexican Institute of Water Technology, where he participated in projects on transfer of irrigation districts, the organization of councils, commissions and committees at basin and aquifer level, characterization of water conflicts, and negotiation processes.

Part IV
International Water Issues

Chapter 12
From the Abundance of Waters to the Scarcity of Studies: Contemplating Hydropolitics in Mexico-Guatemala and Mexico-Belize Borders

Edith F. Kauffer

> *Negotiations on water sharing are intensely political*
> *Sadoff et al. (2008)*

Abstract Facing the paradox of the borders of Mexico with Guatemala and Belize, through which most of Mexico and Guatemala water resources flow, versus the lack of studies about it, how do we articulate local and international dimensions into the political water analysis in the boundary regions that Belize, Guatemala, and Mexico share? This chapter intends to provide some areas of thought regarding this question. Starting from the description of transboundary dimension of water in the studied region, this work proposes a new concept of "hydropolitics" as multiple to analyze diverse international and local dynamics regarding water in this region of multiple borders.

Keywords Mexico · Guatemala · Belize · Transboundary river basins · Borders · Hydropolitics · Water policies

---

A first version of this text was published in Spanish under the title: "De la abundancia de aguas a la escasez de estudios: retos y perspectivas de las hidropolíticas en la frontera México-Guatemala-Belice" as a general introduction to the book edited by Edith F. Kauffer Michel (coord.) *Entre manantiales y ríos desatados: paradojas de las hidropolíticas fronterizas (México-Guatemala)*, México, CIESAS, COLMICH, 2011, pp. 9–31. This text has been modified to be adapted for this work, updated, and summarized in some portions.

E. F. Kauffer (✉)
Centro de Investigaciones y Estudios Superiores en Antropología Social-Sureste,
Chiapas, México

H. R. Guerrero García Rojas (ed.), *Water Policy in Mexico*, Global Issues in Water Policy 20, https://doi.org/10.1007/978-3-319-76115-2_12

12.1 Introduction

In the last three decades, the boundaries that Mexico shares with Guatemala and Belize, traditionally known as the "southern border," have been the object of rich academic work. The so-called southern border went from being considered a forgotten border to a border described and analyzed by diverse disciplines in the social sciences.[1] Consequently, nowadays historical and contemporary phenomena such as colonization processes (De Vos 1993; Rodas Núñez 2014); population issues (Ángeles Cruz 2005; Piedrasanta Herrera 2014a, b), especially migrations (Kauffer Michel 2002; Torras Conangla 2014; Ángeles Cruz 2010; Baltar et al. 2013; Rivera Farfán 2015); historical processes of shaping the borders of Chiapas (Castillo et al. 2006; Guillén 2003; Valdez Gordillo 2006; Fábregas 2015), Tabasco (Vautravers 2005), and Campeche (Torras Conangla 2012); trade relations (Villafuerte 2004); and religious and cultural interactions (Rivera Farfán 2014; Piedrasanta Herrera 2014a, b) have been extensively documented. In this increasingly profuse and detailed literature that recently includes the questioning of the "southern border" notion (Fábregas and González 2014; Kauffer Michel 2013d, 2017), the environmental issue has been understudied in a portion of the Mexican territory characterized by abundance of biodiversity and underground and surface water resources.

Academic work regarding water issues in the Mexico-United States border shows an entirely contrasting reality compared to the situation of the Guatemala and Belize borders. In the northern border, water is a key issue and stands out in five aspects: (1) it is the object of specific publications in both countries with a remarkable increase in the last 20 years; (2) it is an essential aspect in Mexico-United States border and transboundary research; (3) it is significant for the entire American continent; (4) it presents a wide range of studies of various disciplines from human studies to engineering focused on diverse topics and issues; and (5) it holds a special place in the social literature about water in Mexico and in international publications addressing issues of transboundary waters (Kauffer Michel 2011a, b).

The relationship of the southern border with natural resources as well as the sociopolitical dynamics around water in international basins requires further study (Kauffer Michel 2005c). This applies to all scientific disciplines that tend to ignore the existence of borders and the transboundary dimension, even while conducting studies few meters away from the international boundary line on international basins or rivers (Equihua et al. 2006). However, the lack of studies regarding water can be seen especially in Mexican social sciences, which have historically focused their water studies in the center (Boehm 2006), north of the country (Aboites Aguilar 2000a), Distrito Federal (Martínez et al. 2004; Perló and González 2005), and neighboring states of the capital city (Espinosa Henao 2006; Saldívar 2007; Sandré Osorio 2005; Stephan-Otto 2003; Vargas et al. 2006) and to very few exceptions in

[1]Note that most of the contributions regarding the southern border of Mexico are focused on the study of processes involving only the state of Chiapas and don't usually include the three other border states which are Tabasco, Campeche, and Quintana Roo.

Chiapas (Molina 1976). Thus, research on water in Mexico is characterized by an almost total absence of references regarding the bordering states of Chiapas, Tabasco, Campeche, and Quintana Roo, as Aboites Aguilar states (2009) (see Aboites Aguilar 1998; Aboites Aguilar and Estrada Tena 2004; Aboites Aguilar et al. 2000; Ávila García 2002a; Birrichaga 2007; Castañeda González et al. 2005; Durán et al. 2005; Escobar et al. 2008; Kroeber 1994; Meyer 1997).

Recently, some contributions regarding water problems in the Mexican southeast (Aboites Aguilar 2000b) and later regarding the southern border of Mexico have emerged in individual publications (Burguete Cal y Mayor 2000; Birrichaga 2008[2]; Contreras Utrera 2008) or as part of collective works, especially about water in the state of Chiapas (Ávila Quijas et al. 2009; Benez and Kauffer Michel 2012; Contreras Utrera 2009b; García García 2005a, b; García García et al. 2006; Kauffer Michel 2006b, 2009, 2011c, 2012, 2014a, b; Kauffer Michel and García 2003; Kauffer Michel and García 2004; March and Fernández 2003; Mejía González 2011; Mejía González and Kauffer Michel 2008; Rojas Rabiela 2009[3]; Soares 2006; Solís Hernández 2011; Valette 2011; Vera 2005), Tabasco (Jhabvala 2006; Gracia Sánchez and Fuentes Mariles 2004, Kauffer Michel 2013a, b, c) or both states (García García 2013; Kauffer Michel 2005a, 2008, 2013d) and in the two borders, including the Guatemalan and Belizean portions (Kauffer Michel 2005d). Suchiate River has been the focus of very recent studies (Gómora 2013, 2014; Kauffer Michel 2011a; Ordoñez Morales 2011; Santacruz de León 2011a, b) as well as Grijalva river basin. The topic of floodings in Chiapas (Álvarez Gordillo and Álvarez Gordillo 2011) and Tabasco river basins (Capdepont and Marín Olán 2013; Galindo et al. 2013; Kauffer Michel 2013a; Ramos Reyes et al. 2013) has also been studied through collective publications (González and Manse 2014). Mexican regional and interdisciplinary focuses on water issues have been launched (Kauffer Michel and Castillejos 2015) as well as Central American regional projects including the Mexican borders (Kauffer Michel and Medina 2014; Kauffer Michel 2014a, b). Some transboundary river basins remain little studied like Candelaria River (Kauffer Michel 2005b, 2010) or mainly by nonsocial scientists like the Usumacinta River (de La Meza and Carrabias Lillo 2011).

It is worth mentioning contributions on the history of water in Chiapas (Contreras Utrera 2005, 2009a) and Tabasco (Salazar 2013; García García 2013) and the first resource catalog (Sandré Osorio and Kauffer Michel 2014), which includes the states of Campeche, Chiapas, Tabasco, and Quintana Roo, in addition to Yucatan. Among these recent publications, those covering international or transboundary water problems are scarce (Kauffer Michel 2004, 2005c, 2006a, 2011a, b; Santacruz de León 2005, 2006, 2011a, b; Santacruz et al. 2005).

[2]Although in this case just 1 out of 35 presented studies refers to a southern border state; most are dossiers from the north and center part of the country.

[3]In this case, the author makes brief references to the presence of hydraulic works in prehispanic and colonial times in the states of Campeche, Tabasco, Quintana Roo, and a mention to Chiapas.

Facing the paradox of the borders of Mexico with Guatemala and Belize, through which most of water resources of Mexico and Guatemala[4] flow, versus the lack of studies about it, how do we articulate local and international dimensions into the political water analysis in the boundary regions that Belize, Guatemala, and Mexico share? This chapter intends to provide some areas of thought regarding this question. Starting from the description of transboundary dimension of water in the studied region, this work proposes a renewed concept of "hydropolitics" to analyze diverse international and local dynamics regarding water in this region of multiple borders.

12.2 Two "Borders": Exceeding the Hegemonic Notion of Southern Border

This chapter assumes the uncommon position of opposing the hegemonic notion of "southern border," which has allowed over the last two decades to visualize the existence of the area as opposed to the "northern border," still considered nowadays in many political and government circles as the only relevant border in Mexico. This position is based on the fact that it is a contradictory notion to address the issue of transboundary waters and it is a little susceptible to visualize cooperation relationships when diminishing transboundary dimensions. Indeed, "southern border" is a Mexican denomination, since this border is in the case of Guatemala the northern or northwestern border—depending on the referred fragment—and to Belize it stands as the northern border. It also constitutes a regional border that shows the northern limit of Central America. This Mexican denomination excludes those Central American visions of border reality, the existence of "the other side," and the possibility of sharing transboundary positions. It carries the hegemonic vision of more than two decades of research from the Mexican side, which contributes in the reinforcement of economic and academic asymmetries regarding the history of Belize and Guatemala, especially the weakness of existing academic structures.

Besides its excluding character, this notion equalizes the Mexico and Guatemala border with that of Mexico and Belize. Nevertheless, each border has its own history, local characteristics, and relationships, which prevent us from speaking of a single reality (Kauffer Michel 2013d) and lead us to consider the denomination of southern border as inadequate. On these grounds, this chapter will consider two borders and not the southern border of Mexico, when trying to be coherent with the acknowledgment of water flow beyond political boundaries, with the need of looking at or from the other side and while trying to be consistent with the approach of considering political dimensions from local to international settings through the concept of hydropolitics.

[4]Belize is not included in here due to its homogenous situation.

12.3 Transboundary Dimension of Water As a Resource in the Borders Among Mexico, Guatemala, and Belize

The Mexican states of Chiapas, Tabasco, Campeche, and Quintana Roo; the Guatemalan departments of San Marcos, Huehuetenango, El Quiche, and El Peten; and the Belizean districts of Orange Walk and Corozal that comprise the Mexico and Guatemala and Mexico and Belize[5] administrative borders have the advantage of being located in a region particularly rich in water resources as a result of heavy rainfall which could be evidenced by the existence of abundant surface waters in the area.

Thereby on the Mexican side, facing an annual mean rainfall of 740 mm on the national scale, between 1981 and 2010, the bordering states are characterized with the following figures: Tabasco, 2184 mm; Chiapas, 1842 mm; Campeche, 1251 mm; and Quintana Roo, 1267 (Conagua (Comisión Nacional del Agua) 2016). Meanwhile, Belize stands in the 17th place in the world ranking regarding rainfall with 1705 mm (Conagua (Comisión Nacional del Agua) 2008); and in Guatemala, the bordering areas' total rainfall can reach 6000 mm (Dardón Sosa 2002). Regarding per capita water availability, expressed in cubic meters per person per year, the southern border Mexican administrative region (i.e., the one that includes Chiapas and Tabasco, as stated by Conagua from 2010) totaled 18,852 m^3 in 2015 versus 6373 m^3 in the Yucatán Peninsula (that covers the states of Campeche, Quintana Roo, and Yucatán), compared to the national average of 3692 m^3 (Conagua (Comisión Nacional del Agua) 2016). Guatemalan and Belizean availabilities were 8600 m^3 and 61,566 m^3, respectively, in 2007, twice and more than 15 times the one of Mexico (Conagua (Comisión Nacional del Agua) 2008).

These figures undoubtedly show plenty of water in both borders. However, behind this regional and state "natural" abundance, there are specific hidden local realities ranging from scarcity to a glut. Moreover, this annual abundance corresponds in reality to striking contrast between the rainy season, which extends from June to October and the dry season between November and May, meaning that the bordering regions with Guatemala and Belize are not exempt from drought and scarcity episodes and that extreme hydrometeorological phenomena constitute a recurring problem throughout the region.

It is essential to emphasize that despite the abundance conditions of water as a resource, the bordering region is characterized by the highest national shortfalls in access of domestic water and sanitation household services. Whereas in Mexico, the national average reaches 92.5%, the states of Quintana Roo and Campeche are both in the border exceeding such percentage. With 82.6% and 88.8%, respectively, Chiapas and Tabasco (Conagua (Comisión Nacional del Agua) 2016), two states with the highest water abundance in the Mexican Republic, perfectly illustrate the paradox between natural abundance and a lack of access to basic services. In terms of

[5]Situated from west to east.

sewage, compared to an average of 91.4%, Chiapas (84.4%) registers one of the highest shortfalls.

There is no information to evaluate the situation of access to water and sewage services in Belize. In Guatemala, the border region is characterized by high levels of poverty in the four borderline departments: 86.7% of San Marcos population, 78% of Huehuetenango, 81% of El Quiché, and 57% of El Petén are experiencing poverty (Dardón Sosa 2002). The northern border of Guatemala corresponds to peripheral, excluded areas populated by indigenous people[6] that were scenario of an internal armed conflict which lasted more than 30 years. Consequently, the situation of water and sewage services is characterized by a higher shortfall than in the Mexican side. Only 40.7% of the population in the bordering municipalities of San Marcos Department have access to domestic water services and 8.3% to sewage services, while in Huehuetenango Department, the figures represent 50% and 10.4% (Dardón Sosa 2002). In Ixcan municipality, situated in El Quiché, barely 15% of the population has access to domestic water and 8.8% to sewage services.

Water quality constitutes a concerning issue in the region. For instance, Chiapas has just 11 out of 118 municipalities with a wastewater treatment plant in service in 2010 according to a fieldwork. In 2015, although 93 wastewater treatment plants were installed in 122 municipalities (76%) (Conagua (Comisión Nacional del Agua) 2016), interviews evidenced that less than 30 plants were really working (26%). This situation is even more critical in Guatemala, where no urban center conducts wastewater treatments, which are directly discharged into rivers that flow across population centers and carry pollutants downstream. On the other hand, the population is used to dump waste into rivers, and the presence of several waste deposits on the banks of surface currents represents a constant situation in the bordering region, on both sides of the boundary line. Finally, if present, municipal landfills are not established according to environmental norms that prevent soil and water pollution.

As a result of the alarming situation of water quality and sanitation conditions, it is fundamental to note that Chiapas represents the state with the highest child mortality due to diarrheal diseases in the Mexican Republic (43.4 per hundred thousand inhabitants) (Conagua (Comisión Nacional del Agua) 2008), situation that is multiplied in Guatemala with a 3677.46 per hundred thousand inhabitants rate (Dardón Sosa 2002). More recent data about this topic is unavailable.

---

[6]In 1994, in Guatemala (IICA), the Suchiate river basin population presented the following characteristics: 48%, indigenous; 88%, rural; 75%, in extreme poverty; 43% no water services; and 14%, no latrines. In Coatán river basin, 69% of the population was indigenous; 96%, rural; 51%, no water services; 10%, no latrines; and 75% lived in extreme poverty. On the three uppermost tributaries of Grijalva river basin, the situation arises the following indicators: Cuilco river basin, 72% lived in extreme poverty; 95% was rural; 45% lacking, water; and 22%, latrines. Selegua river basin, 53% indigenous; 78%, rural; 39% had no latrines, and 40%, water. Extreme poverty in 75% in Nentón river basin, 93% of the population are indigenous people; 84%, rural; 49%, lacking water; 25%, latrines; and 75% lived in extreme poverty. There is no more updated information of shared river basins in Guatemala. More recent information about bordering areas are not available.

In this context of natural abundance and socio-hydrological shortfalls, what are the different boundary and transboundary dimensions of water resources in the region?

First, three international rivers, i.e., currents with an international border demarcation function, characterize borders among Mexico, Guatemala, and Belize. As Castillo et al. (2006, back cover), correctly states, "making the history of the southern border of Mexico requires a transboundary vision effort of the facts and historical processes that have occurred on both banks of Hondo, Suchiate and Usumacinta rivers." In fact, these three rivers define much of the Mexican border with Guatemala and that of Belize, which totally consist of 1139 km in length: 53% of Mexico-Guatemala border and 87% of Mexico-Belize portion have a fluvial delimitation.[7] Most of Suchiate river serves as an international borderline (77 out of 92 km) between Chiapas and Guatemala,[8] situation that is repeated in the case of Hondo river between Quintana Roo and Belize but representing for Usumacinta river just one third of its length, separating Chiapas from Guatemala.

It is worth noting that those three rivers do not have the same geographic configuration regarding the three countries. Dinar (2008) distinguishes through-border rivers that cross the border from one country to the other one from border-creator rivers or international rivers. Toset et al. (2000) present another three configurations: the upstream/downstream configuration which is equivalent to Dinar's through border, the mixed type when the river crosses a country, functions as a borderline, and flows in a territory of a second country, and the river-boundary configuration when the current flows within a country and establishes the border with another one. This is a basic element to consider in the analysis of conflict and cooperation dynamics regarding international rivers.

The three international rivers in the borders between Mexico and its neighbors present three different configurations. Usumacinta is a mixed river, while Suchiate is a river-boundary configuration in the denomination of Toset et al. (2000). Finally, Hondo river corresponds to none of the configurations proposed, because it includes three states and those consider only two countries; indeed, it crosses the Guatemalan territory, and then it establishes the borderline between Mexico and Belize.

To the international rivers that create "water boundaries" in both studied borders, the existence of territories called transboundary river basins is added: a river basin is an area where surface runoffs gather toward a common exit or convergence point of waters. In the river basins, natural and water resources are closely related in such a way that agricultural, industrial, rural and urban activities conducted upstream, and

[7]Data calculated by Emmanuel Valencia, Laboratorio de Análisis e Información Geográfica (Laige), El Colegio de La Frontera Sur (Ecosur) from a geographic information system.

[8]Note that other authors propose different measures: Santacruz de León (2006) states that Suchiate river is 120 km, of which 84 correspond to borders; Jiménez Castañeda et al. (2006) mention a main riverbed of 79.2 km; Conagua (2007) evokes 75 km of borderline; and CILA (2006) refers 81 km of river border. As a result, probably, of the social construction nature of the basins, and consequently each sources uses delimitation parameters according to work interests. Probably these estimations also vary as a function of tributaries taken into account for its measuring.

the presence of infrastructures have an impact downstream, especially in quality and quantity water availability. A transboundary river basin refers to the physical and geographical reality described above when the catchment is divided by a political border.

Bordering territory between Mexico, Guatemala, and Belize might be divided into six large river basins, called as international, shared, or transboundary. In this chapter, the three adjectives shall be used indistinctively, even though they refer to a similar situation, each one starts from a different scope and emphasizes on different perspectives for water resources management. The "international basin" term refers to a territory that exceeds the jurisdiction of a single nation-state, which implies that its management might be considered through mechanisms that go beyond national bodies. Meanwhile, when talking about "shared basins," emphasis is placed on the idea of cooperation between states for its management. Finally, the use of the term "transboundary" relates to the existence of common spaces that exceed political frontiers and create continuity beyond divisions established by the human being without political connotation.

Until recent times, except for an atlas published in 1987 (CILA), each country used to represent shared basins through boundary basins, that is, delimited basins through each state sovereignty or, in other words, basins truncated by the international division line. Through a binational effort led by Laboratorio de Análisis e Información Geográfica [*Geographic Analysis and Information Laboratory*] (Laige) from El Colegio de la Frontera Sur (Ecosur) on the Mexican side and Sistema de Información Geográfica del Ministerio de Agricultura, Ganadería y Alimentación [*Geographic Information System from Agriculture, Cattle and Food Ministry*] (MAGA) from Guatemala, between 2007 and 2008, there was a joint effort in the development of a consensus map of international basins between Mexico and Guatemala.[9] As a result, a delimitation of six large transboundary basins was obtained, represented on Fig. 12.1.[10]

These basins, located from west to east, are the following: Suchiate river (Mexico-Guatemala), Coatán river (Mexico-Guatemala), and Grijalva basin (Mexico-Guatemala)—known in Guatemala as Cuilco, Selegua, and Nenton river basins—which penetrate into Mexico as three water courses forming Grijalva river in the Mexican territory, Usumacinta river basin (Mexico-Guatemala-Belize) with a

[9]This activity was conducted as a part of the project "Hidropolítica en la frontera México-Guatemala-Belice," financed by Consejo Nacional de Ciencia y Tecnología (Convocatoria SEP-Conacyt de Ciencia Básica) between July 1, 2005 and June 30, 2009. However, it is necessary to establish that this consensus included just the first four international basins (Suchiate, Coatán, Grijalva, and Usumacinta), and it was established among technical groups of geographic information systems and not validated by any political authorities on both states, to which was presented in a meeting with both foreign offices in 2008 but manifested no interest on the subject.

[10]Methodology and characterization of these basins are the object of a thesis (García García 2010) and a publication (García García and Kauffer Michel 2011) that exceed the interest of this chapter. Note that some sources consider five basins instead of six. However, since 2009, Conagua takes as a reference the presence of six transboundary basins.

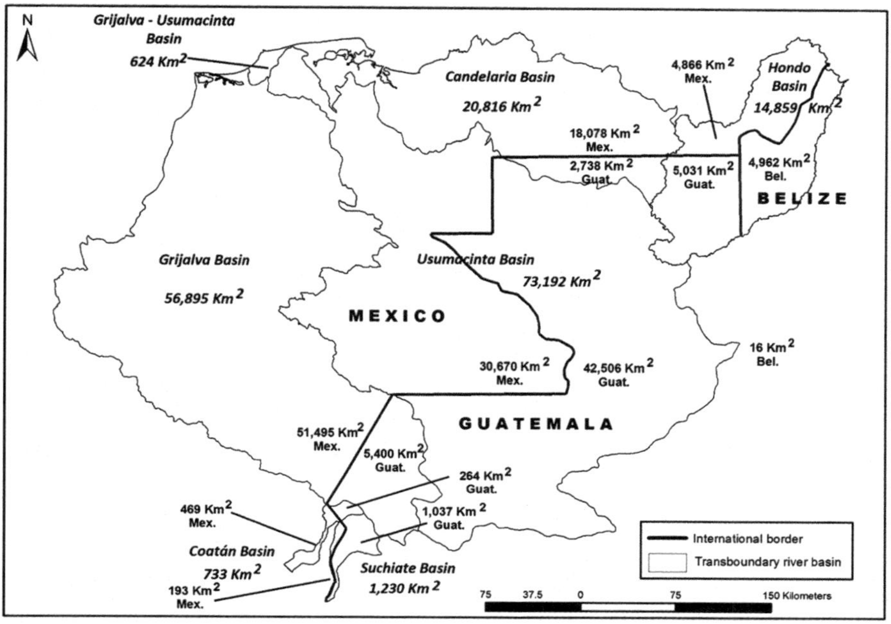

Fig. 12.1 Six transboundary basins on Mexico-Guatemala and Mexico-Belize borders. (Source: Prepared by García García (2010), with the support of Ing. Emmanuel Valencia Barrera, technician of Laige-Ecosur, Unidad San Cristóbal de Las Casas, Chiapas)

very small portion corresponding to Belize, Candelaria river basin (Mexico-Guatemala), and Hondo river which is trinational.

In the transboundary river basins, the upstream situation constitutes a geographic advantage versus downstream locations, as a result of the potential effect of the first over lowermost riparians—individuals or groups—because actions conducted upstream affect water in terms of quality and quantity. This happens in any river basin, apart from its internal or international character. In transboundary river basins, upstream location represents a geographic advantage for the country that holds such a position. However, it is added to other characteristics as political interactions, which could counterbalance such advantage. Two examples of transboundary basins serve to illustrate this situation. Egypt is located downstream in the Nile river basin shared by ten countries, but its condition of regional military power imposed until recently a status quo to the other countries situated upstream, preventing them from making the most of the waters. In fact, Egypt had an exclusive reliance for all necessities on the Nile (Swain 2002) and has even threatened with an armed intervention against its neighbors, in the event of having its interests affected as a result of hydraulic works conducted upstream. This case illustrates that the geographic advantage is not the only dimension involved in terms of international river basins but it also highlights that the situation is able to change as it occurred due to political transformations in Egypt and at regional scale from 2013 (Petersen-Perlman et al. 2017; Salman 2011). The case of Turkey and its neighbors, Syria and Iraq, in the basins of Tigris and Euphrates rivers is another example: as a result of its strategic position and its condition of the US allied regional military power in the region, Turkey has conducted unilaterally hydraulic works that have affected its neighbors situated downstream, stopping or reducing the quantity of water with no influence of neighboring countries on the situation (Mutin 2003) and behaving as a regional hegemon (Conde 2014).

In Mexico-Guatemala and Mexico-Belize borders, Guatemala is located upstream in the six basins, and Mexico and Belize are downstream riparian.[11] However, for many reasons, Guatemala has never used such a strategic advantage to affect the interests of its neighbors. Some of these reasons are lack of economic resources for the construction of major hydraulic works, geographic and topographic conditions that hinder those works, and the location of some parts of this territory within zones affected by the armed conflict, which lasted more than three decades at the end of the twentieth century. However, Mexico, despite its location downstream, during the 1980s, had projected the construction of four major dams in the Usumacinta river basin, which would have flooded part of Guatemalan El Peten. The opposition of the Guatemalan government to this project and the social unrest promoted its abortion in 1989, and this event is the only conflict regarding the water issue registered in the database of the Oregon State University of the Mexico, Guatemala, and Belize border (Transboundary Freshwater Dispute Database (TFDD) 2007).[12] It is important to note that despite its downstream location, Mexico has not experienced any impact from Guatemala regarding water availability, and on the contrary, it has had the intention of unilateral hydraulic works. Moreover, it should be emphasized that Mexico has suffered negative consequences as a result of its downstream location in transboundary river basins regarding water quality due to bacteriological contamination, organic (as a result of indiscriminate pesticide use) and inorganic (especially produced by open-pit mining and intensive agricultural activities such as a chemical spilling from a palm oil plant in 2015 that contaminated La Pasión river in Guatemala); however, this issue does not appear nowadays in the Mexican political agenda.

Finally, the region is characterized for its multiple transboundary surface currents of diverse magnitude that flow from one country to the other, passing over political borders established by humans and making international the water-related issues resulted from the mobility of this resource. Also, in certain points of both borders, there are some surface water bodies like lagoons or wetlands that hold a transboundary character since they are crossed by the international division line.

[11]Note that this statement is not considering the upstream location of Belize in the Usumacinta river basin because of the absence of surface currents and population in the Belize corresponding 16 km^2 of basin, that is, a 0.021% total of its area. In the Suchiate river basin, Guatemala is also situated downstream.

[12]Note that the research of this database included just English language publications, that is, sources with international magnitude events, and reported on that scale but not necessarily reflecting the rich and detailed transboundary local reality.

Thus, the transboundary dimension of the water resources in the region involves diverse realities that can be analyzed from different perspectives. This chapter proposes the notion of hydropolitics to understand diverse political aspects of these water resources.

12.4 Hydropolitics: Looking for an Articulation Between Waters, Politics, and Policies

Contrary to a statement made a few years ago (Kauffer Michel 2004) regarding the scarce use of the concept "hidropolítica" in Spanish and contrasting to the frequent use of "hydropolitics" in English literature about water, politics and policies, and transboundary issues, nowadays it is quite common to read and hear the term "hidropolítica" in Spanish-speaking academic and political circles. However, this frequent use which encompasses at the same time water-related political aspects and international dimensions of water resources does not have a conceptual clarity in most of instances (Cascão and Zeitoun 2010). Several authors fail to define the term and yet use it (López 2008; Oswald 2008; Salman 2011), which is an issue that was already mentioned by Turton (2002) for the English language academic production. Likewise, the frequent use in English and French of an adjective derived from the term "hydropolitics" (*hydropolitical or hydropolitique*) can be recently observed in literature, attesting to its growing success, as De Stefano et al. (2017) constantly refer in a recent paper.

John Waterbury (1979) was the first author in using the term "hydropolitics" in 1979, in his book about the Nile river. He frames the concept around the interaction of hydraulics, water policy, and the results of this relationship. Waterbury's research centers around the analysis of a process involving the conduct of two sovereign states (Egypt and Sudan) in search for their national interest, water-use policies of the Nile river, and their relationship before the international coordination challenge of a shared resource. To do so, the text addresses the following aspects: the natural dynamics of the Nile river and hydraulic interventions that have modified its annual behavior, the relationships between two of the nine riparian states,[13] national policies and politics of both countries, and the impact of development programs in water resources management.

Reviewing traditional literature around the concept "hydropolitics," we can distinguish two major perspectives: on the one hand, the history of conflict and cooperation dynamics regarding transboundary water resources and, on the other hand, the analysis of "the authoritative allocation of values in society with respect to water" from the redefinition proposed by Turton (2002, p. 16). Resuming these two

[13]In 1979, nine states possessed territory in the Nile river basin. Later, as a result of Eritrean independence from Ethiopia, in 1993, they turned into 10 states. In 2011, the secession of South Sudan from Sudan increased the number into 11 states.

approaches, we propose an applicable perspective to the study of hydropolitics in the transboundary river basins among Mexico, Guatemala, and Belize.

12.4.1 Hydropolitics or Conflict and Cooperation Dynamics Around Transboundary Waters

Elhance (1999, p. 3) is the author that best epitomizes this first perspective of hydropolitics study, starting with the definition of hydropolitics as "the systematic study of conflict and cooperation between States over water resources that transcend international borders." This analysis trend of hydropolitics that can be labeled as "international" is nowadays the most consolidated and represented among the specialists on the topic.

This international perspective undoubtedly stems from Waterbury's work. Indeed, despite explicitly mentioning that hydropolitics does not cover international dimension only, Waterbury (1979, p. 87) uses the title of international hydropolitics in a chapter related to the analysis of relations between the two riparian states, giving the impression that this is the main dimension of hydropolitics analysis.

The dynamics of conflict and cooperation related to sharing water proposed by Elhance (1999) as a focal point of the analysis imply the need of paying exclusively attention in state logistics, since states are considered the main players in hydropolitics (Elhance 1999, p. 14) because of their ability to create water conflicts and to promote the means for cooperation. However, the presence of other players and interdependencies with other aspects are acknowledged, but these have no main role on the analysis.

The international perspective of hydropolitics leads to two main debates: a discussion around water scarcity and its effect in the relationships among states and the study of existing and potential conflicts around transboundary waters. Most of the literature about shared waters, in which studies of the Middle East and Asia are predominant, revolves around this type of issues (Kauffer Michel 2004).

The vision of hydropolitics proposed by Elhance (1999, p. 15) considers geography as crucial for the resulting conflict and cooperation processes in relation to international basins. Physical geography (topography, meteorology, hydrology, and geology) is directly involved in water availability, in the interdependence between riparian states and the potential uses of waters. Economic geography has an impact on water demand and the responses toward its needs. Finally, political geography defines conflict and cooperation dynamics among countries in interaction with a series of other elements that turns the study of hydropolitics into a complex field.

Making an assessment about cooperation and conflict dynamics regarding hydropolitics from six case studies of international river basins, Elhance (1999, pp. 236–242) demonstrates the main elements leading to noncooperation: sovereignty of states, nationalism and pressure of national interest groups, difficulty of assigning economic value to water which involves divergent interests, and lack of information and available technology. However, he also mentions favorable trends toward cooperation (Elhance 1999, pp. 242–247): the fact that states are not necessarily looking for an armed conflict to solve their differences, international political and economic changes favoring cooperation, accumulated experience in water international agreements, emergence of new technologies for water uses, and the existence of the international convention of 1997—convention of the law of non-navigational uses of international water courses—which defines the rights of international water courses use for purposes other than navigation and was ratified by 35 countries in August 2014 to become an international conventional law for the signatory states.

Thus, the international analysis perspective of hydropolitics is focused in the interaction of various favorable dynamics of cooperation among countries in the field of shared waters and in the local, national, and international elements that promote conflicts among state players.

12.4.2 Hydropolitics As an Adjective: Recent Proposals Regarding Political Dimension

More recent contributions (Turton 2008; Wolf 2007) ponder around the international perspective of hydropolitics and use the *hydropolitical* adjective to create new concepts that enable the analysis of different aspects about cooperation and conflicts dynamics in transboundary waters.

Based on the definition of hydropolitics as "the ability of geopolitical institutions to manage shared water resources in a politically stable sustainable manner, i.e., without tensions or conflicts between political entities," that is, from the traditional and international analysis perspective, Wolf (2007, p. 4) proposes two new concepts: hydropolitical resilience and hydropolitical vulnerability. Hydropolitical resilience refers to the ability to face changes and hydropolitical vulnerability to the emergence of disputes around shared waters. Conflicts emerge in function of two elements: velocity of physical change and the institutional capacity to cope with it. The latter refers, according to Wolf (2007, p. 4), to favorable international relationships, to the presence of treaties or institutions for the management of transboundary waters that are the same or even more important than the change in itself. Thus, institutional capacity is a resilience factor, while vulnerability is the result of changes and institutional weaknesses toward them. Consequently, a river basin that tends to be oriented toward hydropolitics resilience would include international agreements and institutions on shared waters, a history of cooperation in this area, favorable political relationships, and economic development conditions. On the contrary, a trend toward hydropolitical vulnerability would be the result of brutal environmental changes, population growth or an economic asymmetric growth, unilateral projects of magnitude development, absence of institutional capacity, bad international relationships, and climatic variability which is particularly characterized by alternating floods and droughts (Wolf 2007, p. 5).

Another analysis proposed by Turton (2008, pp. 35–38) incorporates the concept of hydropolitical complex in reference to South Africa. This author starts with recognition of the diversity of basins and riparian states around economic and military capabilities and especially in various situations regarding their dependence, in some basins, of waters from other countries and, for some states, of waters from a certain basin in particular. The notion of hydropolitical complex leads to two additional notions, pivotal states and basins and impacted states and basins that are used to refer to states and river basins. Turton (2008, p. 37) explains that pivotal states have an economic development and a high dependence degree of shared river basins for water supply. Impacted states share a sharp dependence on waters from the same river basin in pivotal states but find themselves unable to negotiate an equidistribution with the first.

Pivotal basins are situated close to the water availability limit, because all available water is distributed among diverse uses (basin closure concept) and is essential for pivotal states. Meanwhile, impacted basins are integrated with a pivotal state and an impacted state. The second state is not under equity conditions regarding distribution of shared waters (Turton 2008, p. 38). Finally, the concept of hydropolitical complex constitutes an intriguing framework for analysis regarding international relationships in the transboundary basins that Turton (2008) develops for a group of basins in southern Africa, demonstrating an asymmetries game among states.

In both proposals, the political dimension is a key factor for the analysis. The formulation by Wolf (2007) refers to institutions interacting through a set of hydrological, climatic, economic, demographic, and diplomatic elements. Under the perspective offered by Turton (2008), power relationships are articulated through water availability within the scope of a certain basin.

While these approaches to hydropolitics renew the international perspective and highlighted the political dimension of the concept, they still focus their analysis in states as exclusive players of hydropolitics. From traditional hydropolitics literature (Magrin 2016) to current analysis (Salman 2011; De Stefano et al. 2017), the state-centrism remains one of the major characteristics (Allouche and Daoudy 2010). Interstate relations approaches are today considered as a narrow perspective by critical hydropolitics (Sneddon and Fox 2006) and by constructivist hydropolitics (Julien 2012). Recent proposals argue for the necessity to go beyond state-state relations (Thomas 2017; Sneddon and Fox 2006) and include non-state actors in water governance and hydropolitics analysis (Mirumachi and Chan 2014) because water politics and transboundary issues are multi-scalar and multi-stakeholders (Menga 2016). Developing these proposals with other types of players that share transboundary waters remains a pending and highly relevant task for Mexico and Guatemala and Mexico and Belize borders.

12.4.3 Hydropolitics and Water Conflicts in Mexico

Patricia Avila (Ávila García 2002b) is the first person in using the term "hidropolítica" in Spanish in Mexico. This conceptualization emerges from an international perspective, but her main contribution consists in the adjustment of the concept for its use in water local problems in Mexico. Avila (Ávila García 2002b) explicitly starts from Elhance (1999) and Maury (2003) texts—who in turn bases his work on Elhance as well—to create a definition of hydropolitics understood as "the manifestation of tensions arisen from control and management of an increasingly scarce and strategic resource," and which refers to "water use as a political resource and local power source." In this conceptualization effort, Maury (2003) and Avila (Ávila García 2002b) make a distinction between hydropolitics and water policies, being the first result of the second one, that is, hydropolitics is derived from the consequences of a "critical situations set" associated to water policies (Maury 2003). Recently, several authors have taken up again this conceptualization to analyze diverse local conflicting situations related to water in Mexico (de Alba 2007; Rojas 2008).

The conceptualization of the "hydropolitics" term proposed by Avila García (2002a, b) for the Mexican case is focused on the analysis of conflicts surrounding water, and it is interesting in this aspect. Although it helps to translate the previously described international vision to local aspects of water management in Mexico, it does not permit the analysis of diverse dimensions involved in Mexico-Guatemala and Mexico-Belize borders from local to international since they are focused on the conflict. Our position is based on the premise that although the conflict represents a fundamental dimension of politics, this is situated, above all, because of its social regulation function, as defined by Leca (1973). Therefore, the hydropolitics conceptualization proposed by Avila is far too narrow for the apprehension of interactions among local and international dimensions of the hydropolitics study and for the analysis of its complexity.

12.4.4 Hydropolitics According to Turton: Extending the Concept

Based on the complexity of hydropolitics dynamics, Turton (2002) proposes to extend the "hydropolitics" concept as a result of a set of reasons. First, hydropolitics cannot be solely limited to the analysis of state players and should involve other players with an important role in diverse aspects. Derived from this first aspect, Turton mentions that it is necessary to consider interactions among state and non-state players to the extent that Elhance's definition does not include all political interactions regarding water.

For that purpose, Turton (2002, p. 16) uses the notion of "politics" imbedded in the term hydropolitics and to the definition of politics proposed by David Easton as "the authoritative allocation of values in society." Politics is a dynamic process and on track, where the main focus is the allocation of values through laws and policies, authoritatively exercised (i.e., by an authority). This implies a decision process favoring certain groups, which leads to challenging values and legitimacy of the authority. Therefore it is relevant to understand the who, what, when, where, and how in the analysis of politics.

The definition of hydropolitics resulted from these elements is as follows: "the authoritative allocation of values in society with respect to water" (Turton 2002, p. 16). It suggests an opening of the concept to a wide range of aspects beyond conflict and cooperation processes among states in relation to transboundary waters. Turton goes no further on the definition but explains two basic aspects of the extension on the hydropolitics field. The first one is the scale of hydropolitics studies: unlike the traditional vision, which places them exclusively in the international scope, Turton (2002, p. 239) proposes a multiscale approach of the hydropolitics field, that is, from individual to international, through the following scales: family, community, city, province, national, regional, or international basin. The second element of Turton proposal focuses in the opening of potential topic scopes that the notion of "hydropolitics" involves, which, in the case of South Africa, include dimensions linked to the local context.

Turton (2002) proposes a matrix format to cross diverse scales with thematic scopes that builds three groups: legal-institutional, social, and economic. This configuration enables the location of the richness of thematic scopes related to hydropolitics and its diverse analysis scales. Some scopes involve different scales and others only center on several scales of analysis. For instance, the thematic scope "poverty" involves all scales of social group analysis. Likewise, the scope of international rights is located solely between the scales of river basin and international, because it does not apply to the others.

The fundamentals of Turton's proposal are, precisely, the redefinition of the "hydropolitics" concept, which permits its expansion toward less reduced aspects than those proposed by the international perspective prevailing on the literature of the subject.

12.4.5 Hydropolitics and Policy Interactions Regarding Water: A Proposal Toward Water and Transboundary River Basins

Water issues in the borders among Mexico and Guatemala and Mexico and Belize have created no open conflicts among riparian countries or any relevant bilateral or trilateral cooperation actions. There is no treaty regarding shared waters nor actions for managing transboundary basins because of a lack of interest to include the topic in the political agenda of states, which is dominated with migration issues and problems of security linked to drug trafficking and illegal actions. Thus, on one side, the state interactions regarding water are extremely limited, while on the other side, there is a multitude of local relationships surrounding water. In fact, water is an everyday issue for communities settled on both sides of international division lines, as well as an interest point for non-state players that pretend to have an influence over the issue of international river basins, like nongovernmental organizations, academic groups, or epistemic communities.[14] Taking into consideration of the above, it is difficult to apply the international perspective of the hydropolitics analysis proposed by Elhance (1999) to the reality of the borders among Mexico and Guatemala and Mexico and Belize.

Finally, water issues in the six transboundary basins, representing a 167,727 km^2 territory, involve dimensions that are not related with the presence of an international division line but with characteristics of the local context, among which we can mention, as an example, the presence of indigenous groups in large areas of transboundary river basins territory, belonging to different political systems, administrative division, national and local institutions, geophysical conditions of the environment, incidence of extreme meteorological phenomena, and territorial representations. Consequently, the analysis of hydropolitics in the borders between Mexico-Guatemala and Mexico-Belize shall necessarily emerge from a proposal which permits to widely involve the realities of political interactions regarding water, that is, its international aspects—which are not reduced to the relationships among states—and local and national dimensions to further analyze interrelationships between these two components. In such a way, the proposal herein is part of the continuity of the "hydropolitics" concept that Turton (2002) extends and emerges from a definition of hydropolitics as *the political interactions regarding water in different scales and pertinent specific topics* for the study of water resources in the borders between Mexico and its southern neighbors.

12.4.6 Conclusion: From One Hydropolitics to Multiple Hydropolitics

Starting from the previously proposed definition of hydropolitics and the complexity of water problems in the borders among Mexico, Guatemala, and Belize that we have pointed out, our proposal is to extend the study of hydropolitics in the region by articulating two major scales: international and local.

As a result of the diversity and complexity of the analysis of water-related political problems in the region, along with diverse visions and perspectives resulted from different disciplinary approaches as political sciences, anthropology,

[14]Concept developed by the constructivist theory of international relationships that Haas (1992, p. 3) defines as a network of professional with a certain expertise and renowned abilities in a particular matter and formulates an articulated approach around a relevant issue of policy derived from understanding its sphere of specialization.

sociology, economy, and history, we have considered convenient to talk about various hydropolitics rather than a single hydropolitics. Talking about multiple hydropolitics then emphasizes our approach from different visions, processes, and phenomena of political interactions regarding waters that are intertwined and characterized by a certain plurality and heterogeneity besides those common elements.

As an example, we are able to find similar realities in faraway places on both sides of the border that repeat themselves through space and time. Other interactions depend on a specific history, political context, or even past and present water policies of each country of each administrative subdivision and, occasionally, of each place. On the other hand, there are transboundary local relationships between inhabitants that may be opposed to international interstate dynamics of water aspects and transboundary basins. There are also external players present with intent of approach that promotes action strategies from local inhabitants under the modality of conflicts or collaboration actions. The essence of politics, regarding the relationship between conflict and regulation, converges in the heart of multiple hydropolitics, from a combination of a bordering and a transboundary perspectives and as a continuum from local to international water issues.

As the most recent proposals from international literature upon hydropolitics argue from a critical and constructivist approaches, transboundary water issues are characterized by fluidity (Cascão and Zeitoun 2010; Hussein and Grandi 2017), and they must enlarge to a diversity of scales and actors (Thomas 2017) to include social dynamics (Julien 2012) and ecosystems (Mirumachi and Chan 2014) and enable an "alternative imaginings and associated praxes" of river basins (Sneddon and Fox 2006, p. 198). As a matter of fact, hydropolitics relates with state-state interactions but mainly depends on "what societies make of it" (Julien 2012, p. 45) as fieldwork in the mentioned borders indicates (Kauffer Michel 2014a, b).

References

Aboites Aguilar, L. (1998). *El agua de la nación. Una historia política de México (1888–1946)*. Mexico: CIESAS.

Aboites Aguilar, L. (2000a). *Demografía histórica y conflictos por el agua. Dos estudios sobre 40 kilómetros de historia del río San Pedro Chihuahua*. Mexico: CIESAS.

Aboites Aguilar, L. (2000b). Optimismo nacional: Geografía, ingeniería hidráulica y política en México. In B. von Mentz (Coord.), *Identidades, Estado nacional y globalidad. México, Siglos XIX y XX* (pp. 95–152). Mexico: CIESAS.

Aboites Aguilar, L. (2009). *La decadencia del agua de la nación. Estudio sobre desigualdad social y cambio políticos en México. Segunda mitad del siglo XX*. Mexico: El Colegio de México.

Aboites Aguilar, L. y Estrada Tena, V. (Comp.). (2004). *Del agua municipal al agua nacional. Materiales para una historia de los municipios en México 1901–1945*. Mexico: CIESAS-AHA-CONGUA-COLMEX.

Aboites Aguilar, L., Gardida, D. B., González, R. C. y Cortez, B. E. S. (2000). *Fuentes para la historia de los usos del agua (1710–1951)*. Mexico: CIESAS-CONAGUA.

Allouche, J., & Daoudy, M. (2010). L'hydropolitique et les relations internationales. *Dynamiques internationales*. 1–4.

Álvarez Gordillo, G. D. C., & Álvarez Gordillo, L. M. (2011). La vulnerabilidad social en Motozintla, Chiapas y el huracán Stan. In E. F. Kauffer Michel (Coord.), *Entre manantiales y ríos desatados: paradojas de las hidropolíticas fronterizas (México-Guatemala)* (pp. 217–240). Mexico: CIESAS, COLMICH.

Ángeles Cruz, H. (Coord.). (2005). *La población en el Sureste de México*. Tapachula: ECOSUR-SOMEDE.

Ángeles Cruz, H. (2010). Las migraciones internacionales en la frontera sur de Mexico. In F. Alba, M. Á. Castillo, & G. Verduzco (Eds.), *Los grandes problemas de México. Migraciones internacionales. T-III* (pp. 437–480). Mexico: El Colegio de Mexico.

Ávila García, P. (Ed.). (2002a). *Agua, medio ambiente y desarrollo en el siglo XXI*. Zamora/Morelia/Jiutepec: El Colegio de Michoacán-Secretaría de Urbanismo y Medio Ambiente-IMTA.

Ávila García, P. (2002b). Cambio global y recursos hídricos en México: hidropolítica y conflictos contemporáneos por el agua. In *Instituto Nacional de Ecología*. http://www.ine.gob.mx/dgioece/cuencas/descargas/cambio_global_y_rec_hdricos_mex.pdf. Accessed 26 Feb 2009

Ávila Quijas, A. O., Gómez Serrano, J, Escobar Ohmstede A., & Sánchez Rodríguez M. (Coord.). (2009). Negociaciones, acuerdos y conflictos en México, siglos XIX y XX. Agua y tierra. Mexico/Zamora/Aguascalientes: CIESAS-Colmich-Universidad Autónoma de Aguascalientes.

Baltar, E., Marroni, M. D. G, & Villafuerte Solís, D. (Coords.). (2013). *Viejas y nuevas migraciones forzadas en el sur de México, Centroamérica y el Caribe*. Mexico: Universidad de Quintana Roo, SITESA.

Benez, M. C., & Kauffer Michel, E. F. (2012). Percepciones y cosmovisiones en torno a la calidad del agua de los manantiales en una comunidad indígena: El Crucero, municipio de San Juan Chamula, Chiapas. In L. Daniel Murillo (Ed.), *Culturas del agua y cosmovisión india en un contexto de diversidad cultural* (pp. 123–151). Jiutepec: IMTA.

Birrichaga Gardida, D. (Coord.). (2007). *La modernización del sistema de agua potable en México 1810–1950*. Toluca: El Colegio Mexiquense.

Birrichaga, G. D. (2008). *Agua e industria en México. Documentos sobre impacto ambiental y contaminación 1900–1935*. Mexico: El Colegio Mexiquense-CIESAS.

Boehm Schoendube, B. (2006). *Historia ecológica de la Cuenca de Chapala*. Zamora/Guadalajara: El Colegio de Michoacán-Universidad de Guadalajara.

Burguete Cal y Mayor, A. (2000). Agua que nace y muere. Sistemas normativos indígenas y disputas por el agua en Chamula y Zinacantán. México: UNAM.

Capdepont, J. L., & Marín O. P. (2013). Las inundaciones de 2007 y 2008 en Tabasco: efectos en el sector productivo agropecuario. Un acercamiento estadístico. In E. F. Kauffer Michel (Coord.), *Cuencas en Tabasco: una visión a contracorriente* (pp. 167–189). Mexico: CIESAS/RISAF.

Cascão, A. E., & Zeitoun, M. (2010). Power, hegemony and critical hydropolitics. In A. Earle, A. Jägerskog, & J. Öjendal (Eds.), *Transboundary water management. Principles and practice* (pp. 27–42). London: Earthscan/SIWI.

Castañeda González, R., Escobar Ohmstede, A., & Andrade Galindo, J. A. (Comps.). (2005). *Desastre económico o debilidad federal en los primeros gobiernos posrevolucionarios*. Mexico: CIESAS-AHA-CONAGUA-UAEM.

Castillo, M. Á., Ribot, M. T., & Olivera, M. V. (2006). *Espacios diversos, historia en común. México, Guatemala y Belice: la construcción de una frontera*. Mexico: Secretaría de Relaciones Exteriores.

CILA (Comisión Internacional de Límites y Aguas). (1987). *Atlas físico de las cuencas de los ríos internacionales entre México y Guatemala*. Mexico: CILA.

CILA (Comisión Internacional de Límites y Aguas). (2006). *Estado del río internacional Suchiate, después del paso del ciclón tropical "Stan" en octubre de 2005*. Mexico: Presentación en powerpoint. CILA.

Conagua (Comisión Nacional del Agua). (2007). *Estadísticas del Agua 2007*. Mexico: SEMARNAT.

Conagua (Comisión Nacional del Agua). (2008). *Estadísticas del Agua 2008*, Mexico.

Conagua (Comisión Nacional del Agua). (2016). *Estadísticas del Agua 2016*, Mexico.

Conde, G. (2014). El agua entre Turquía, Siria e Iraq. *Regions and Cohesion, 4*(2), 81–100.

Contreras Utrera, J. (2005). Proyecto hidráulico para el abasto de agua potable en la Ciudad de Comitán, Chiapas. Finales del siglo XIX y principios del siglo XX, *Boletín del Archivo Histórico del Agua, Mexico*, 10(mayo-agosto), 38–50.

Contreras Utrera, J. (2008). *Entre la higiene y la salubridad. El abasto de agua en los principales centros urbanos de Chiapas, 1880–1949, tesis de doctorado*. Bilbao: Universidad del País Vasco.

Contreras Utrera, J. (2009a). *La manzana de la discordia: el conflicto por el agua del río Sabinal y sus afluentes, Estado de Chiapas (1909–1920). Historia Judicial Mexicana: La propiedad* (pp. 1–19). Suprema Corte de justicia de la Nación: Mexico.

Contreras Utrera, J. (2009b). Suministro de agua, higiene y conflictos sociales en los centros urbanos del departamento del Soconusco, Chiapas, 1876–1945. In Ávila Quijas, A.O., Gómez Serrano, J., Escobar Ohmstede, A., & Sánchez Rodríguez (Coords). *Negociaciones, acuerdos y conflictos en México, siglos XIX y XX. Agua y tierra*. (pp. 317–362). Mexico: CIESAS-Colmich-Universidad Autónoma de Aguascalientes.

Dardón Sosa, J. J. (Coord.). (2002). *Caracterización de la frontera Guatemala-México: aporte para su caracterización*. Quetzaltenango: FLACSO.

Contreras Utrera, J. (2009b). Suministro de agua, higiene y conflictos sociales en los centros urbanos del departamento del Soconusco, Chiapas, 1876–1945. In Ávila Quijas, A.O., Gómez Serrano, J., Escobar Ohmstede, A., & Sánchez Rodríguez (Coords). *Negociaciones, acuerdos y conflictos en México, siglos XIX y XX. Agua y tierra*. (pp. 317–362). Mexico: CIESAS-Colmich-Universidad Autónoma de Aguascalientes.

De Alba, F. (2007). La geopolítica del agua en México. La oposición entre la hidropolítica y el conflicto socio-político. Los nuevos rostros de las "luchas sociales", en Interações. *La revista internacional del desenvolvimiento local, 8*(1), 95–112.

de La Meza, J., & Carrabias Lillo, J. (2011). *Usumacinta: bases para una política de sustentabilidad ambiental*. Mexico: IMTA-Natura y Ecosistemas Mexicanos A.C.

De Stefano, L., Petersen-Perlman, J. D., Sproles, E. A., Eynard, J., & Wolf, A. T. (2017). Assessment of transboundary river basins for potential hydro-political tensions. *Global Environmental Change, 45*, 35–46. https://doi.org/10.1016/j.gloenvcha.2017.04.008.

De Vos, J. (1993). Las fronteras de la frontera sur. Reseña de los proyectos de expansión que figuraron la frontera entre México y Centroamérica. Villahermosa: UJAT-CIESAS.

Dinar, S. (2008). *International water treaties. Negotiation and cooperation along transboundary rivers*. London/New York: Routledge.

Durán, J. M., Sánchez, M., & Escobar, A. (Eds.). (2005). *El agua en la historia de México. Balance y perspectiva*. Zamora: El Colegio de Michoacán.

Elhance, A. P. (1999). *Hydropolitics in the third world. Conflict and cooperation in International River basins*. Washington, DC: United States Institute of Peace Press.

Equihua, M., Benítez, G., Mñoz, L., Medina, A., Álvarez, J. L., Pulido, M. T., Plestina, R. e Acosta, I. (2006). Bosques y agua en el sur de México: un balance general, In D. Villafuerte Solís and X. Leyva Solano. *Geoeconomía y geopolítica en el área del Plan Puebla-Panamá* (pp.171–203), México: CIESAS.

Escobar Ohmstede, A., Rodríguez, M. S., & Gutiérrez Rivas A. Ma. (Coords.). (2008). *Agua y tierra en México*. (2 volúmenes). Zamora/San Luis Potosí: El Colegio de Michoacán-El Colegio de San Luis.

Espinosa Henao, Ó. M. (2006). Sociedad y agua en Zacualpan de Amilpas. In *Una aproximación entre territorio, comunidad, organización social y conflictos, en torno a una cultura del agua*. Mexico: CEDRSSA- Cámara de Diputados.

Fábregas, P. A. (2015). La frontera sur de México. In A. Hernández & A. E. Campos Delgado (Coords.), *Líneas, límites y colindancias. Mirada a las fronteras desde América Latina* (pp. 249–278) Mexico: El Colegio de la Frontera Norte/CIESAS.

Fábregas Puig, A., & González, R. P. (2014). La frontera México-Guatemala, Guatemala-México: 1983–2013. *Frontera norte, 26*(SPE 3), 7–35.

Galindo Alcántara, A., Acosta, S. R. y Hernández, A. M. (2013). Peligro, vulnerabilidad y riesgo por fenómenos hidrometeorológicos en la cuenca baja del Usumacinta. In E. F. Kauffer Michel (Coord.), *Cuencas en Tabasco: una visión a contracorriente* (pp. 137–166). Mexico: CIESAS/ RISAF.

García García, A. (2005a). La cuenca endorreica de San Cristóbal de Las Casas, Chiapas: entre la gestión local y la nacional. In S. Vargas y E. Mollard (Coords.), *Problemas socio-ambientales y experiencias organizativas en las cuencas de México* (pp. 219–244). Mexico: SEMARNAT, IMTA, IRD y CONACYT.

García García, A. (2005b). *La gestión del Agua en la Cuenca Endorreica de San Cristóbal de las Casas, Chiapas, México*. Tesis de maestría, San Cristóbal de Las Casas Universidad Autónoma de Chapingo.

García García, A. (2010). *Instituciones y pluralismo legal: la hidropolítica en la cuenca transfronetriza del río Grijalva*, tesis de doctorado, El Colegio de la Frontera Sur.

García García, A., & Kauffer Michel, E. F. (2011). Las cuencas compartidas entre México, Guatemala y Belice: un acercamiento a su delimitación y problemática general. *Frontera Norte, 23*(45), 131–161.

García García, A. (2013). Las inundaciones fluviales históricas en la planicie tabasqueña: un acercamiento integral de largo aliento en la perspectiva de cuencas hidrógraficas. In E. F. Kauffer Michel (Coord.), *Cuencas en Tabasco: una visión a contracorriente* (pp. 61–99). Mexico: CIESAS/RISAF.

García García, A., Kauffer Michel, E. F. y Quezada, A. M. (2006). El agua doméstica en San Cristóbal de Las Casas, Chiapas: Entre la gestión local, la centralización y la privatización (1935–2004). In V. V. García, D. Soares Moraes, A. de la Rosa Regalado, & Á. Serrano Sánchez (Coords.), *Gestión y Cultura del Agua* (pp. 119–143). Mexico: IMTA, COLPOS and SEMARNAT.

Gómora Alarcón, J. (2013). *Hidropolítica del río Suchiate: los conflictos en la cuenca transfronteriza México-Guatemala desde mediados del siglo XX. Tesis de licenciatura en geografía.* Mexico: Universidad Nacional Autónoma de México.

Gómora Alarcón, J. (2014). La ribera del río Suchiate, territorio fronterizo en extinción. Conflictos generados por la abundancia del recurso hídrico. *Pueblos y fronteras digital,* 9(17), 59–77. http://www.redalyc.org/pdf/906/90630737005.pdf

González Espinosa, M., & Manse, M. C. B. (Coords.). (2014). *Montañas, pueblos y agua. Dimensiones y realidades de la cuenca Grijalva.* Mexico: El Colegio de la Frontera Sur, Juan Pablos Editor.

Gracia Sánchez, J., & Fuentes Mariles Ó. A. (2004). La problemática del agua en Tabasco: inundaciones y su control. In B. Jiménez y L. Marín (Eds.). El agua en México vista desde la academia (pp. 177–185). Mexico: Academia Mexicana de Ciencias.

Guillén, D. (2003). Redimensiones de una frontera largamente olvidada; Chiapas 1973–1993, *Frontera Norte*, 15(julio-diciembre), 121–149.

Haas, P. M. (1992). Introduction: Epistemic communities and international policy coordination. *International Organization, 46*(1), 1–35.

Hussein, H., & Grandi, M. (2017). Dynamic political contexts and power asymmetries: The cases of the Blue Nile and the Yarmouk Rivers. In *International environmental agreements: Politics, law and economics*. Dordrecht: Springer. https://doi.org/10.1007/s10784-017-9364-y.

Jhabvala, F. (2006). Gestión de agua en Tabasco. In D. Barkin (Coord.), *La gestión del agua urbana en México. Retos debates, bienestar.* Guadalajara/Mexico: Universidad de Guadalajara-Universidad Autónoma Metropolitana.

Jiménez Castañeda, A. A., Sánchez, J. G., Ramos, J. G., Hernández, J. E. M. F., Domínguez, R., Espinosa, J. G. L., Romero, J. O., & Franco, V. (2006). *Anteproyecto de la rectificación integral del río Suchiate y la rehabilitación del bordo de la población de Hidalgo*. Mexico: Facultad de Ingeniería, UNAM.

Julien, F. (2012). Hydropolitics is what societies make of it (or why we need a constructivist approach to the geopolitics of water). *International Journal of Sustainable Society, 4*(1/2), 45. https://doi.org/10.1504/IJSSOC.2012.044665.

Kauffer Michel, E. F. (Ed.). (2002). *Identidades, migraciones y género en la frontera sur de México*. San Cristóbal de Las Casas: El Colegio de La Frontera Sur.

Kauffer Michel, E. F. (2004). El concepto de hidropolítica en la frontera sur de México. In B. Graizbord y J. A. Alejandre (Coords.), *El futuro del agua en México* (pp. 181–195). México/Guadalajara/Los Angeles: Universidad de Guadalajara-El Colegio de México- UCLA Program on Mexico- PROFMEX/Casa Juan Pablos.

Kauffer Michel, E. F. (2005a). El consejo de cuenca de los ríos Usumacinta y Grijalva: los retos para concretar la participación y la perspectiva de cuencas. In S. Vargas & E. Mollard (Coords.), *Problemas socio-ambientales y experiencias organizativas en las cuencas de México* (pp. 195–218) Mexico: SEMARNAT, IMTA, IRD y CONACYT.

Kauffer Michel, E. F. (2005b). Hidropolítica de la cuenca del río Candelaria. In J. A. Benítez (Coord.), *Sistema de Información Geográfica de la Cuenca del Río Candelaria* (pp. 48–49) Campeche: EPOMEX-Universidad Autónoma de Campeche.

Kauffer Michel, E. F. (2005c). Hidropolítica. ¿Un concepto útil para entender la problemática del agua en la Frontera México-Guatemala-Belice? In E. Kauffer Michel (Ed.), *El agua en la frontera México-Guatemala-Belice* (pp. 45–57). Tuxtla Gutiérrez: UNACH-ECOSUR-The Nature Conservancy.

Kauffer Michel, E. F. (Ed.). (2005d). *El agua en la frontera México-Guatemala-Belice*. Tuxtla Gutiérrez: UNACH-ECOSUR-The Nature Conservancy.

Kauffer Michel, E. F. (2006a). El agua en la frontera sur de México: una aproximación a la problemática de las cuencas compartidas con Guatemala y Belice", *Boletín del Archivo Histórico del Agua*, 11, mayo-agosto, 22–29.

Kauffer Michel, E. F. (2006b). La ley de aguas nacionales frente a las prácticas indígenas: ¿Una historia de desencuentros?. In D. S. Moraes, V. V. García, Á. S. Sánchez, & de la Rosa Regalado, A. (Coords.), *Gestión y Cultura del Agua* (pp. 215–236). Mexico: IMTA, COLPOS and SEMARNAT.

Kauffer Michel, E. F. (2008). Comités de cuenca en Chiapas y Tabasco: entre participación endeble y riesgo de politización. In D. Soares, S. Vargas y M. R. Nuño (Eds.), La gestión de los recursos hídricos: realidades y perspectivas, Mexico: Instituto Mexicano de Tecnología del Agua, Universidad de Guadalajara and SEMARNAT.

Kauffer Michel, E. F. (2009). Donde el agua de la nación tiene dueños: la relación entre agua y tierra en los Altos de Chiapas. In Ávila Quijas, A. O., Gómez Serrano, J., Escobar Ohmstede, A., & Sánchez Rodríguez (Coords). *Negociaciones, acuerdos y conflictos en México, siglos XIX y XX. Agua y tierra*. (pp. 401–437). Mexico: CIESAS-Colmich-Universidad Autónoma de Aguascalientes.

Kauffer Michel, E. F. (2010). Hidropolítica del Candelaria: del análisis de la cuenca al estudio de las interacciones entre el río y la sociedad Ribereña. *Relaciones, 31*(124), 187–226.

Kauffer Michel, E. F. (2011a). Agua y territorio en la cuenca binacional del río Suchiate: apropiaciones encontradas de un espacio sujeto a inundaciones. In E. F. Kauffer Michel (Coord.), *Entre manantiales y ríos desatados: paradojas de las hidropolíticas fronterizas (México-Guatemala)* (pp. 111–143). Mexico: CIESAS, COLMICH.

Kauffer Michel, E. F. (2011b). De la abundancia del agua a la escasez de estudios: retos y perspectivas de las hidropolíticas en la frontera México-Guatemala-Belice. In E. F. Kauffer Michel (Coord.), *Entre manantiales y ríos desatados: paradojas de las hidropolíticas fronterizas (México-Guatemala)* (pp. 9–31). Mexico: CIESAS, COLMICH.

Kauffer Michel, E. F. (Coord.). (2011c). *Entre manantiales y ríos desatados: paradojas de las hidropolíticas fronterizas (México-Guatemala)* (pp.111–143). Mexico: CIESAS, COLMICH.

Kauffer Michel, E. F. (2012). La política del agua en Chiapas frente a los repertorios indígenas: de la ignorancia a la yuxtaposición de sistemas normativos. In D. Murillo Licea (Ed.), *Culturas del agua y cosmovisión india en un contexto de diversidad cultural* (pp. 33–58). IMTA: Jiutepec.

Kauffer Michel, E. F. (2013a). Introducción. Las cuencas en Tabasco: un análisis histórico y político a contracorriente de espacios indundados. In E. F. Kauffer Michel (Coord.), *Cuencas en Tabasco: una visión a contracorriente* (pp. 101–132). Mexico: CIESAS, RISAF.

Kauffer Michel, E. F. (2013b). Represas en la cuenca transfronteriza del río Usumacintta: ¿un conflicto crónico?. In E. F. Kauffer Michel (Coord.), *Cuencas en Tabasco: una visión a contracorriente* (pp. 101–132). Mexico: CIESAS, RISAF.

Kauffer Michel, E. F. (Coord.). (2013c). Cuencas en Tabasco: una visión a contracorriente. Mexico: CIESAS, RISAF.

Kauffer Michel, E. F. (2013d). De las indefiniciones a las demarcaciones inacabadas: repensar las fronteras fluviales y terrestres entre Mexico, Guatemala y Belice, *Liminar, Estudios sociales y humanísticos*, año 11, vol XI, núm 2, julio-diciembre, pp. 70–81.

Kauffer Michel, E. F. (Coord.). (2014a). Cuencas en Chiapas. La construcción de utopías en cascada, Mexico: CIESAS.

Kauffer Michel, E. F. (2014b). Conflits et coopération dans les bassins versants transfrontaliers en Amérique Centrale et au sud du Mexique: du Lempa à l'Usumacinta. *Regions and Cohesion, 4* (2), 30–53.

Kauffer Michel, E. F. (2017). Entre rigidez política (hacia el conflicto) y fluidez hídrica (hacia la paz): Las fronteras de agua de México con Guatemala y Belice. *Revista de Paz y Conflictos, 10* (1), 61–86.

Kauffer Michel, E. F., & Castillejos, D. E. (Coords.). (2015). *De Chiapas a la Penísula de Yucatán: intersticios hídricos*. Mexico: UNACH, RISAF.

Kauffer Michel, E. F. y García, A. G. (2003). Mujeres en los comités de agua del Estado de Chiapas: elementos para entender una participación con segregación genérica. In E. Tuñón Pablos (Coord.), *Género y Medio Ambiente* (pp. 295–322). Mexico: El Colegio de la Frontera Sur, SEMARNAT, Plaza y Valdés.

Kauffer Michel, E. F. y García, A. G. (2004). Aguas sucias para trabajar, agua limpia para tomar: transformaciones en torno al agua en comunidades tzotziles de los Altos de Chiapas. In F. Peña (Coord.), *Los pueblos indígenas y el agua: desafíos del siglo XXI* (pp. 109–138). Colombia: El Colegio de San Luis-Instituto Mexicano de Tecnología del Agua, WALIR y SEMARNAT.

Kauffer Michel, E. F., & Medina, L. (2014). Entre conflictos y cooperación: pensar las cuencas transfronetrizas a la luz de sus actores. *Regions and Cohesion, 4*(2), 1–9.

Kroeber, C. B. (1994). El hombre, la tierra y el agua. In *Las políticas en torno a la irrigación en la agricultura de México* (pp. 1885–1911). Mexico/Jiutepec: IMTA-CIESAS.

Leca, J. (1973). Le repérage du politique. *Projet, París, 71*, 11–24.

López Ramírez, A. (2008). *Conflictos socioambientales en América Latina. Hidropolítica de los cursos de agua internacionales*. Conferencia, 14 de abril. Instituto de Estudios Latinoamericanos, Viena.

Magrin, G. (2016). The disappearance of Lake Chad : History of a myth. *Journal of Political Ecology, 23*, 204–222.

March, I. J. y Fernández, J. C. (2003). La gran cuenca del río Usumacinta: una contradicción regional. In P. Ávila (Ed.), Agua, medio ambiente y desarrollo en el siglo XXI: México desde una perspectiva global y regional (pp. 117–134). Zamora: El Colegio de Michoacán-Secretaría de Urbanismo y Medio Ambiente-SEMARNAT-IMTA.

Martínez, O., Concepción, V. L. M., Quiñones Castillo, A. M., López Hernández, R. I., Rendón, G. A. O., & Montesillo, J. L. (2004). *Gestión del agua en el Distrito Federal*. Mexico: UNAM.

Maury, R. G. (2003). Hidropolítica y conflictos por el agua en el Mediterráneo: el caso del medio Oriente. In P. Ávila (Ed.), *Agua, cultura y sociedad en México* (pp. 387–396). Zamora/Jiutepec: El Colegio de Michoacán- IMTA.

Mejía González, L. (2011). La participación pública en la política del agua potable en la cabecera municipal de El porvenir, Chiapas: del marco legal a la práctica local. In E. F. Kauffer Michel (Coord.), *Entre manantiales y ríos desatados: paradojas de las hidropolíticas fronterizas (México-Guatemala)* (pp. 189–216). Mexico: CIESAS, COLMICH.

Mejía González, L., & Kauffer Michel, E. F. (2008). Historia de una descentralización fracasada: la política del agua potable en El Porvenir, Chiapas. In D. Soares, S. Vargas, & M. R. Nuño (Eds.), *La gestión de recursos hídricos: realidades y perspectivas* (pp. 343–374). Mexico: Secretaría de Medio Ambiente y Recursos Naturales, Instituto Mexicano de Tecnología del Agua, Universidad de Guadalajara.

Menga, F. (2016). Reconceptualizing hegemony: The circle of hydro-hegemony. *Water Policy, 18* (2), 401–418.

Meyer, M. C. (1997). *El agua en el suroeste hispánico. Una historia social y legal 1550–1850.* Mexico/Jiutepec: IMTA-CIESAS.

Mirumachi, N. y Chan, K. (2014). Anthropocentric hydro politics? Key developments in the analysis of international transboundary water politics and some suggestions for moving forward. *Aquatic Procedia, 2*, 9–15. https://doi.org/10.1016/j.aqpro.2014.07.003.

Molina, V. (1976). *San Bartolomé de Los Llanos: una urbanización frenada.* México: CIESAS-INAH.

Mutin, G. (2003). Le Tigre et l'Euphrate de la discorde. *VertigO, La revue en sciences de l'environnement, 4*(3), 1–10.

Ordoñez Morales, C. E. (2011). Frontera y economía informal en el área d elos puentes sobre el río suchiate de Guatemala y México. In E. F. Kauffer Michel (Coord.), *Entre manantiales y ríos desatados: paradojas de las hidropolíticas fronterizas (México-Guatemala)* (pp. 59–94). Mexico: CIESAS, COLMICH.

Oswald, U. (2008). Cambio climático, conflictos e hidropolítica en México, Conferencia, XLIV Reunión Nacional de Investigación Pecuaria y II Reunión Nacional de Innovación Agrícola, 3–8 de noviembre, Merida.

Perló, C. M. y González, R. A. (2005). ¿Guerra por el agua en el Valle de México? Estudios sobre las relaciones hidráulicas entre el Distrito Federal y el Estado de México. Mexico: UNAM-Fundación Friedrich Ebert Stifung.

Petersen-Perlman, J. D., Veilleux, J. C., & Wolf, A. T. (2017). International water conflict and cooperation: Challenges and opportunities. *Water International, 42*(2), 105–120. https://doi.org/10.1080/02508060.2017.1276041.

Piedrasanta Herrera, R. (2014a). La frontera del noroccidente guatemalteco: de su emergencia a las dinámicas globales. In I. R. Nuñez, E. F. Kauffer Michel, C. R. Farfán, R. T. Conangla y R. P. Herrera. *Más que una línea: historia y dinámicas en la frontera Guatemala-México, Cahier-Cuadernos CEMCA, Serie FabricaMig* (pp. 10–15). Mexico: CEMCA, CNRS, Ambassade de France, ANR, Número 7, octubre.

Piedrasanta Herrera, R. (2014b). Territorios indígenas en frontera: los Chuj en el período liberal (1871–1944) en la frontera Guatemala-México. *Boletín Americanista, 69*(2), 69–78.

Ramos Reyes, R., Barba Macías, E., & Velásquez Mazariegos, S. (2013), Delimitación de microcuencas en Balancán, Tabasco, México. In E. F. Kauffer Michel (Coord.), *Cuencas en Tabasco: una visión a contracorriente* (pp. 190–206). Mexico: CIESAS, RISAF.

Rivera Farfán, C. (2014). Creencias divergentes en territorios comunes. Religiosidades en la frontera Guatemala-México. *Boletín americanista, 69*(2), 97–108.

Rivera Farfán, C. (2015). Trabajadores migrantes en la frontera sur de méxico. Caracterización del tarbajo temporal centraoamericano en el Soconusco. In A. Hernández & A. E. Campos Delgado (Coords), *Líneas, límites y colindancias. Mirada a las fronteras desde América Latina* (pp. 249–278). Mexico: El Colegio de la Frontera Norte, CIESAS.

Rodas Núñez, I. (2014). Las expediciones arqueológicas y las poblaciones del frente pionero en el Usumacinta medio guatemalteco. La recolonización contemporánea de la zona fronteriza. *Boletín americanista, 69*(2), 33–54.

Rojas, J. P. (2008). La hidropolítica en Jalisco en los albores del siglo XXI: tres escenarios de conflicto y negociación política en torno al proyecto público de la presa de Arcediano. *Estudios Sociales Nueva Época, Universidad de Guadalajara, Guadalajara, 2*, 103–135.

Rojas Rabiela, T. (2009). Las obras hidráulicas las épocas prehispánica y colonial. In *Comisión Nacional del Agua. Semblanza histórica del agua* (pp. 9–25). Mexico: Conagua.

Sadoff, C., Greiber, T., Smith, M., & Bergkamp, G. (Eds.). (2008). *Share- Managing water across boundaries*. Gland: IUCN.

Salazar Ledesma, F. (2013). Las cuencas fluviales de Tabasco como recurso teórico-metodológico en el estudio de la organización territorial prehispánica y española de los siglos XVI y XVII. In E. F. Kauffer Michel (Coord.), *Cuencas en Tabasco: una visión a contracorriente* (pp. 27–60). Mexico: CIESAS, RISAF.

Saldívar, V. A. (2007). *Las aguas de la ira: Economía y cultura del agua en México, ¿sustentabilidad o gratuidad?* Mexico: UNAM.

Salman, S. M. A. (2011). The new state of South Sudan and the hydro-politics of the Nile Basin. *Water International, 36*(2), 154–166. https://doi.org/10.1080/02508060.2011.557997.

Sandré Osorio, I. (2005). *Documentos sobre posesión de aguas de los pueblos indígenas del Estado de México, siglos XVI al XVIII*. Mexico/Zinacantepec: CIESAS-AHA-CONAGUA El Colegio Mexiquense.

Sandré Osorio, I., & Kauffer Michel, E. (Coord.). (2014). *Documentos para la historia del agua en el sureste de México (1890–1993). Catalógo de documentos del Archivo Historico del Agua*. México: CIESAS, RISAF, CONAGUA, AHA, Conacyt.

Santacruz de León, G. (2005). La cuenca del río Suchiate: los potenciales problemas ambientales asociados la agua. In S. Vargas & E. Mollard (Coords.), *Problemas socio-ambientales y experiencias organizativas en las cuencas de México* (pp. 298–315). Mexico: SEMARNAT, IMTA, IRD, CONACYT.

Santacruz de León, G. (2006). Los potenciales conflictos sociales por el uso del agua en la frontera México-Guatemala: la cuenca del río Suchiate. *Boletín del Archivo Histórico del Agua, año, 11* (mayo-agosto), 30–34.

Santacruz de León, G. (2011a). Aletargamiento gubernamental en la gestión de los recursos hídricos: el caso de la cuenca del río Suchiate. In E. F. Kauffer Michel (Coord.), *Entre manantiales y ríos desatados: paradojas de las hidropolíticas fronterizas (México-Guatemala)* (pp. 95–110). Mexico: CIESAS, COLMICH.

Santacruz de León, G. (2011b). Problemática ambiental y conflictos sociales en torno al uso del agua en la cuenca del río Suchiate. In E. F. Kauffer Michel (Coord.), *Entre manantiales y ríos desatados: paradojas de las hidropolíticas fronterizas (México-Guatemala)* (pp. 35–58). Mexico: CIESAS, COLMICH.

Santacruz, G., Santacruz, E. y Santacruz, E. E. (2005). Abundancia y despilfarro del agua: una visión desde el ejido Once de Abril, municipio de Unión Juárez, Chiapas. In E. F. Kauffer Michel (Ed.), El agua en la frontera México-Guatemala-Belice (pp. 353–371). Tuxtla Gutiérrez: UNACH-ECOSUR- The Nature Conservancy.

Sneddon, C., & Fox, C. (2006). Rethinking transboundary waters: A critical hydropolitics of the Mekong basin. *Political Geography, 25*(2), 181–202. https://doi.org/10.1016/j.polgeo.2005.11.002.

Soares, D. (2006). Acceso, abasto y control del agua en una comunidad indígena Chamula en Chiapas, Un análisis a través de la perspectiva de género, ambiente y desarrollo. *Región y Sociedad, Hermosillo, XIX, 38*, 25–50.

Solís Hernández, M. G. (2011). Gestión local del agua por la Coordinación de la Zona norte de San Cristóbal de Las Casas, Chiapas: entre la autoorganización y las p'rcticas. In E. F. Kauffer Michel (Coord.), *Entre manantiales y ríos desatados: paradojas de las hidropolíticas fronterizas (México-Guatemala)* (pp. 169–188). Mexico: CIESAS, COLMICH.

Stephan-Otto, E. (Coord.). (2003). *El agua en la cuenca de México. Sus problemas históricos y perspectivas de solución (2 tomos)*. Mexico: Asociación Nacional de Investigadores de Xochimilco, A.C.

Swain, A. (2002). The Nile river initiative: Too many cooks, too little broth. *SAIS Review. The Johns Hopkins University Press, 22*(2), 293–308.

Thomas, K. A. (2017). The river-border complex: A border-integrated approach to transboundary river governance illustrated by the Ganges River and Indo-Bangladeshi border. *Water International, 42*(1), 34–53. https://doi.org/10.1080/02508060.2016.1247236.

Torras Conangla, R. (2012). La tierra firme de enfrente. In *La colonización campechana sobre la región de Los Ríos (siglo XIX)*. Mexico: UNAM.

Torras Conangla, R. (2014). Los refugiados mayas yucatecos en la colonización de El Petén: vicisitudes de una frontera. *Boletín americanista, 69*(2), 15–32.

Toset, H. P. W., Gleditsch, N. P. y Negre, H. (2000). Shared rivers and interstate conflict. *Political Geography, Elsevier, 19*, 971–996.

Transboundary Freshwater Dispute Database (TFDD). (2007). *Oregon State University*. http://www.transboundarywaters.orst.edu/. Accessed 3 Feb 2009.

Turton, A. (2002). Hydropolitics: The concept and its limitations. In A. Turton y R. Henwood (Eds.), *Hydropolitics in the developing world: A southern African perspective* (pp. 13–19). Pretoria: African Water Issues Research Unit.

Turton, A. (2008). The southern African Hydropolitical complex. In O. Varis, C. Tortajada, & A. K. Biswas (Eds.), *Management of Transboundary Rivers and Lakes* (pp. 21–79). Berlin: Springer.

Valdez Gordillo, M. E. (2006). *Desencuentro y encuentro de fronteras: El Petén guatemalteco y el Sureste mexicano: 1895–1949*. Tuxtla Gutiérrez/San Cristóbal de Las Casas: UNICACH-UNICH.

Valette, C. (2011). ¿Descontrol, gestión, o instrumentalización? El caso de las inundaciones en San Cristóbal de Las Casas, Chiapas, México. In E. F. Kauffer Michel (Coord.), *Entre manantiales y ríos desatados: paradojas de las hidropolíticas fronterizas (México-Guatemala)* (pp. 147– 168) Mexico: CIESAS, COLMICH.

Vargas, S., Soares, D., & Guzman, N. B. (Eds.). (2006). *La gestión del agua en la cuenca del río Amacuzac: diagnósticos, reflexiones y desafíos*. Jiutepec: IMTA-Universidad Autónoma del Estado de Morelos.

Vautravers, T. G. (2005). *Estudio comparativo de la frontera Tabasco, México- El Petén, Guatemala*. Villahermosa: Universidad Juárez Autonóma de Tabasco.

Vera Cartas, J. (2005). Participación, consejos de cuenca y política hidráulica mexicana: el caso de la costa de Chiapas. In S. Vargas y E. Mollard (Coords.), *Problemas socio-ambientales y experiencias organizativas en las cuencas de México* (pp. 276–297). Mexico: SEMARNAT, IMTA, IRD y CONACYT.

Villafuerte Solís, D. (2004). La frontera sur de México. In *Del TLC México-Centroamérica al Plan Puebla-Panamá*. México, UNAM: Plaza y Valdés.

Waterbury, J. (1979). *Hydropolitics of the Nile Valley*. Syracuse: Syracuse University Press.

Wolf, A. T. (2007). Hydropolitical vulnerability and resilience: Series introduction. en United Nations Environment Programme, *Hydropolitical Vulnerability and Resilience along International Waters*. Latin America and the Caribbean, Corvallis, United Nations Environment Programme-Universidad Nacional de Costa Rica-Oregon State University.

Edith F. Kauffer is a Professor and Researcher at CIESAS (Centro de Investigaciones y Estudios Superiores en Antropología Social)-Sureste, Mexico, and Resident of IMéRA (Institute for Advanced Studies) at Aix-Marseille University (2016–2017). She was former coordinator of the researchers' water network on the border between Mexico, Guatemala, and Belize (RISAF) (2003–2016); co-coordinator of the Comparative Research on Regional Integration and Social Cohesion Consortium (RISC) working group on Management of Strategic Resources, Environment, and Society. She is also an associate editor of the journal *Regions & Cohesion* (Berghahn Journals) and is the representative of CIESAS at the consultative council of Diplomatic Historical Documentary Archive of the Mexican Minister of Exterior Relations. She conducts research on transboundary waters and river basins, water policy and politics, hydropolitics, gender and water in Southern Mexico and Northern Central America, and also recently on transboundary river basins in the Mediterranean.

Chapter 13
The US-Mexico Border: Institutional Weaknesses in Meeting the Growing Demand for Water

Vicente Sánchez Munguía

Abstract Water has been a central issue on the bilateral relationship between Mexico and the United States since the border between both countries was established. Despite the conflicting character of their relationships, the countries have reached agreements to secure access to this natural resource in their shared watersheds. We can see very established and institutional capabilities face the critical issues and achieve solutions through negotiations, primarily on surface water, while an equivalent agreement has not occurred on groundwater. But it is likely that the bargaining capacity of Mexico be affected if Mexican water authorities fail in efficient water management, in the context of shared river basins with the United States. At the same time, that effectiveness to solving conflicts between competing users for the same resource fails also.

Keywords Water management · Transboundary basin · Conflict · Cooperation · Asymmetry

13.1 Introduction

"Water is one of the most widely shared resources on the planet, and the most vital for human survival after oxygen. It has a capacity to unite people that share a source of water, or to incite conflict among them as they compete for it" (UNEP 2002). However, despite the potential conflict that represents the water scarcity, the ordinary so far have been not the water wars predicted by some media but agreements and treaties between countries to ensure access to the resource and share responsibility for their quality (Wolf et al. 2005). The World Water Council has pointed out that in the twenty-first century, societies must face six challenges related to water: shortages, lack of access, deterioration in its quality, awareness of decision-makers

V. Sánchez Munguía (✉)
El Colegio de la Frontera Norte, Tijuana, Mexico
e-mail: vsanchez@colef.mx

H. R. Guerrero García Rojas (ed.), *Water Policy in Mexico*, Global Issues in Water Policy 20, https://doi.org/10.1007/978-3-319-76115-2_13

of decisions and the public, dwindling financial resources, and the fragmentation of its management (Becerra Pérez et al. 2006).

On the other hand, there is no doubt that the very challenges are in spaces of relative scarcity and growing demand, where governments and societies will have to undertake great efforts to improve management leading to great operational efficiency and standards available, in addition to taking care the water resource. In this sense, it is important to note that the future in some regions of the world is related with the need to taking alternatives regarding better distribution and access to water. Then this natural resource is undoubtedly the limiting factor of any aspiration and form of sustainable development in the long run (Pombo and Wright 2012).

Some of these regions also face the challenge of being part of an international regime of shared waters, because the river flows through more than one country or is in the boundaries between various countries. We are speaking of 263 international watersheds containing more than 60% of the total flow of the rivers on the planet and 40% of the world's population (Wolf et al. 2008).

Mexico shares water in its border rivers with neighboring countries, both in the Northern and in the Southern international frontiers. The rivers are part of the international boundary, and water driven by these streams is subject to a shared regime of management with the United States in the Northern border. Groundwater also crosses political boundaries, although not subject to any international regulation for sustainable use and the protection of its quality.

At the same time, Mexico faces particularly relevant conditions related with short water supply in two-thirds of the country's surface, due the aridity and scarce rainfall conditions. Water authorities recognize the deterioration of surface waters quality and the depletion of 17% of its aquifers (Becerra Pérez et al. 2006). Of course, the US-Mexico border is in the middle of desert and semiarid regions, where water is the scarcest resource. Moreover, water streams consist of very small water flows compared with those in Mexico's southern border with Guatemala or Belize.

The main problems related with water in the US-Mexico border are scarcity because of the climatic conditions and great demand represented by the growth of population in two border sides. The area is one of the most populated of the world (Granados-Olivas et al. 2010) with little more than 14 million people in 2010 (Castro et al. 2015), which represents a great challenge to ensure water access for domestic consumption and economic activities underpinning it and at the same time maintain the relationship of cross-border cooperation.

In general, social and economic growth in Mexican borderlands has occurred without considering the conditions of water scarcity. Population and economic activity are concentrated in areas of high shortage, not only in country regions with low rainfall but also in areas where shortages were not perceived as a problem at the beginning of urban growth or the establishment of irrigation agriculture (Becerra et al. 2006). Now we have an induced shortage due negligent water policy but also by lack of control and planning of growth in some regions, especially on the US-Mexico border.

In this sense, both countries the United States and Mexico have registered a considerable growth of its population during second part of the twentieth century along the common border, and that growth has occurred in the absence of consideration of water supply conditions. The Mexican border population has grown in recent decades at a pace much higher than recorded by the American side of the border (Castro et al. 2015) but also with strong lags in the provision of public services (Kelly 2002). The absence of planning and urban control on the Mexican side impacts water quality of streams that cross the border from South to North and contribute to tensions between the two countries.

However, the Mexico and US governments are open to cooperation to solve such problems and have been making efforts to achieve goals on the intelligent use of these shared resources. It also requires a binational understanding "over the system to achieve a correct cross-border management of these water resources, have a better planning for the sustainable use of groundwater and address issues on which there may be disputes and conflicts" (Granados-Olivas et al. 2010).

Given the water stress conditions, due to shortages and an increasing demand for water in the border between the two countries, also by increasing interdependence and cross-border ties that have been created over time in different dimensions that include the business, culture, and families connected to both sides through the border, it is important to create trust and cross-border communication to facilitate understanding and cooperation between both countries. Most relevant is to consider and understand the difference between both countries in their levels of development. That is evident in the way in which societies on both sides of the border have agreed to the provision of water services and how water management is organized thereof. Both countries have different legal frameworks and, then, different water management culture and rules that are difficult to converge. Thus, binational or transboundary water planning and water management are so difficult to do within oriented governance frameworks to achieve efficiency and protect the resource addressing principles of environmental kind or integrated management in the context of the watershed (see Browning Aiken 2007).

This chapter discusses the main challenges on the shared water in the border between Mexico and the United States. The solution of problems that there arise necessarily requires the binational cooperation and efficient working of the institutions related to the water management, where the Mexican side must make a great effort to meet those challenges because of its institutional weaknesses.

13.2 The Progressive Construction of the Institutional Management Framework for Transboundary Basin

To understand the complex relationships between Mexico and the United States related in terms of shared waters in the border between the two countries, it is necessary to take into account the way in which the normative framework of the binational scope between Mexico and the United States has been built to establish arrangements, agreements, and bargaining access to water from the shared basins across the common border, which has clearly had a progressive and incremental character through the years. Since the priority at the beginning of common neighboring were territorial limits, the first legal instrument was the *Guadalupe Hidalgo Treaty* or *Treaty of Peace, Friendship and Limits* signed on February 2, 1848, once the war of intervention had ended, and Mexico recognized the loss of more than half of its territory ceded to the United States, thus securing the border between the two countries. Only a few decades later, in 1882, the first of six conventions sought to address and negotiate the issues related to water within the border limits. This framework was built closely related to the dynamics of territorial occupation and the pressure on water distribution according to the principles of ownership related to the claim of users and the rhythm of human settlements on the borders of the watersheds, especially in the head rivers located in the United States, where the intensive colonization resulted in claims of rights and intense negotiations between the riverine states pushed by interest groups emerging around the water use of cross-border basin.

We must take into account that it was only possible to agree upon the international treaty between the two countries once the seven riverine states in the Colorado River Basin in the United States signed the Compact of Santa Fe in 1922, in which Mexico was recognized as the eighth user of the waters of that river and agreed to grant it an assignment of 1,233 Mm^3 which was expanded in the 1944 treaty, after an intense and prolonged fight of Mexico to achieve higher volumes of water (Castro et al. 2015), whereas the traumatic history of the relationship between the two countries, especially regarding the origin of the border, means that the beginning of relations between the two countries was marked by a mutual distrust of the respective governments, taking also into account the fascination and fear that the United States has always caused among Mexican elites, which is expressed in the ambivalence of the Mexican government regarding the northern neighbor. For many years the Mexican border remained distant and almost empty, because the dominant thought in the country stated that Mexico's best defense against the United States was separation via a desert. In some way, the vast territory in the north of Mexico was seen as hostile and distant space. The sentences "so far from God and so close to the United States" or "between Mexico and the United States, the desert" reflect the kind of thinking in the Government of Porfirio Díaz. These thoughts reflect a fear to deal with Mexico's north neighbor and its expansionist spirit.

Justo Sierra, the ideologist of the political regime, led by Porfirio Díaz, that had clear opinions, considered that due to the extension of the vast territory inherited from the Spanish Crown, Mexico had inherited also the northern part of a territory rich in some areas but sterile and not likely to be populated in others (Sierra 1977).

Governments emerging from the beginning of the last century Mexican Revolution opted for a different idea, especially in the postrevolutionary stage last century when cast agricultural and irrigation policy arose as a chance to take advantage of the waters of the major basins for productive purposes and attraction of settlers from other regions of the country. The opening of the irrigation districts as of the Mexicali Valley in the lower basin of the Colorado River, after the expropriation of land from American companies and the opening of the lower Río Grande/Bravo basin irrigation district, both among the largest in the country, boosted economic activity in those areas of the border, at the same time becoming a source of pressure for the signing of an international treaty with the United States on access to and distribution of water in transboundary basins.

Earlier, in 1906, the two countries had reached an agreement to ensure the water in the basin of the Rio Grande/Bravo to agricultural users of the Valley of Juarez in Chihuahua, but it lacked a similar agreement to water down the river and on the Colorado and Tijuana Rivers also. In 1944 the *Treaty for the Use of the Waters of the Rivers Colorado and Tijuana and the Rio Grande* was signed, as it is said, once the seven riverine states bordering the Colorado have agreed and signed the *Compact of Santa F*e in 1922. The Convention of 1889 laid the basis for agreeing upon the definitive boundaries between the two countries and gave rise to the International Commission of Limits. This bureaucratic body was responsible for maintaining boundaries. The Treaty of 1944 gave broad attributions to this binational Commission, which was renamed as International Border and Water Commission (IBWC or CILA in México). It is composed of two sections, one for each country, working in a joint and coordinated manner. In fact, the mandate to the Commission goes beyond the pure care pursuant to the provisions of the Treaty, since it is responsible for operational infrastructure of storage and distribution of water corresponding to the two countries, each section does in its country.

13.3 The Binational Water Management in an Asymmetric Relations Framework

There is frequent reference to the asymmetric nature that characterizes relations between Mexico and the United States (Ojeda 1983; Fagen 1983; Fernandez de Castro and Dominguez 2001; Mumme 2015), although we seldom specify the contents of these asymmetries. Assuming the existence of such asymmetric relationship, generated by the differences in the levels of development of the two countries, we can see the specific conditions and type of priorities in each country, regarding the demands of water services from shared watersheds. Of course, asymmetries go much further and are represented in the power of the United States in broader contexts and in different dimensions, such as the economic, political, and military, in addition to the influence exerted by the United States at the international level and in multilateral bodies that function as forums for discussion of international and global problems that affect the planet. To better understand, we must consider both the Treaty of 1906, which determined the volume of water that Mexico would get to meet the water demand for the Juarez Valley, and the Treaty of 1944 that were signed under the asymmetric conditions described.

However, it is noteworthy the diplomatic work undertaken by Mexico to obtain the United States acceptance and the way in which government agents of Mexico addressed the issues of access to water in the two main basins, through bold actions to put pressure to United States government to take a more active role with the border states of that country. Mexican government found allies in farmers of Texas of the lower basin of the Rio Grande that exerted pressure in favor of the treaty because they looked for guaranteed water access in the Rio Grande/Bravo that irrigates their land (see Walsh 2004). Regarding shared water and transboundary management of it, the first aspect refers to volumes assigned by the international treaty, establishing fixed volumes and terms of delivery in annual cycles; the second aspect relates to the operation and maintenance of the storage infrastructure, control, and water delivery that has been built with shared resources within a framework of cooperation and understanding between the two countries.

In the Colorado River Basin, the United States delivers to Mexico 1850 Mm^3 of water per year, and in the Rio Grande/Bravo Basin, Mexico delivers to the United States 432 Mm^3 of water yearly. This is an exchange that is complemented by the delivery of 70 Mm^3 of water from the United States to Mexico in El Paso-Ciudad Juarez area as part of the agreement to the binational Convention of 1906. The United States has complied fully and timely the annual delivery of water volumes agreed in the Treaty, but Mexico has struggled to meet its commitment in the Rio Grande/Bravo Basin corresponding to the section between Forth Quitman and the mouth of the river in the Gulf of Mexico. Problems are officially attributed to recurrent drought periods in the area comprising the basin and in particular in the Conchos River basin, main tributary in the middle part of the Rio Grande/Bravo Basin, although it is likely that there is also another type of factors, external to the climatic conditions of the region, associated to the institutional performance and the urban economic and social dynamic that has taken place in recent decades in the Mexican border area.

Furthermore, although there seems to be always some relative advantage of the country and the water users are upstream versus downstream found in an international relationship that flows through water basin shared and where the water is scarce, the benefits would be realized in unilateral actions in any impact on the country that is downstream, causing discomfort, tension, or conflict between countries sharing the water of the basin. In the case of Mexico and the United States, such situations have not been absent through time, but this is not common, as this would make the relationship between the two countries more contentious.

The shared water between the two countries in these watersheds translates into an exchange where these supposed advantages become also evident to Mexico which also occupies a position upstream in some cases, such as the Tijuana River and in some way in the Rio Grande/Bravo Basin, where the Conchos River contribute most to the water flow to distribute according to the International Treaty of 1944 regime.

Disparity between the two countries in favor of the United States is a determinant of unilateralism expressed at various times in the history of mutual relations in shared water issues, but it has mentioned that in specific negotiations concerning the financing of priority water projects and sanitation for the border, Mexico has been largely favored under items associated with the differences in development between the two countries (Fernández 2006).

In this sense, under the condition of asymmetry, upstream has not necessarily been a factor of advantage for Mexico, with very particular exceptions, as in the case of the Tijuana River, where from the priorities of the San Diego community affected by the pollution of the waters of this river and the reduced availability of resources by Mexico to finance the sanitation infrastructure necessary to correct the problem it has involved a greater burden for the United States in the supporting the projects (Fernández 2006), although in exchange, wastewater treatment facilities are located in US territory, and they are operated by personnel in that country, i.e., it is a relative advantage for Mexico.

Something similar seems to have happened with the technical project of modernization and improvement of the districts of irrigation of the Conchos River. This basin provides the water involved in the Treaty for delivery to the United States in the basin of the Rio Grande, but the problems to meet water deliveries committed to the United States have required a considerable investment for the infrastructure modernization of the irrigation district. For the implementation of this project aimed to improve efficiency and water savings that will ensure the timely delivery of the volumes set out in the international treaty and avoid new conflict events between the two countries in the future, the support of the American Government on the necessary financial resources for the implementation of the mentioned project was necessary.

Because of the asymmetric nature of the relations between the two countries, a clear dependence of Mexico from the United States at the border is observed, and it acquires a sense of reality in the distribution and management of water by Mexican authorities, less ability to respond to emerging situations, and a lower capacity for long-term planning, increasing the asymmetry and dependence for Mexico.

Additionally, asymmetries are not just about the great differences between both countries in the areas mentioned. More important is institutional strength of them, defined as the ability of governments to implement regulations and enforce in their respective areas of jurisdiction, which is very closely related to levels of institutions legitimacy in different levels of government. The binational water planning is very important for both countries to care and protect this scarce resource, since it is not just access to water for different uses possible or permitted, but also the management and control of the activities connected to the use of water of the basins within the jurisdictional area of each country.

Better management and control of activities connected to the uses of water are needed to guarantee the water access now and in the future, but in some cases the scarce resource is not only water but management skills too. Decisions taken by government to affect the use, care, and protection of natural resources are taking place more frequently because access to water for various uses possible is being difficult, and here it requires water authorities with effective capacity of planning and implementation of policies and projects. However, in the case of Mexico, not only more management capabilities are required but also a real opening for social participation. While in law and in the public discourse this talk of such participation, is still very limited and less considered by decision makers. There is a great difference between Mexico and the United States in this aspect too (See Browning-Aiken et al. 2007).

Then, it is the ability of government authorities to control and guide the processes of land use and management of border space, mainly those government agencies directly responsible for leading the policy and management of water, as well as other government agencies linked to water resources, as it would be the case of environment and agriculture agencies, of local governments on urban water, sanitation and drainage services operation, and in urban planning, management and control of the land use.

In this regard, it is important to point out that the institutional scheme of management of the water in Mexico corresponds to a centralized model, where federal government has the power, and until recently state and municipal governments have tended to have a greater role albeit limited. Regional decentralization through the instances created for such purposes, as the watershed councils and committees, is still under the control of federal water authorities, although they are defined as spaces of participation of water users. So far, some of the technical committees of underground waters (COTAS in Spanish acronym) seem to have the best results, also in some of the organizations of irrigation district users, whose functions are to operate programs of delivering volumes of water for irrigation.

In other words, the two countries are very far from the latest trends of multilevel water management that can be more efficient and achieve better results in a model of integrated water border cooperation (Maganda 2008), and Mexico requires much greater efforts in terms of increased and effective decentralization of water management, but also the substantial improvement of the capacities of local governments, both financially and in management skills.

At this regard, Mumme (2015) has pointed out that despite the institutional reforms associated with the North American Free Trade Agreement, the water allocation regime between Mexico and the United States has not been touched upon and agreements such as those achieved in 2012 to allocate water for environmental use in the Colorado River delta area acquire a marginal character in the framework of a power structure highly resistant to changes, so the author points out the need to be cautious in theorizing about multilevel governance and collaborative watershed management, in relation to the specific case of the US-Mexico relationship in the water issue. Previously, in 2010, the same author had mentioned that the drought-related problems in the Rio Grande/Bravo Basin, were more of a governance problem than one of hydrological type (Mumme 2010), where weakness derives not only from the allocation regime emerged from the Treaty, but also from the lack of integration and coordination between the binational institutions responsible for water management at the border.

If border communities in Mexico are more inclined to improve the quality of water and wastewater services that conservation issues facing broader phenomena and impact as climate change and global warming (Browning Aiken 2007), they are indicative of the gaps between the communities of the two countries and the kind of priorities that arise, but also the delay of Mexican institutions regarding operating conditions of water services and more, the involved authorities accountable to the issues raised, where binational convergence toward a more integrated model sounds good, but it is not easy in the short and medium term.

Consequently, although the asymmetries are given by the aforementioned elements, which constitute the most obvious differences between the two countries and upon which Mexico has no chance of control, there are also elements that correspond to the country at a disadvantage. Mexico must work to overcome it, as a condition to try to bridge the gap and better address the specific issues of the bilateral agenda, in this case concerning shared water basin and groundwater. It is not expected that countries are in the same conditions to address the issues on the agenda, but Mexico must strengthen institutional capacities in the border region, and transfer more responsibility to states and municipalities, and at the same time incorporate more opening to community involvement for working within a framework of cross-border relations and achieve better results in proper water use.

In recent decades governments of both countries have promoted the construction of a new institutional framework aimed at cooperation, and this has been the result of pressure from border communities and environmental problems arising from changes in the border economy and trade integration proposals which facilitated the mobility of goods across the border. In a broader sense, we can see that despite the asymmetries, the governments of the two countries have managed to give sense and meaning to the institutional framework which through bargaining overcomes conflicts and strengthens cooperation.

In fact, since four decades ago, both countries have been impulse the international border economy in a progressive interconnected sense. Outcomes of this process are an asymmetric interdependence between two countries with strong ties between the cross-border areas since the NAFTA signing. Ties between the two countries at the border are very complex. Oscar J. Martinez (1994) coined the concept of asymmetric interdependence with which he recovers the functionality of cross-border relationships especially in the economic field; Ganster and Lorey (2008) regained the discussions about the different profiles of cross-border relationships in different parts of the world, but the relations between Mexico and the United States in the border are unique, since it is a relation of integration between very unequal countries. But there is a great gap in water governance in Mexican side, because social participation is limited by government institutions, although public discourses tell us another thing. Then, binational policies fail in integration sense because differences in institutional development between both countries.

13.4 The Water Institutions Facing the Social and Economic Dynamics of the Border

As shown by the Minutes of the International Boundary and Water Commission (IBWC), from the late 1970s of the past century, the border between Mexico and the United States had serious problems in terms of the stream's water quality that cross the international limits and was the cause of discomfort among neighboring communities and governments of the two countries.

In successive IBWC records related to the agreements to identify the specific type of problem and the way to solve at each location of the border, the diagnosis shows the recurrent dynamics of population growth on the Mexican side and the lack of infrastructure services necessary to avoid the problems of water pollution and the risks of environmental and public health of the border populations on both sides.

That is, the quality of boundary waters in this perspective began to deteriorate from Mexico's inability to support the growth of the population and establishment of assembly industries in the border, with the necessary infrastructure. This situation occurred despite the IBWC as an institution itself responsible for implementing the international treaty, which in Article 3 establishes the obligation of countries to address sanitation preferably for water crossing the border.

We must consider that an important part of the problems related to water on the border between Mexico and the United States, or at least those that have generated greater binational tension, is due to the structural changes that the Mexican border area experienced in its economic dynamism and demographics in a very short time.

The first economic boost of the border in the past century occurred from the opening of the irrigation districts in the lower basin of the Colorado River and lower Rio Grande/Bravo in the two countries. Intense competition to secure rights to water mainly for agricultural use, especially in expectation of an eventual agreement between the two countries and the claim of water based on the volumes actually used (Walsh 2004), was recorded in that period.

Agriculture became the main water user sector and farmers, in the irrigation districts, in possession of this natural resource rights. Agriculture turned into an important economic activity, at least for the first part of the twentieth century, so that the water rights were also assigned to the production sector. However, the International Treaty of 1944 between the two countries designated as priority use municipal consumption over agricultural and other urban uses of water.

In the mid-1960s, the Mexican Federal Government boosted a program of opening up to foreign investment in the border, to facilitate the establishment of assembly plants and promote employment for individuals displaced from agriculture or deported from the US government after the Second World War. The border cities began an industrialization period and dynamic population growth. The urbanization processes without planning by local governments, in a trend that prevailed for decades, were fueled also by subsequent problems in the Mexican economy. People expelled from affected regions of the country moved toward the US border looking for employment.

The alternative of employment launched by the Mexican government at the border, through the Border Industrialization Program, became in long term as the basis of the economic growth in that area of Mexico and, in the greater factor of attraction, demographic and urban growth of major cities at the border. Notwithstanding that the impulse came from the Mexican Federal Government, there was no planning capacity and control of urban growth by federal or local government. The new settlements took place in the absence of basic services, resulting in severe environmental impacts. The deficit in services infrastructure, especially of piped water, drainage, and treatment as well as the lack of control over discharges of water from industrial plants, became apparent in the pollution of surface water bodies, including cross-border flows (Kelly 2002).

However, the opening of the irrigation districts, and other activities promoted by the government at the border, where the population settled in the different cities located there grew, had been promoted without considering the aridity and low rainfall that characterize the region and availability of water resources essential to sustain the needs of the people and productive activities.

The type of environmental externalities and public health arising from the conditions described above generated critical situations, tension, and conflict between the two neighboring countries in the early 1980s, especially in Baja California, where cities like Tijuana and Mexicali lacking sanitation infrastructure discharged sewage waters into streams and rivers crossing the border, prompting protests from the Southern California communities receiving polluted water from Mexico.

The regional economy in the border acquired a more urban face, oriented on the Mexican side by government programs to the industry, trade, and services as main activities. Both sides of the border tended to be increasingly interdependent. Despite those changes, agriculture continued to be the main user of water, although in the midst of scenarios of greater competition for structural change pointed out, in where the growth of the water demand for other urban uses has acquired considerable weight and is a source of major concern for the authorities.

On the specific issue of water, it is worth to mention that since 1979 a hygienist policy between Mexico and the United States began, led by the IBWC. As instructed by Presidents Carter and Lopez Portillo in the sense that within the framework of the existing agreements between the two countries will identify the problems and the measures that could be taken for a permanent solution of the border sanitation (IBWC, Minute 261). With Minute 261 of IBWC began the sanitation problem diagnosis in all the Mexican border cities and recommendations on measures that should be taken for a solution in each one of the border towns that were identified by common agreement by the two sections of the IBWC.

The Commission recommended to solve the problems of sanitation of the Rio Nuevo in Mexicali, lower California-Calexico, California, issued on the basis of the international treaty which in the last paragraph of Article 3 refers to the responsibility of the two countries to address problems of sanitation of water crossing the border, presidential instructions of 1979 and the Minute 261 which incorporated such instructions, pledging to make the diagnosis of the problems referred to those presidential instructions and recommendations on solutions in each case.

The Minutes of the IBWC show that the origin of the sanitation problem is located on the Mexican side, and it is directly associated to accelerate growth of the population from the 1960s onward and the lack of coverage of services related to water and drainage. Of course, the solutions that are the subject of recommendations in the Minutes to agree in each case refer to the urgency of investing in infrastructure to prevent untreated water from crossing the border and maritime boundaries; also it must be free of toxicants, in clear reference to the discharge of industrial wastewater into sewers without treatment.

However, it was necessary that the two countries sign a new treaty to make possible a real breakthrough in the solution of problems related to cleaning the border waters, in order for the IBWC to diagnose problems and make the relevant recommendations. In 1983, the Presidents Reagan and De la Madrid signed the La Paz Agreement, consisting of a series of annexes related to environmental problems that were taking place on the border between the two countries.

Commitments relating to the binational cooperation to resolve environmental problems caused by rapid growth, without order and lack of control originated from the dynamics of urban and industrial growth at the border, were established in this agreement. The agreement goes beyond a hygienist vision that had started with the IBWC, since it not only refers to the remediation of situations caused by untreated sewage that had become a source of conflicts most recurrent among the governments of the two countries, but the elimination of the sources of water pollution, air and soil in the border area, which was defined for the first time as the area of up to 100 kilometers on both sides of the land and sea border line dividing between Mexico and the United States of America (Semarnat 2010). The objective of the agreement is to preserve a "healthy environment for economic and social welfare in the long run, of present and future generations of each country, as well as the international community" (Semarnat 2010).

The signing of this agreement opened an avenue of binational cooperation widely favorable to Mexico, especially because in those years it was going through a severe financial crisis and faced serious restrictions to deal with the problems defined on the diagnosis elaborated by the IBWC about border problems. The problems associated with the water quality at the border were defined within a broader environmental framework, facilitated through cooperation with the environmental agency of the United States (EPA) channeling resources in support of projects implemented on the Mexican side.

In practice, the La Paz Agreement was a widely significant step for binational cooperation opportunities, and 10 years later, in 1993, was created the Commission of Border Environmental Cooperation (BEC), and the North American Development Bank and the Free Trade Agreement or NAFTA were signed this same year too (Thompson 2009; Mumme and Collins 2014). One of the priority issues on the agenda of these new institutions was the infrastructure water services, which have exceeded by far the backlog that had in the early 1980s in water coverage and drainage, as well as treatment. The binational cooperation has been crucial for

Mexico's border cities that have passed the lag in water service and drainage coverage, and some of its water authorities are among the best qualified according to efficiency indicators in the country (Hufbauer and Schott 2005; Lutz Ley and Salazar Adams 2011). Although it is likely that greater efforts are required to achieve water of adequate quality to be incorporated to the reuse and enter into patterns of substitution leading to the saving of water.

New binational institutions came to join the efforts initiated by the IBWC around problems related to the quality of the boundary waters. Thus, although the Minute 295 of September 19, 1996, refers to the system of sewage from Naco, Sonora, the IBWC's Minute refers for the first time to the existence of the BEC, noting that the agreement by which the governments of Mexico and the United States created this new commission established in its article III, section 6, Chapter I, the mandate that BEC and IBWC must cooperate for solving problems (IBWC, Minute 295).

From the Minute 261 of September 1979 and up the Minute 311 of February 20, 2004, related to sanitation problems and solutions agreed between the two national sections of IBWC, all of them relating to situations originating in Mexico are indicative of the structural weaknesses of the Mexican institutions responsible for different aspects related to the water quality, urban planning, and controlling in the cities of the border with the United States.

At the same time, there are 16 Minutes of the IBWC for the same period, corresponding to agreements related on the operation and maintenance of the water infrastructure under IBWC responsibility to secure the water supply to each country but also on emerging measures against problems associated with lack of water derived from extreme drought in the Rio Grande/Bravo Basin that affected Mexico dramatically, as Minute 293 agreed on October 4, 1995, in terms of "measures emergent cooperation to meet the water needs of the Mexican towns located along the Rio Grande down the 'La Amistad Dam'." The agreement allowed Mexico to hold water corresponding to the United States in binational storage dams up to 100 million cubic meters, because of the presence of a season of extreme drought in the basin of the Conchos River.

The Minute 307 signed on March 16, 2001, authorized to partially cover the water deficit from Mexico to the United States in the basin of the Rio Grande in the drought period. The delay in Mexico led to conflicts over water between the two countries, with pressure from Texan farmers, American congressmen, and the Government of Texas. The US Federal Government exerted pressure to the Mexican government in order to pay for water. The solution according to the Minute was the authorization of the Mexican government to send water not corresponding to runoffs of the streams included in the Treaty of 1944, generating controversy in Mexico also and criticism by congress representatives and water users of the State of Tamaulipas.

The Minute 308 signed on June 28, 2002, set out allocations of water to the United States in the last year of the 5-year cycle which was concluding, in order to pay off the existing deficit. At the same time, municipal consumption needs of the Mexican cities located on lower Rio Grande/Bravo were defined, so they tried to cover the debt of water to the United States by Mexico but at the same time to ensure the supply needs of the population located on the Mexican side. Also, it is important to mention the interest in modernizing irrigation districts located in this basin in Mexican side, in order to achieve a more efficient water use and saving it regarding to deliver the volumes of water which the treaty assigns to the United States.

In fact, the Minute 309 refers to the volume of water that will be saved with the modernization and automation of the irrigation districts in the Rio Conchos and the methodology to driving water to the Rio Grande/Bravo for delivering the share of the United States. Volumes up to 396 million cubic meters of savings resulting from the above measures are estimated, but also the National Water Commission undertakes to keep track of the efficiencies achieved, the volumes extracted from dams located in basin[1].

Also, there are at least successive Minutes 317, 318, and 319 relating to the conduct of studies aimed at developing recommendations for "identifying innovative measures and cooperation between the two countries for the conservation, storage, increased supply, and environmental protection" in the Colorado River Basin (Minute 317). The Minute 318 signed on December 17, 2010, refers to the adjustment of water delivery schedule of the United States to Mexico in the Mexicali Valley because of the damage experienced by the infrastructure of the irrigation district 014 in April same year from an earthquake. Finally, the Minute 319 was signed on November 20, 2012, as an extension of the measures agreed in Minute 318 but also as a broader cooperation agreement related to the storage and management of water deliveries in the lower Colorado River Basin (Sánchez-Munguia and Cortez-Lara 2014).

These are all cases of cooperative actions and decisions taken promptly to situations that have arisen due to the changes experienced by the space of the frontier and its dynamics of population and economy but also with issues related to weather conditions affecting water availability across the border between the two countries.

13.5 Conclusion

The border between Mexico and the United States is in an arid and semiarid region, with low rainfall and distant from the national averages in both countries. At the same time, the population located on both sides of the border has grown considerably, although the Mexican side has had a much higher rate for several decades, from the federal government policies aimed to boosting the border economy, resulting in a growing demand for water to meet the needs of people to create a viable economy.

Scarce surface water sources are shared under an international regime agreed between Mexico and the United States, which has worked relatively well so far,

[1]However, it is important to note that in this as in other river basin in Mexico, the main problem is the absence of water authority in exercise of its power to order water withdrawals or limited ability to do so.

although the two countries have been involved in some conflicts related to extreme droughts in recent years.

Some of the problems related to water in the border and the relationship between the two countries are related with the inadequacies and omissions since the signing of the International Treaty. The issue of the water quality that countries would receive was excluded; the lack of a clear definition of the concept of extraordinary drought, mentioned in the international treaty, and in the procedure declared by the parties involved; the separation between surface water and groundwater, with no mention of transboundary aquifers, the same can be said of promised fixed volumes of water without sufficient considerations on possible changes in future climate conditions, and which have required different formulas in each basin regarding the possibility of a reduction of water available for delivery of the compromised volume from one country to another.

Those issues are recurrent to events in which countries face emergencies resulting from scarce rainfall seasons, and low water volumes in the storage dams. Nevertheless it is important to point out that the problems arising from such situations are regularly addressed by the Commissioners of the two countries at the IBWC, and generally solved.

However, as we have seen throughout this chapter, the binational relationship on the water issues has been characterized in recent decades by a broad cooperation, where the focus has been the resolution of problems arising from a relationship between countries with very different levels of development and therefore with very differentiated available resources and management capabilities to face challenges. Although the convergence of the two countries to a much more comprehensive water policy and integrated water management for a reasonably sustainable use of the resource is desirable, it is not an easy thing; if the gaps are so large in the institutional capacities of performance, they require consistent guidelines aimed at that purpose by Mexico.

Even so, the persistent ability to reach agreements between the two countries on issues of water management and conflict resolution in different moments is worth noting, derived from the asymmetrical condition between them, availability of water, and institutional omissions or failures due to the lack of more integrated water policies in the border area. Above all, Mexico needs to improve their capabilities in water planning, management, and collection but also in law enforcement.

References

Becerra Pérez, M., et al. (2006). Los conflictos por el agua en México. Diagnóstico y análisis. *Gestión y Política Pública, XV*(1), 11–143.

Browning Aiken, A., et al. (2007). Climate, water management, and policy in the San Pedro Basin: results of survey of Mexican stakeholders near the U.S.-Mexico border. *Climate Change, 85* (3–4), 323–341.

Castro, J. L., et al. (2015). La relación binacional y la gestión del agua en la frontera México-Estados Unidos: lecciones del pasado, desafíos del futuro. In María Concepción Martínez Omaña (Coord.), *Agua e Historia* (pp. 315–346), Instituto Mora, México.

Fagen, R. R. (1983). The politics of the United States-Mexico relationship. In C. W. Reynolds & C. Tello (Eds.), *U.S.-Mexico relations. Economic and social aspects* (pp. 331–347). Stanford University Press.

Fernández, L. (2006). *Wastewater management across an international border*. Paper presented at the 6th meeting on game theory and practice, Zaragoza, Spain, 10–12 July.

Fernández de Castro, R., y Domínguez, J. I. (2001). *¿Socios o adversarios? México-Estados Unidos hoy*. México: Océano.

Ganster, P., & Leroy, D. E. (2008). *The U.S.-Mexican border into the twenty-first century*. London/New York/Toronto/Plymouth: Rowman & Littlefield Publishers.

Granados-Olivas, A., et al. (2010). Future solutions: Research needs in the Mexican section of the Rio Grande/bravo watershed. *Journal of Transboundary Resources, 1*, 147–157.

Hufbauer, G. C., & Schott, J. J. (2005). *NAFTA revisited: Achievements and challenges*. Washington, DC: Peterson Institute for International Economics.

International Boundary and Water Commission. http://www.ibwc.gov/Treaties_Minutes/Minutes.html. visited 20/02/2013.

Kelly, M. E. (2002). Water management in the binational Texas/Mexico Rio Grande/Rio Bravo Basin, in human population and fresh water resources: U.S. cases and international perspectives, *Bulletin Series* Num. 107 Yale School of Forestry & Environmental Studies. http://environment.research.yale.edu/documents/downloads/0-9/107Front.pdf. visited 02/10/2013.

Lutz Ley, A. N., & Salazar Adams, A. (2011). Evolución y perfiles de eficiencia de los organismos operadores de agua potable en México. *Estudios Demográficos y Urbanos, 26*(3., El Colegio de México), 563–599.

Maganda, C. (2008). *The impact of regional integration on transboundary water management: Politics, planning and public participation in water governance*. Paper, UNESCO international on water scarcity, global change, and groundwater management responses. http://www.groundwater-conference.uci.edu/files/Alphabetical/2008_conf_ Maganda,%20C.pdf. visited 10/11/20112.

Martinez, O. J. (1994). *Border people: Life and society in the U.S.-Mexico borderlands*. Tucson: University of Arizona Press.

Mumme, S. P. (2010). Environmental governance in the Rio Grande watershed: Binational institutions and the transboundary water crisis. *Journal of Transboundary Water Resources, 1*. http://www.wrri.nmsu.edu/publish/journal_of_transboundary/Mumme.pdf.

Mumme, S. P. (2015). Scarcity and power in US–Mexico transboundary water governance: Has the architecture changed since NAFTA? *Globalizations*, published on line el 8-01-2016. http://www.tandfonline.com/action/showCitFormats?doi= 10.1080%2F14747731.2015.1129710

Mumme, S., & Collins, K. (2014). The La Paz agreement 30 years on. *The Journal of Environment Development, 23*(3), 303–330.

Ojeda, M. (1983). The future of relations between Mexico and the United States. In C. W. Reynolds & C. Tello (Eds.), *U.S.-Mexico relations. Economic and social aspects* (pp. 331–330). Stanford University Press.

Pombo, A., & Wright, R. (2012). Chapter 7: Water perspective in Western U.S.-Mexican border: Future conflict? In E. Lee & P. Ganster (Eds.), *The U.S.-Mexican Border Environment: Progress, and Challenges for Sustainability, Southwestern Consortium for Environmental Research and Policy, SCERP Monograph Series No. 16* (pp. 187–200). San Diego: State University Press.

Sánchez-Munguia, V., & Cortez-Lara, A. A. (2014). Binational cooperation processes for transboundary waters in the Colorado river basin: Minute 319 of the international boundary and water commission between the U.S. and Mexico. *International Journal of Water Resources Development, 31*(1), 1–11.

Semarnat (Secretaria del Medio Ambiente y Recursos Naturales), Convenio de La Paz. http://www.semarnat.gob.mx/temas/internacional/norte/Paginas/ConveniodelaPaz.aspx. Visited 15/02/2013.

Sierra, J. (1977). *La evolución política del pueblo mexicano*, Lumen, Buenos Aires, primera edición.

Thompson, O. N. (2009). *Binational water management: Perspectives of local texas officials in the U.S.-Mexico border region*. https://digital.library.txstate.edu/handle/10877/3758

United Nations of Environmental Program. (2002). *Atlas of international fresh water agreements, UNEP*. http://www.unep.orgblications/search/pub_details_s.asp?ID=115. visited 20/03/2013.

Walsh, C. (2004). Aguas Broncas: The regional political ecology of water Conflict in the Mexico-US Borderlands. *Journal of Political Ecology, 11*, 45–60.

Wolf, A., et al. (2005). Gestionando conflictos por el agua y la cooperación. En Michael Renner et al (2005) La Situación del Mundo 2005: Redefiniendo la seguridad mundial. Informe anual del Worldwatch Institute sobre el progreso hacia una sociedad sostenible, Barcelona, Icaria, Centro de Investigación para la Paz, pp. 115–178.

Wolf, A., et al. (2008). Healing the enlightenment rift: Rationality, spirituality and shared waters. *Journal of International Affairs, 61*(2, summer), 51–73.

Vicente Sánchez Munguía holds a PhD in Contemporary Latin American Studies at the Ortega y Gasset University Institute of Madrid; and PhD Studies in Advanced Public Management at Complutense University of Madrid; also a Diploma in constitutional law and political science at the Center for Constitutional Studies of Madrid. He is a professor and researcher at El Colegio de la Frontera Norte (Northern Border College) in Tijuana (Mexico) and is a member of the National System of Researchers (SNI). His research work is about the water relationship between Mexico and the United States at the common border, the water management in northern of Mexico, and the border area specifically. He has published articles, books, and chapters of books on water management and conflict and cooperation in this border region.

Part V
Present and Future Challenges

Chapter 14
Climate Change and Vulnerability of Water Resources in Mexico: Challenges for Basin Management

Eduardo Ríos Patrón, Daniel Iura González Terrazas, and Ignacio Daniel González Mora

Abstract In this chapter we will review the importance of water resources, the development of the Mexican public policy (within the framework of water management), the risk to which availability is subject (in the face of climate change scenarios), and how, in front of a differentiated approach by basins and ecosystems, can be a model, from actions of adaptation and conservation of the functionality of the basins, for the reduction of water vulnerability and a crosscutting element of public policies in Mexico, which allows to adapt to climate change and meet the international commitments that Mexico has signed.

Keywords Water resources · Vulnerability · Basin management · Climate change

14.1 Introduction

Water is and has been the fundamental basis for life and development, considering that it is necessary to support biogeochemical processes, which preserve the ecosystems that are a source of direct and indirect benefits for human societies. It is also a necessary input for the functioning of cities as well as industrial processes and food production. However, the uncertainty of a relative scarcity level regarding its existence, use, and renewal represents a major challenge toward the larger

E. Ríos Patrón (✉)
Mexican Network of Hydrographic Basins (Remexcu), Ministry of Environment and Natural Resources (SEMARNAT), Morelia, Michoacán, Mexico
e-mail: eduardoriospatron@gmail.com; eduardo.rios@semarnat.gob.mx; coordinacion@remexcu.org

D. I. González Terrazas
National Institute of Ecology and Climate Change (INECC), CdMx, Ciudad de México, México
e-mail: daniel.iura@inecc.gob.mx

I. D. González Mora
Mexican Network of Hydrographic Basins (Remexcu), World Wildlife Fund (WWF México), CdMx, Ciudad de México, México
e-mail: coordinacion@remexcu.org; igonzalez@wwfmex.com

H. R. Guerrero García Rojas (ed.), *Water Policy in Mexico*, Global Issues in Water Policy 20, https://doi.org/10.1007/978-3-319-76115-2_14

demand of the vital liquid, as a consequence of an increasing population and an increase in the levels of production and new satisfiers. Additionally, climate change and its differentiated impacts, which intensify year after year, are enhanced by an inadequate management of the territory in the basins of Mexico.

Reducing impacts on natural and human systems due to changes in the availability of the water resource as a result of climate change and territory management is a national challenge that aligns with recent global agendas that have emerged as a recognition of the global priority of accelerating and guiding the actions of nations toward a sustainable development. This way, the 2030[1] Agenda emerges, with 17 sustainable development objectives, as well as the Paris Agreement[2] intended for preventing the planet from increasing the average global temperature in more than 2 °C from the reduction of greenhouse gases (GHG); and if it cannot be done, there would be huge economic, social, and environmental impacts.

Besides this international commitments signed by Mexico and which are systematically reflected in the national public policies and regulatory framework; there is the constitutional recognition that "everyone has the right to water access, disposal and sanitation for personal and domestic consumption, in a sufficient, safe, acceptable and affordable manner,"[3] which, in front of a climate change context, represents a huge commitment of the state for achieving the effective fulfillment of such human right.

This huge commitment implies the relevance of the design and implementation of integral and regional public policies that in a synergic manner and, from scientific information in an effective framework of participation and governance, consider (i) attention of vulnerability to climate change from the strengthening of local and national capabilities; (ii) integration and strengthening of the preservation policy to increase resilience from the functional maintenance and improvement of water resource production zones; (iii) the strengthening of early warning systems (SAT – Sistema de Alerta Temprana) and the integrated risk management; and (iv) the insertion of the basin approach in public policies from the development of norms,

[1]In September 2015, more than 150 heads of state and government met in the Sustainable Development Summit in New York, EU at UN headquarters, where the 2030 Agenda was approved. This Agenda includes 17 universal application objectives that, since January 1, 2016, govern the efforts of the countries to achieve a sustainable world in 2030 (Source: http://www.un.org/sustainabledevelopment/es/la-agenda-de-desarrollo-sostenible/).

[2]The Paris Agreement was adopted on December 12, 2015, in the 21st, session of the Conference of the Parties to the United Nations Framework Convention on Climate Change (UNFCCC) held in Paris, France. The Paris Agreement came into force on November 4, 2016. One hundred ninety-five out of 197 parties of the UNFCCC signed the Agreement, and 169 have already ratified it (Source: http://unfccc.int/paris_agreement/items/9444.php).

[3]The recognition of this human right was added by a Decree, published in the Official Gazette of the Federation on February 8, 2012, in the sixth paragraph of Article 4 of the Political Constitution of the United Mexican States (Source: https://www.juridicas.unam.mx/legislacion/ordenamiento/constitucion-politica-de-los-estados-unidos-mexicanos#10539).

programs, and projects linking supply and demand of water environmental services under climate change scenarios and anticipate actions for their maintenance over time.

Climate change is evident and unequivocal, and its consequences are an unprecedented challenge for the design and implementation of public policies for water resources management in Mexico. This challenge implies adapting and generating governance conditions to changes in water availability, as well as against an increase in extreme meteorological events and its consequences, from the acknowledgement of basins functionality and the relation of hydrological environmental services (SAH) with the local and regional economic, social, and environmental dynamics. Maintaining the SAH implies the improvement of human well-being and the reduction in the vulnerability of human and natural systems toward hydrometeorological phenomena and must come from (i) inserting the basin integral approach in the planning and implementation of actions for the preservation and restoration of ecosystems, (ii) inserting the design and effective use of early warning systems, and (iii) exploiting the synergies of coordination platforms posed by existing laws on water and climate change, in order to face the consequences of climate change in the basins in Mexico and to reduce the vulnerability by collaborating and articulating efforts in different scales and levels.

14.1.1 Water Resources in Mexico

Only 0.77% of total water in the world is estimated as accessible to human beings as fresh water (CONAGUA 2016); nevertheless, it is subject to high pressures as a result of a continuous increase in the demand, contamination, and climate change. Facing these pressures in quality and quantity of the vital liquid, the World Water Development Report 2017 concludes that "it would be unthinkable not to take advantage of the opportunities of a better management of waste water" (WWAP 2017), and obviously, this better management implies an integrated management approach considering the basin as a convenient territory for water management resources.

Mexico, within its 1959 million km^2 continental surface, has 1471 watersheds (INEGI-INE-CONAGUA 2007) integrating a 633 thousand kilometer network of rivers and streams. Among those, 1389 are exorheic basins flowing into some point of its 11,122 kilometers of coastline; 77 are endorheic and 5 arheic in the Yucatan Peninsula, sharing 8 basins with neighboring countries, 4 with Guatemala (Grijalva-Usumacinta, Suchiate, Coatan, and Candelaria), 1 with Belize and Guatemala (Hondo River), and 3 with the USA (Bravo, Colorado, and Tijuana). It has 653 aquifers that supply 38.9% of the volume for consumptive uses, 105 of those are overexploited, 32 with the presence of saline soils and brackish water, and 18 with seawater intrusion (CONAGUA 2016).

According to data from the National Water Commission, Mexico receives around 1,449,471 million square meters of water as rainfall. From this water, 72.5% is evapotranspired back to the atmosphere, 21.2% drains through rivers or streams, and the remaining 6.3% naturally infiltrates to the subsoil recharging aquifers. Taking into account output (exportations) and input (importations) water flows with neighboring countries, the country has 446,777 million cubic meters of renewable freshwater, data estimated for 2015, giving an availability of 3592 cubic meters per inhabitant per day.

However, there are large temporary and spatial variations in the territory, requiring an analysis per basin to characterize and to identify the particular situation of the water resource, since in the northern and central regions of the country, there is less water availability, more contribution to the gross domestic product (GDP), and larger population, while the southern area of the country presents a larger water availability, lesser contribution to the GDP, and less population. Besides, it is worth noting that each basin presents unique socioecosystemic characteristics, and hence the management should be differentiated and reflected in water and climate change policies.

In 2015, a volume of 266,559 million cubic meters was in concession, 180,895 of which were for non-consumptive use (which does not affect the volume), that is, for hydroelectric use, and 85,664 million cubic meters for consumptive uses (the difference between extracted and discharged volume while conducting an activity). The water source of the latter was 37.9% of underground origin and 61.1% of superficial origin (CONAGUA 2016). Available water in the country requires satisfying the population growing demands of about 120 million inhabitants and a globally integrated economy with strong pressure on production and consumption in a climate change context.

14.1.2 Climate Change and Vulnerability: Commitments of Mexico

The Agreement in the United Nations Framework Convention on Climate Change reached in Paris, France, in December 2015 recognizes climate change as "an urgent and potentially irreversible threat to human societies and the planet" (Naciones Unidas 2015).

It is evident and unequivocal that the planet has presented a 0.85 °C increase in average temperature[4] that the Arctic ice extension in the North Pole is being lost between 3.5% and 4.1% each decade,[5] that glaciers have decreased at a rate of

[4]Measure conducted from 1850 to 2012.

[5]Measure conducted from 1972 to 2012; this means that just in the last 10 years, 3.5 million km^2 have been lost.

275 giga tons per year,[6] and that the sea level has globally increased 0.18 m.[7] Likewise, the evidence is compelling, and carbon dioxide (CO_2), methane (CH_4), and nitrogen oxide (N_2O) have demonstrated an increase of 40, 150, and 20%, respectively, since 1750.[8] At the same time, an increase in the industrial activity has caused that almost 30% of carbon dioxide is absorbed by oceans causing acidification (Conde 2016; IPCC2013). It is also alarming to confirm year after year a new maximum temperature record; the global mean temperature for the period between January and September 2017 was 0.47° ± 0,08 °C warmer than 1981–2010 mean (estimated in 14.31 °C), which represents a temperature increase of around 1,1 °C from the beginning of the preindustrial period (OMM 2017). These changes in the global climate system have important and differentiated local impacts, especially in Mexico, which is a country that due to its geographic location and social conditions is especially vulnerable to climate change.

Climate change in Mexico is also evident and unequivocal, and its impacts are already being noted. Thus, according to data from the National Institute of Ecology and Climate Change (INECC), "in a little more than 100 years terrestrial and marine surface temperatures have increased in all the territory, however, in certain areas in the north of the country, changes have been greater, ranging from 1.2 and 1,5 °C above its historical average." The global warming observed is joined by an increase in the number of extreme hot days and a decrease in extreme freezing days and frost. All this is added with the impact of a growing number of extreme hydrometeorological phenomena, as tropical cyclones and hurricanes. Between 1970 and 2013, 10 out of 22 category 3 cyclones in the Saffir-Simpson scale that affected the Pacific and Atlantic coasts of Mexico occurred in the last 12 years. In the case of droughts, so far this century, five important events have occurred between 2000 and 2003, in 2006, between 2007 and 2008, in 2009, and between 2010 and 2012. In some cases, drought has been so severe that it has affected large extensions in the country, as in 2011, when 90% of the territory was affected.[9] Sea level has also been raised in many coastal areas in Mexico. Seventeen sites were studied in the Gulf of Mexico and the Pacific, between 1950 and 2000, standing out the elevations observed in Ciudad Madero, Tamaulipas (up to 9.16 millimeters per year) and Guaymas, in Sonora (4.23 millimeters per year)" (INECC 2015).

According to INECC in Mexico, for the near horizon 2015–2039, annual temperatures 2 °C above in the north, and between 1 and 1.5 °C, are estimated in most of the territory. On average, the forecast is an average reduction between 10 and 20% in rainfall (INECC 2015); these scenarios require action and the development of

---

[6]Measure conducted from 1993 to 2009. A giga equals 10^9 or 1,000,000,000 (thousand million).

[7]Measure conducted from 1901 to 2010. At a 1.7 mm/year rate in 1901–2010, 2.0 mm/year between 1971 and 2010, and 3.2 mm/year in 1993–2010.

[8]Percentage of increase of CO_2, CH_4, and N_2O from 1750 to 2011.

[9]From this situation 74% of the territory presented severe, 47% extremely, and 22% exceptional drought. See http://smn.cna.gob.mx/es/climatologia/monitor-de-sequia/monitor-de-sequia-en-mexico.

comprehensive public policies that allow us to reduce damage from an effective management of the water resource.

Within the commitments derived from the Paris Agreement, an adaptation component to climate change is included. The non conditional commitments of Mexico to 2030 are to (i) reduce the high vulnerability of at least 160 municipalities considered as very vulnerable to climate change, (ii) reach a zero deforestation rate and strengthen actions for the protection and restoring of ecosystems, and (iii) generate prevention and early warning systems before extreme hydrometeorological events.

14.2 Response Strategies to Changes in Water Availability Associated to Climate Change

Climate change can be interpreted either as a compelling problem that threatens biodiversity and world population or as a factor that aggravates the effects of other existing threats. In this sense, climate change affects the provision of water resources in the ecosystems, by modifying rainfall and temperature patterns. For instance, with the reduction of forests and the increase of urban land use, the eco-hydrological functionality of basins is modified, and the water provision for society and ecosystems is reduced in a faster manner in the context of climate change (Ríos et al. 2016).

Impacts on rainfall and temperature variations have an effect on the components of the hydrological cycle, in runoff, evaporation and infiltration, as well as in the intensity and frequency of extreme hydrometeorological phenomena and their consequences as droughts and floods.[10] The strategies for responding to water availability changes associated to climate change must be designed to ensure the water balance of the basin, either by improving infiltration processes and reducing flood, by decreasing runoffs velocity, or by recovering the catchment functionality of land ecosystems and lotic and lentic systems. To reach the improvement of hydrologic functionality, it is important to have the information of rainfall and temperature scenarios, as well as their impact in the water balance components for proper actions and decision-making. It is essential to consider differentiated strategies before such scenarios of water abundance or scarcity, with drought or flooding risks, or with water increase or decrease for the development of economic activities in the basins of Mexico. Strategies should also link – from the knowledge of water balance considering different rainfall and temperature scenarios – trends in water resource use or demand for the activities of the basin and the actual offer, in order to determine impact mitigation strategies in an advanced and concerted manner, through a participation process promoting present and future governance.

[10] Idem.

Mexico, particularly, has established international commitments with important challenges in climate change adaptation, which emerge from three differentiated strategies (INECC 2015):

- Adaptation of the social sector to climate change
- Adaptation based on ecosystems
- Adaptation of productive systems and strategic infrastructure

Integrated management of basins implies an understanding of a certain region functioning, its ecosystems, and socioeconomic processes that take place in there. In this manner, anticipated and well-informed strategies can be developed to implement measures of adaptation, considering possible vulnerabilities of the population and ecosystems (Lampis 2013). Planning, based on scientific evidence, may contribute in the reduction of vulnerability in front of local impacts of climate change, ensuring decision-makers have the ability and information needed for the prevention and adaptation to changes associated to this phenomena.

Strategies for the reduction of water vulnerability to climate change imply a search for sustainable development, which, "cannot be achieved without a world with safe water, A safe water world appropriates the productive power of water and minimizes its destructive force" (GWP 2017). In a climate change context, water security implies (i) having clean, sufficient, and affordable water for a healthy life and responding to the human right for water access and sanitation; (ii) protecting the community from flooding, droughts, landslides, erosion, and water-related diseases; (iii) addressing environmental protection and the negative effects of a bad management; and (iv) eliminating the fragmenting responsibility of an integrated management of water resources in all sectors. Integration must be the core of the strategy and should be supported in good governance.

14.2.1 Water Resource as an Integrating Axis in Territorial Management: An Approach from Basin Management

A way to understand that water resources management is subject to territorial management is through the basin approach. This involves making explicit the relation among territories and users of high and low basins, considering all generated impacts resulted from activities in the high basin toward the lower side, and recognizing the services that ecosystems provide to the basin (Sanjurjo et al. 2017). In these territories, there is a spatial and temporary interrelation and interdependence among the biophysical surrounding (aquatic and terrestrial ecosystems, crops, water, biodiversity, geomorphological and geological structure), appropriation modes (technology and/or markets), and institutions (social organization, culture, regulations, and/or laws) (Cotler et al. 2013).

Hydrographic basins are considered the basic territorial unit for planning and managing natural resources and as a spatial and temporary dimension for the adaptation to climate change (Biestroek et al. 2009; SEMARNAT-INECC 2013).

The watershed approach is considered both in the National Waters Law (LAN), in the acknowledgement of water resources management units, in the preservation of the hydrologic cycle in the territory, and in water availability and governance, and in the General Law on Ecological Balance and Environmental Protection (LGEEPA) to guarantee the sustainability of economic activities through official Mexican norms in environmental issues, for the sustainable achievement of natural resources, as well as to prevent contamination, fundamentally. But its articulation is an area of opportunity in the Mexican regulations. Basin has been relevant in water planning and management issues, but not in the regulation of territorial management from the perspective of terrestrial and aquatic ecosystems in permanent interaction, which provide ecosystemic services needed for the well-being of human societies. In this sense, water has been treated as a material resource and not as an asset that results from the functionality and integrity of terrestrial and aquatic socio-ecosystems.

14.3 Analysis of Demand and Supply of Hydrological Environmental Services (SAH)

A way to change the wrong approach of considering water as a resource with no territorial connection is through the incorporation of the supply and demand analysis of hydrological environmental services, identifying in detail the contribution of hydrological and ecosystemic processes in water "production" for socioeconomic activities in a basin facing different scenarios of climate change. This necessarily involves the implementation of the basin approach, the functional link of territorial management, as well as the zone prioritization for its preservation and protection.

This approach of linking supply and demand of water resources, from areas where physical, chemical, and biological processes are involved for the functioning of the hydrological dynamics of a basin, includes reducing risks and maintaining and improving the quality and quantity of water in a climate change context, so it should be a common approach at the local level for conducting the management of water resources. Fulfillment requires the strengthening national and local capabilities, from entities that plan, participate, and coordinate the water policy and climate change policy.

14.3.1 Evolution of the Instruments of Environmental Policy in Mexico that Promote Conservation and Sustainable Use of the Hydrological Environmental Services (SAH)

In Mexico, environmental protection is directly bonded with the social function notion of private property, substituting property as an absolute right: it is expressed in the 1917 Constitution, where the principle of use and conservation of natural resources is introduced, particularly in constitutional Article 27, Section 3, where the foundations for the territorial planning of the nation are mentioned:

> the necessary measures shall be declared to order human settlements and to establish proper provisions, uses, reserves, the purpose for lands, waters, and forests, in order to execute public works and to plan and regulate funds, conservation and improvement, and growth of population centers; to preserve and restore the ecological balance

The first manifestations of territorial planning considering water resource appear in the postrevolutionary times for infrastructure reconstruction, with a boost on agriculture, developing an integrating vision of regional planning (Calderon 2001).

In 1947, during the presidential administration of Miguel Alemán, and seeking an impulse on the agricultural and industrial development of the country, the first hydrographic basin commission was created, the Tepalcatepec River basin, dependent of the Ministry of Water Resources in charge of planning and conducting the necessary works for the integral development of natural resources in that basin. However, efforts were minimal in relation to the industrial-urban tax model of the following 30 years, when industrial growth policies began around Mexico City and regional economy poles. In the administration of Luis Echeverría (1970–1976), the so-called specific regional planning begins (Calderón 2001; Gasca 2009), based on the generation of sectoral productivity policies, the socioeconomic planning strategies on territorial basis boosted decentralization efforts, and Gasca (2009) refers to this period as "the golden age" of planning. However, recurring crises from de la Madrid administration (1982–1988) caused by the mobility of oil prices, the flight of capital, economic slowdown, and high inflation forced the Mexican state to sign international loans, deregulate planning processes, and privatize key sectors, actions that initiated a phase of neoliberal policies (Gasca 2009). Decentralized efforts of the public administration were blocked; there was a dismantling process of the institutional structure and programs focused on territorial planning, a loophole that remained in subsequent administrations.

Nevertheless, in the evolution of the environmental policy framework,[11] triggered by an evident degradation of natural systems and an increase in population, it is from the last quarter in the twentieth century that the development of territorial planning instruments oriented to evaluating the satisfaction degree of the social demand from

[11] In Mexico, this has been widely described by experts in judicial (Brañes 2004; Azuela 2006a, b; Quintana 2009) and administrative systems (Rosete 2006).

the evaluation of offer provided by the environment from a territorial perspective, in terms of goods, services, and spaces, began, the latter with the sole purpose of promoting a sustainable development.

Regarding environmental services[12] to consolidate this approach in laws and Mexican normativity, instruments whose objective is to maintain functional interrelations of the landscape elements that intervene in the offer of these services through conservation, restoration, protection, and sustainable use of environmental policies have been materialized. Among the policy instruments that address the continuity of environmental services in space and time are: the National Development Plan (PND), and within the same, Sectoral Programs like the one of Environment and Natural Resources (PROMARNAT) and the Water National Program (PNH), environmental General Laws and deriving programs with a direct effect on the territory. In this sense, the incorporation of the basin approach is a milestone in Mexican environmental policy[13] as one of the main strategies to comply with two of the PROMARNAT 2013–2018 objectives, strengthening integral and sustainable management and guaranteeing access to the population and ecosystems and recovering basins and landscapes functionality through conservation, restoration, and sustainable appropriation of the natural heritage, objectives that are directly related to the safeguard of environmental services demanded by the population, ecosystems, and productive activities. PNH 2014–2018, based on basins and aquifers, proposes the objectives of strengthening integrated and sustainable water management, increasing water security toward droughts and floods, and strengthening water supply and access to drinking water, sewerage, and sanitation services.

Regarding General Laws, in terms of synchrony of precepts and operational instruments with territorial incidence for the conservation of environmental services, the General Law of Ecological Balance and Environmental Protection is worth noting (LGEEPA 1998). In the absence of clear legal rules and norms of the rural territory use, and specifically of conservation-minded territory, in 1987 the constitutional Article 27 was amended by adding "preserving and restoring the ecological balance." Subsequently, LGEEPA was published explicitly incorporating in a Chapter the "Sustainable development of water and aquatic ecosystems."

LGEEPA focuses on environmental planning, through diverse modalities of its ruling instrument, the Territorial Ecological Ordering (OET); inducing or regulating out of the population centers, land uses to protect the environment

[12]Regarding the definition of the National Waters Law, it is understood that social interest benefits are generated or emerge from hydrological basins and its components, such as climate regulation, hydrological cycle conservation, erosion control, flooding control, aquifer recharge, runoffs maintaining in quality and quantity, soil formation, carbon capture, purification of water bodies, as well as conservation and protection of biodiversity.

[13]Conceptual and methodological framework which considers characteristics related to the functionality of hydrographical. Especially (i) the relation among users and high and low basin territory, (ii) cumulative impacts devastated by rivers and streams, and (iii) the consideration of water quantity, quality, and temporality as an indicator of territorial management (PROMARNAT 2013–2018)

and preserve, restore and appropriate natural resources in a sustainable manner, basically by conducting productive activities and locating human settlements. In this sense, it is the first time in the history of Mexican environmental policy that the national development goals are harmonized, environmental legislation and a bonding instrument with territorial incidence, represented by the OET in its local feature, to safeguarding balance between supply and demand of environmental services, especially those related to water resources. However, facing the climate change threat, the short-term challenge is to adequate the design of this instrument considering (i) thresholds or limits in ecosystems use, (ii) relations of hydrographic connectivity between environmental services supply and demand zones for the prioritization of conservation zones, and (iii) possible impacts related to pattern changes of the climate that might compromise the functionality of ecosystems and, as a consequence, the associated environmental services.

14.4 Adaptation and Management of Basins: A Strategy to the Reduction of Vulnerability to Climate Change Impacts on the Water Resource

The most immediate effect of climate change is an increase in temperature and alteration of the hydrological cycle with the consequent reduction in rainfall but also an increase in extreme hydrometeorological phenomena.

Undeniably, adapting to climate change is to protect water sources and to maintain water availability for future generations. The impact of climate change on natural water sources can be characterized by an alteration in water quality (e.g., by pollutants, temperature, or dissolved oxygen), water quantity, and the pattern of its cycles (in the regular periods of drought and floods). Worldwide, alterations in the seasonal pattern of water might be the most important consequence for human beings and other species. Unfortunately, this might be also the most uncertain variable according to diverse general circulation models. Hence, water policy should be focused in sub-annual changes, for instance, per season or per month. On the other hand, uncertainty should be no excuse for non-acting (Mathews and Le Quesne 2009).

In Mexico, water availability in the 757 water basins or water management units is published every year and considers no sub-annual or monthly periods, which constitutes an opportunity area for a better integrated management of water resources. In this sense, the Mexican official norm of availability (NOM-011-CONAGUA-2015) and the Mexican norm of ecological flow determinations (NMX-AA-159-SCFI-2012), whose purpose is the ecological conservation of basins and which can be used to know inter-annual periods of variation, shall be harmonized. Publishing the annual availability of all basins in the country has been a breakthrough in water management in Mexico, yet allows non-sufficient detail to have an adaptive management up to the challenge of climate change.

Adaptation to climate change is the process that allows reducing the vulnerability of population, ecosystems, and its environmental services through the adjustment in human and natural systems as a response to projected or actual stimuli. Examples of adjustments include the reduction in the consumption of water to compensate low rainfall indexes, movement of a certain industry in an area with a drought growing trend toward a more humid region, or the modification in the morphology of the urban current to compensate for larger and more frequent flooding. Perhaps the biggest threat of climate change for freshwater aquifers is the interaction between those relatively "traditional" problems, like excessive extraction or habitat fragmentation, and those promoted by climate alterations, like the most frequent droughts (Matthews and Le Quesne 2009). In this sense, the vulnerability analysis to climate-associated problems is the first step toward triggering adaptation processes through the identification of:

- Who or what is vulnerable?
- What makes it vulnerable?
- Where in the territory is that vulnerability expressed?

Thus, in order to reduce climate change vulnerability through adaptation actions, basin integral management must be the approach that transcends Mexican management public policies to maintain ecosystemic services and reach its co-responsible compensation under participation and governance schemes. Under this principles, water management should be based upon instruments as regulated areas and their corresponding regulations and water reserves (Arts. 38, 39, and 41 from LAN and 73–76, 78, and 79 from LAN Regulation), which permit the establishment of administrative actions to those who do not respect the guidelines for a sustainable management of the resource. In this manner, authority acts can be executed to promote and monitor the application of adaptation measures to climate change in water management and its availability in hydrological basins.

14.4.1 Vulnerability Analysis of Populations, Economic Activities, Infrastructure, and Natural Capital to Changes in Water Availability

As a result of its physiographic, bioclimatic, and social characteristics, Mexico is a country particularly vulnerable to climate change. Mexican territory is distributed in the interface between Neartic and Neotropical ecozones; it has direct influence from two oceans, the Gulf of Mexico and the Pacific Ocean, with 11.122 km of coastline and 150 municipalities with coastal front, which constitutes approximately 21% of the continental surface in the country. It also presents four out of five main types of climate, dry (type B), humid-temperate (type C), humid-cold (type D), and polar (E); thus heterogeneous climatic conditions have been developed which can be translated into high endemism and ecosystem particularities that could be sensitive to changes in the hydrological response associated to climate change. Regarding human population, in 1950, a little less of 43% of the population in Mexico lived in urban areas, in 1990 it was 71%, and in 2010 this figure increased to almost 78% (INEGI 2010). This concentration in urban centers trend has been joined with a significant services demand, especially in water supply for urban, agricultural, and industrial public use. In the context of climate uncertainty, degradations of soils and forests, and high social inequity, the relation between supply and demand of the water resource is seriously compromised. Therefore, it is essential to identify differential vulnerabilities in the territory for the design of targeting and prioritization strategies of actions to safeguarding in time and space, the offer of the water resource from ecosystems.

The analysis of climate change vulnerability identifies (i) the main problems in the territory derived from climate, (ii) objects that will or potentially will be affected by these problems, (iii) intrinsic (of the object) and extrinsic (of the territory) characteristics that increase or reduce the impact of previously identified problems, and (iv) the relevant unit of analysis according to the problem and the vulnerable object.

In this sense, the analysis of the vulnerability in a context between supply and demand of environmental services must locate in space and identify the relations of the landscape elements that intervene in the functional relations emerged from these services, as well as the consumption strategies of the demanding objects. In this context, vulnerability to climate change is understood as "the degree of susceptibility or inability of a system to face the adverse effects of climate change and, especially, to climate variability and extreme phenomena. Vulnerability will depend on the exposure of a system to the character, magnitude and speed of climate change, and its sensitivity and adapting capabilities. Thus, vulnerability is a function of three components" (IPCC 2007).

$$Vulnerability = f\,(Exposition + Sensitivity - Adaptive\ Capacity).$$

Exposition: Refers to the type and degree (or nature), to which a system is exposed as a result of significant climate variations (climatic hazards or climatic stressors are considered) (IPCC 2007).

Sensitivity: Degree in which a system may become affected by a particular problem associated to climate variability or climate change. The effects can be direct (i.e., change in crops yield in response to rainfall variations) or indirect (i.e., potential impacts caused by a greater frequency of coastal flooding's) (IPCC 2007).

Adaptive capacity: Set of capacities, resources, and institutions of a country or region that will allow the implementation of effective adaptation measures for reducing impacts resulted from a particular problem associated to climate variability or climate change (IPCC 2007).

Based on this conceptual and operational framework of the vulnerability evaluation, operational variables of the exposed component should be related to climate variables that affect the interrelation of biotic and abiotic factors of the landscape involved in the supply of water environmental services, both in flooding regulation as in the provision of surface water, as an example. Sensitivity variables must be designed to explain why a climate event turns into a problem, both in the vulnerable object level and in the territory where this object is developing, for instance, an increase in urban population and the condition of primary vegetation in the supply basins, respectively, if the identified problem is water availability for public urban consumption.

In this sense, the evaluation of the vulnerability is the first step toward the design and approach of actions to reduce current and potential impacts of climate change; it is directly framed in the monitoring and evaluation processes of the implemented actions by different levels of government and of the civil society organized through the guiding question: the implemented adaptation actions are efficient in diminishing the vulnerability of an object and of a specific problem related to climate. One of the main challenges is to incorporate in the regulation of the territorial planning in Mexico the results of the vulnerability analysis under an environmental services supply and demand approach, in the current context and under climatic uncertainty scenarios.

14.5 Targeting Actions to Reduce Vulnerability to Changes in Water Availability: Examples

One of the distinctive elements in our country is the wide diversity in social, cultural, and environmental, and in terms of the vulnerability degree to climate change, such diversity and special characteristics of the territory and basins are interrelated to a trend in the demand of natural resources and its management, especially the water resource. All this has an influence on the risk of suffering damages or not satisfying the demand from the reduction of supply due to changes in rainfall and/or temperature or as a result of an increase in the water resource demand or a combination of both.

Thus, one of the main challenges for Mexico is to build the capacities to prevent and react in an anticipate manner to water scarcity or abundance and in particular before drought and flooding scenarios. These capacities, which are an element of vulnerability, must be constructed in different spatial and temporal spheres and emerge from the design and implementation of normative, planning, and prioritization instruments, some of which are mentioned in this chapter, as well as an active participation of social stakeholders in the basin, understanding their functional integration from the watershed to the lake or the sea.

There are cases of targeting of actions that have a water resources supply and demand approach and its relation and anticipation of scenarios of rainfall and temperature decrease or increase, and which may generate a participative methodological basis to promote, from scientific information and watershed management; adaptations to climate change to reduce the vulnerability of the water resource toward climate change.

14.5.1 *Instruments for Action Planning and Prioritization to Reduce the Vulnerability to Climate Change*

Main instruments are specifically included in the General Law for Climate Change, issued on June 6, 2012, and initiated with the creation of the National System for Climate Change.

14.5.1.1 National System for Climate Change

The National System for Climate Change is created from the General Law for Climate Change to act as a permanent mechanism of concurrency, communication, collaboration, coordination, and agreement about the national policy for climate change and to promote its transversal application in the short, medium, and long term among the authorities of the three levels of government, in their respective competence scope, coordinating the federation, state, and municipal efforts for conducting actions toward the adaptation, mitigation, and reduction of vulnerability, to cope with the adverse effects of climate change, promoting concurrence, connection, and coherence of programs, actions, and investments of the federal government, states, and municipalities, with the National Strategy and the Special Program on Climate Change (LGCC 2012). The System is formed by:

- Interministerial Commission on Climate Change (CICC), integrated by 14 Federal State Secretariats
- Climate Change Council, as a permanent CICC consulting entity, integrated by members with recognized merits and experience in climate change from social, private, and academic sectors
- National Institute of Ecology and Climate Change (INECC), agency that coordinates the Climate Change policy
- Congress
- States and Municipalities

This National System for Climate Change will use public policy instruments especially:

- Planning: National Strategy on Climate Change, Special Program on Climate Change that will integrate the National Policy on Climate Change, as well as State Programs on Climate Change that will side with the National Policy
- Financial: Emissions Trading and the Climate Change Fund
- Informative and regulating: the inventory of national and state greenhouse gases effect emissions, National Registry of Emissions, Information System on Climate Change, and Mexican Official Norms

14.5.1.2 Information System on Climate Change (SICC)

It is one of the information instruments which integrate the National System on Climate Change,[14] and it is in charge of the National Institute of Statistics and Geography, in accordance with the provision of the Law of the National System of Information, Statistics and Geography, that must generate, with the support of governmental agencies, a set of key indicators that will address at least the following issues as stated in Article 77 of the General Law on Climate Change:

I. Emissions from the national inventory, state inventory, and registry
II. Projects for the reduction of emission of the Registry or those who participate in the agreements of which Mexico is a party
III. Atmospheric conditions of the national territory, weather forecasts in the short-term and long-term projections and characterization of climate variability
IV. Vulnerability of human settlements, infrastructure, islands, coastal areas and river deltas, economic activities, and environmental effects, attributable to climate change
V. Average elevation of sea level
VI. Estimation of costs, in a given year, attributable to climate change that will be included in the Ecological Net Domestic Product (PINE)
VII. Soil quality, including carbon content
VIII. Protection, adaptation, and management of biodiversity

14.5.1.3 Atlas of Water Vulnerability to Climate Change

It is a key instrument for planning, prioritization of actions, and development of public policies for reducing the water resource vulnerability in Mexico toward climate scenarios. This Atlas was created by the Mexican Institute of Water Technology (IMTA) in 2010 and was updated in 2015, considering social, agriculture, and hydrometeorology sectors (quality of water, superficial drain off, and extreme events) (Arreguín Cortés 2015).

Particularly, the Atlas of Water Vulnerability to Climate Change emphasizes the relevance of sustainable management of aquifers, preventing its overexploitation and promoting its natural or induced recharge, since it could be a strategic reserve for the improvement of the degree of adaptability during drought. And as a part of the Atlas results, in the vulnerability issue of water resource, the challenge on the current generation is stressed to ensure that adaptation proposals are environmentally sustainable – with due respect to the ecologic expenditure in rivers and preventing overexploitation of aquifers – since an overexploited aquifer may generate sanitation problems to society and a possible collapse in economic activities (Rivas 2015).[15]

[14]SICC is available online at http://gaia.inegi.org.mx/sicc2015/

[15]The Atlas, in its updated 2015 version, is available at https://www.imta.gob.mx/biblioteca/libros_html/atlas-2016/

14.5.1.4 System of Early Warning of Hydrometeorological Risks

Climate change in Mexico has been accompanied by human loses and high economic and social costs. Just between 2001 and 2013, those affected by meteorological phenomena nearly reached 2.5 million people, and the economic costs amounted 338.35 million pesos (INECC 2015).

These systems are meant to protect the population and mitigate damages caused by hydrometeorological phenomena. An effectiveness improvement in the application of these systems toward the greatest and most intense incidence of hydrometeorological phenomena resulted from a modification in the global and regional climate system is one of the main challenges that Mexico faces, a highly vulnerable country because of its geographic location and social vulnerability.

Improvement in the basin functionality from integral management increases the resilience to suffering damages by reducing downpours, improving infiltration processes and generating participation processes to respond to risks and adapt to them. This functionality improvement with a work per basin along the implementation of early warning systems against hydrometeorological risks will allow synergistic results to prevent damage and to adapt in an effective manner from the social, ecosystems, and protection of productive systems and infrastructure.

These systems are composed of four components (CENAPRED 2017) that should be integrated in a coordinated way for the functioning of the system: (i) prior knowledge and identification of meteorological phenomena associated risks, to take measures in the preparation and self-protection, (ii) measurement and monitoring system to conduct forecasts or science-based risk warnings, and (iii) plans of response or contingency to know what to do against the impact of perturbing phenomena and public alert issuing with clear and precise information. In Mexico, there are early warning systems for tropical cyclones and tsunamis, from the National Meteorological Service, which is an agency of the National Water Commission (CONAGUA) that also operates a Drought Monitor.

The **Drought Monitor** acquired its national character by issuing fortnight drought maps, since February 2014. This important early warning system is based in obtaining and interpreting diverse drought indexes or indicators as the Standard Precipitation Index (SPI) which quantifies the precipitation deficit or surplus conditions (30, 90, 180, 365 days), Rain Anomaly in Percent of Normal (30, 90, 180, 365 days), Satellite Vegetation Health Index (VHI) which measures the degree of stress in vegetation through observed radiance, Leaky Bucket Soil Moisture Model CPC-NOAA which estimates soil moisture through a one layer hydrological model, Normalized Difference Vegetation Index (NDVI), Mean Temperature Anomaly, and Water Availability Percentage in dams in the country and input from local experts. This indexes are deployed in layers through a geographic information system (GIS), and by consensus, drought-affected regions are determined according to a scale of intensities ranging from abnormally dry (D0), moderate drought (D1), severe drought (D2), extreme drought (D3), up to exceptional drought (D4) (SMN 2017).[16]

In Mexico, the development of effective flooding early warning systems associated with Flooding Risks Atlas and in simulators through rainfall monitoring and simulators of hydrologic behavior in the basins is essential. Likewise, it is very important to strengthen the Early Warning Systems on the local level, in order to reduce vulnerability against hydrometeorological risks.

14.5.2 Initiatives and Actions to Reduce the Vulnerability to Climate Change

Strategies of adaptation to climate change based on ecosystems have been developed in Mexico as pilots in order to develop and refine methodological schemes of measurement and implementation that allow the development of synergistic strategies of preservation and restoration actions, associated to climate change mitigation actions, and that permit the strengthening of the national policy of climate change adapting in Mexico. Consider some cases.

14.5.2.1 Interinstitutional Platform for the Attention of Coastal Wetlands[17]

One of the most important ecosystems in Mexico is the coastal wetlands; they are transition ecosystems between terrestrial and aquatic environments and constitute transitory or permanent flooding areas. Their environmental services include, among others, being the support of a characteristic biodiversity, reproduction and feeding site of numerous species of ecological and fishing interest, contribution to water supply, carbon catchment, and absorbing the impact of extreme hydrometeorological events, all of which give them a growing social and economic value now and in the context of climate change. The biological richness and strategic environmental services that wetlands provide place them among the most significant natural resources of more than 11,000 kilometers of coastline in the country.

As part of a conservation, restoration, and sustainable management strategy of coastal wetlands in Mexico, a Digital Platform of Institutional Coordination was designed to access, visualize, and analyze information from the federal government agencies on the subject of coastal wetlands. The objective is to establish a coordination of efforts that

---

[16]Information of Drought Monitor is available at http://smn.cna.gob.mx/es/climatologia/monitor-de-sequia/monitor-de-sequia-en-mexico

[17]INECC-CONAGUA. Priorización de cuencas hidrológicas para la atención de humedales costeros. Coordinación interinstitucional para la Atención de Humedales Costeros. Disponible en: http://sigagis.conagua.gob.mx/atencion_humedales/

converge in time and space the actions from diverse government institutions, academic institutions, and civil society organizations that are committed to the preservation of wetlands under the approach of basin integral management.

INECC along with CONAGUA conducted a prioritization to geographically locate potential areas of joint intervention, to coordinate efforts of institutions with attributions for the attention of coastal wetlands. A territorial vision was considered when incorporating the cartographic base of hydrological basins as prioritization unit. Prioritization considers three main criteria:

i. Ecological importance of coastal wetlands and the anthropic pressure on them
ii. Exposition to current extreme events and vulnerability to climate change
iii. Current environmental policy instruments on coastal wetlands

This methodology makes it possible to conduct different basin prioritizations depending on the objective and the institutional and financial natural resources for the attention of wetlands included. For instance, wetlands with many instruments of attention may be considered as a priority in areas of high exposition to climate events, or of high ecological value and need to reinforce the implementation of conservation policy instruments.

Results of this prioritization identify basins with coastal wetlands of high ecological value, with high exposition to tropical cyclones, with those municipalities most vulnerable to climate change and where federal efforts meet for its management. The complementarity of rehabilitation activities of forest masses in high grounds, sustainable productive activities in the middle grounds, and wetland protection in low grounds of basins ensure the supply of wetlands environmental services, essential for the vulnerability reduction through mitigation promotion, triggering of adapting processes against climate change and preserving biodiversity.

Of the 304 hydrological basins with coastal wetlands, 46 present priority category according to the considered criteria, 19 correspond to the Gulf of California region, 11 to the Gulf of Mexico, 5 to Northeast Pacific, 10 to Tropical Pacific, and 1 to Caribbean Sea region.

14.5.2.2 Action Plans for Basin Management (PAMIC)

One of the main challenges in territorial planning is the implementation of instruments that incorporate the territorial dynamics of hydrographical basins. As a response to this need, PAMICs were created; their objective is to promote the functional connectivity of the territory by providing information to focus conservation programs for important areas of water environmental services. Currently, it is used to identify better areas for the payment of environmental services of CONAFOR, in the implementation of preservation actions the National Commission of Protected Natural Areas (CONANP) and diverse projects with an environmental approach operated by regional funds.

PAMICs are an instrument of territorial planning designed by the Directorate of Hydrological Environmental Services and Adaptation of the Institute of Ecology and

Climate Change (INECC). Its objective is to promote the functional connectivity of hydrographical basins through the identification of territorial relations between supply and demand areas of SAH. Allowing, through cartographic products, to focus on actions for the conservation, rehabilitation, and sustainable use of high potential SAH areas and which are hydrographically related to areas with a high demand for those services. The instrument has three components; one analytical, one relational, and another participative. The first one corresponds to the geographic modeling and identification of high potential areas for the supply of superficial waters and preservation of soils, that is, supply priority areas of SAH. The relational component identifies water uses and volumes and licenses and determines the hydrographic relation between supply and demand areas of SAH for the identification of priority intervention areas. The participative component covers the local traditional knowledge, the community and institutional capacities for implementing actions focused on the conservation of natural capital, and the adequacy of productive practices in priority supply areas of SAH.

In the context of climate uncertainty, the proposal of implementing conservation actions in PAMICs prioritizes areas with a high supply of SAH and with a potential change in the vegetation structure considering the projections of climate change, contributing in the reduction of the population vulnerability and productive activities of basins.

Up to date, in the "Conservation of Coastal Basins in the Context of Climate Change" (C6) framework, the PAMICs that have been developed on the coastal basins are: Tuxpan, Jamapa, and Antigua Rivers, in the state of Veracruz; the basin systems that flow into Bahia de Banderas, in Jalisco; and San Pedro and Baluarte River basins, in Nayarit and Sinaloa.

C6 operates with a grant from Global Environmental Facility (GEF) managed by the World Bank. Three public institutions (the Institute of Ecology and Climate Change, CONANP, CONAFOR) and a private entity, the Mexican Fund for the Preservation of Nature (FMCN), collaborate. The architecture of the project coordination constitutes an innovative component that seeks the initiation of collaboration processes and synergies in the territory for promoting the integrated management of coastal basins, preserving their biodiversity, triggering adaptation processes, and contributing in the mitigation of climate change. All the actions in the C6 framework seek to contribute in the recovery of basins functionality and the maintenance of ecosystem services of regulation and provision.

14.5.2.3 National Program Against Drought (PRONACOSE)

PRONACOSE, coordinated by CONAGUA, emerges at the beginning of the administration in 2014 as a planning or programming scheme of actions or preventive and mitigation measures against droughts, which, as mentioned before, began its fortnight monitoring through the National Meteorological Service of the National Water Commission (SMN-CONAGUA) since February 2014. These measures or actions inserted in programs according to basin councils, or cities, or metropolitan areas establish a pathway toward a severe drought program. There are already programs for Preventive Measures and Drought Mitigation (PMPMS) in 26 basin councils as well as in 13 major cities or metropolitan areas. Consistent with CONAGUA (2015), these programs seek to:

- Guarantee water availability required to ensure health and life of the population: public, domestic, urban, and rural supply
- Prevent or minimize the negative effects of drought on the environment, especially on ecological flow regimes
- Minimize negative effects on economic activities, according to prioritization of uses established in the water legislation and in water programs

This program necessarily links the measures established in the PMPMS, in basin approach and with the analysis of supply and demand of hydrological environmental services, in the face of scenarios of low water availability due to meteorological drought seeking to avoid socioeconomic drought.

14.5.2.4 National Program of Water Reserves (PNRA)

LAN establishes that for the ecological protection, which includes preservation or restoration of vital ecosystems, and for the preservation of wetlands, the legal solution is the declaration or decree of total or partial reserves of national waters as ecological reserves. Article 41 of that Law establishes that the Federal Executive may declare by decree total or partial reserve of national waters in order to "guarantee minimal flows for the ecological protection, including preservation or restoration of vital ecosystems."

In order to promote water reserves, the National Program of Water Reserves (PNRA) was created, a powerful private-public initiative[18] for strengthening water management for the conservation and adaptation to climate change (BID 2015), based upon the model developed by Alliance WWF-Fundación Gonzalo Rio Arronte, I.A.P. (FGRA) and other partners since 2005 and whose objectives are:

- Establishing a national system of water reserves
- Demonstrating its benefits as an instrument, guarantor of the functionality of the water cycle and its environmental services
- Strengthening the application of the Mexican norm of ecological flow in the country

CONAGUA and the Alliance WWF-FGRA led the study to establish the feasibility of hydrological basins in Mexico, in order to implement water reserves. Results show that 189 basins are feasible to be decreed if the necessary studies are completed, and there would be water available for the environmental protection of

[18]The main partners of PNRA are the National Water Commission, Binational Commission of Protected Natural Areas, WWF Mexico, FGRA, and Inter-American Development Bank.

55 Ramsar sites (41% of the total) and for 97 ANP (50% of the total) (Conagua 2011; González Mora et al. 2014).

14.6 Conclusion

Integrate in an effective manner, the water resources supply and demand approach in the context of climate change in the governance, coordination and participatory spaces of the water and climate change policy, is an unprecedented challenge for the operation of the Mexican environmental normativity. The basic legal and normative framework exists but the interaction and functionality of attributions and the institutions acting in the complex systems that basins mean, not yet. As it is reaffirmed in this chapter, it is an urgent need to make it from a socioecosystemic perspective based on the approach of integral basin management that will translate into a systematic reduction in the vulnerability against impacts, both in human and in natural systems.

However, facing the problem of water resources in the future, with the principles, approaches, and tools of public policy discussed in this chapter in an interdisciplinary environment, intersectoral coordination, with the formation of the required capacities, will allow Mexico to fulfill non-conditioned commitments in climate change adaptation subscribed in the Paris Agreement. Solutions imply a paradigm shift in the design and implementation of public policies integrating a common vision of the territory for all sectors, stakeholders, and levels, as well as the strengthening of national, local, and basin of Mexico inhabitants' capacities.

References

Arreguín Cortés, Felipe, I Coord. (2015), *Atlas de vulnerabilidad hídrica en México ante el cambio climático*, Felipe I. Arreguín Cortés, Mario López Pérez, Olivia Rodríguez López y Martín José Montero Martínez, coordinadores Jiutepec, Mor., Instituto Mexicano de Tecnología del Agua.

Azuela de la Cueva, A. (2006a). Distancias y disonancias del ordenamiento territorial en la legislación mexicana. En Javier Delgadillo (Compilador), Política territorial en México, hacia un modelo de desarrollo basado en el territorio (467–492). SEDESOL y UNAM.

Azuela de la Cueva, A. (2006b). El ordenamiento ecológico del territorio en México: génesis y perspectivas. SEMARNAT.

Banco Interamericano de Desarrollo (BID). (2015), *Programa Nacional de Reservas de Agua en México: experiencias de caudal ecológico y la asignación de agua al ambiente.* BID, WWF México.

Biestroek, R., Swart, R., & van der Knapp, W. (2009). The mitigation-adaptation dichotomy and the role of spatial planning. *Habitat International, 33*, 230–237.

Brañes, R. (2004). *Manual de derecho ambiental mexicano.* Fondo de Cultura Económica y Fundación Mexicana para la Educación Ambiental.

Calderón, M. (2001). *Lázaro Cárdenas del Río en la cuenca del Tepalcatepec-Balsas. En La Tierra Caliente de Michoacán*, Coordinador editorial. José Eduardo Zárate. El Colegio de Michoacán y el Gobierno del Estado de Michoacán. Zamora, Michoacán, pp. 233–264.

Centro Nacional de Prevención de Desastre (CENAPRED). (2017). Sistemas de Alerta Temprana, Contenido de curso de la Escuela Nacional de Protección Civil. En: http://www.cenapred.gob.mx/es/documentosWeb/Enaproc/curso_Alerta_Temprana.pdf

Comisión Nacional del Agua, CONAGUA. (2011). *Identificación de reservas potenciales de agua para el medio ambiente en México* (85 pp). México: Semarnat/Conagua.

Comisión Nacional del Agua, CONAGUA. (2015). Programa Nacional de Combate a la Sequía (PRONACOSE). https://www.gob.mx/conagua/acciones-y-programas/programa-nacional-contra-la-sequia-pronacose-programas-de-medidas-preventivas-y-de-mitigacion-a-la-sequia-pmpms-para-ciudades Consultada el 20 de noviembre de 2017.

Comisión Nacional del Agua, CONAGUA. (2016). *Estadísticas del Agua en México. Edición 2016*, Capítulo 2. Situación de los recursos hídricos, página 31, Capítulo 3. Usos del agua, 70, 71 Capítulo 8. Agua en el mundo, página 195.

Conde, C. (2016). Conferencia: Bases conceptuales y legales de cambio climático y adaptación en México, 21 de septiembre de 2016. *Ciclo de videoconferencias: Manejo de cuencas y cambio climático en México*, Cecadesu/Semarnat/Red Mexicana de Cuencas Hidrográficas.

Cotler, H., Galindo, A., González, I. D., Pineda, R. F. y Ríos, E. (2013). *Cuencas hidrográficas. Fundamentos y perspectivas para su manejo y gestión.* Cuadernos de Divulgación Ambiental (32 pp). México: Cecadesu/Semarnat y Red Mexicana de Cuencas Hidrográficas (Remexcu).

Gasca Zamora, José. (2009). Geografía Regional, La región, la regionalización y el desarrollo regional en México. México: Instituto de Geografía, UNAM.

Global Water Partnership, GWP. (2017). *What is water security.* http://www.gwp.org/en/The-Challenge/What-is-water-security/. Consultado el 20 de noviembre de 2017.

González Mora, I. D., Salinas Rodríguez, S. A., Guerra Gilbert, A., Sánchez Navarro, R. y Ríos Patrón, E. (2014). *Ríos libres y vivos. Introducción al caudal ecológico y reservas de agua.* CECADESU-SEMARNAT, Alianza WWF México y Fundación Gonzalo Río Arronte, I.A.P. y Remexcu.

INEGI. (2010). *Censo de Población y Vivienda 2010.* México.

INEGI-INE-CONAGUA. (2007). Mapa de cuencas hídricas de México (escala 1: 250 000). En: *Cuencas Hídricas de México, escala 1:250,000.* INE 2003, autores Priego A.G., Isunza E., Luna N. y Pérez J.L.

Instituto Nacional de Ecología y Cambio Climático, INECC. (2015). *Compromisos de mitigación y adaptación ante el cambio climático para el periodo 2020–2030.* Gobierno de la República, En: http://www.inecc.gob.mx/descargas/adaptacion/2015_indc_esp.pdf

Intergovernmental Panel on Climate Change, IPCC. (2007). Summary for policymakers. In M. L. Parry, O. F. Canziani, J. P. Palutikof, P. J. van der Linden, & C. E. Hanson (Eds.), *Climate change 2007: Impacts, adaptation and vulnerability. Contribution of Working Group II to the Fourth Assessment Report of the Intergovernmental Panel on Climate Change* (pp. 7–22). Cambridge: Cambridge University Press.

Intergovernmental Panel on Climate Change, IPCC. (2013). *Fifth assessment report (AR5). WGI The physical science basis.*

Lampis, A. (2013). La adaptación al cambio climático: el reto de las dobles agendas. Postigo.

Ley General de Cambio Climático (LGCC), 6 de junio de. (2012). Capítulos V artículos 38 al 97. Cámara de Diputados del H. Congreso de la Unión. México. En: http://www.diputados.gob.mx/LeyesBiblio/ref/lgcc.htm

LGEEPA. (1998). Ley General de Equilibrio Ecológico y Protección al Ambiente 2017. Cámara de Diputados del H. Congreso de la Unión. México. Disponible en: http://www.diputados.gob.mx/LeyesBiblio/pdf/148_240117.pdf Fecha de consulta: 31 de octubre de 2017.

Matthews, J. L. y Le Quesne, T. (2009). *Cómo adaptar la gestión hídrica. Guía para hacer frente al cambio climático*, Serie Seguridad Hídrica de WWF, Vol. 3. Alianza WWF-Fundación Gonzalo Río Arronte, I.A.P. WWF México.

Naciones Unidas/Convención Marco sobre el Cambio Climático. (2015). Aprobación del Acuerdo de París. Disponible en. http://unfccc.int/resource/docs/2015/cop21/spa/l09s.pdf

Organización Meteorológica Mundial, OMM. (2017). *El año 2017 va camino de ser uno de los tres años más cálidos, con registros de tiempo extremo sin precedentes.* 6 de noviembre de 2017. Boletín de prensa número: 06112017 En: https://public.wmo.int/es/media/comunicados-de-prensa/el-a%C3%B1o-2017-va-camino-de-ser-uno-de-los-tres-a%C3%B1os-m%C3%A1s-c%C3%A1lidos-con

Quintana, J. (2009). *Derecho ambiental mexicano, lineamientos generales.* Distrito Federal: Porrúa.

Ríos Patrón, E., González Mora, D., González Terrazas, D., Merino Pérez, M., Mokondoko Delgadillo, P., & Ortiz Paniagua, F. (2016). *Adaptación al cambio climático, fundamentos desde el manejo de cuencas.* CECADESU-SEMARNAT, Remexcu, INECC y Alianza WWF México y Fundación Gonzalo Río Arronte, I.A.P.

Rivas, I. (2015). *Efectos del cambio climático en el recurso hídrico de México (agua superficial).* En: *Atlas de vulnerabilidad hídrica en México ante el cambio Climático*/Felipe I. Arreguín Cortés, Mario López Pérez, Olivia Rodríguez López y Martín José Montero Martínez, coordinadores, Jiutepec, Mor., Instituto Mexicano del Tecnología del Agua.

Rosete, F. (2006). *Semblanza histórica del ordenamiento ecológico territorial, perspectiva institucional: México*, Instituto Nacional de Ecología, SEMARNAT.

Sanjurjo Rivera, E., González Mora, I. D., Ortiz Paniagua, C. F., & Ríos Patrón, E. (2017). *Economía y medio ambiente. Reflexiones desde el manejo de cuencas.* CECADESU-SEMARNAT, Red Mexicana de Cuencas Hidrográficas, Red de Socioecosistemas y Sustentabilidad, Alianza WWF-Fundación Gonzalo Río Arronte, I.A.P.

Secretaría de Medio Ambiente y Recursos Naturales (SEMARNAT). (2013). Programa Sectorial de Medio Ambiente y Recursos Naturales (PROMARNAT) 2013–2018, Gobierno de la República. México, 138 p.

Secretaría de Medio Ambiente y Recursos Naturales/Instituto Nacional de Ecología y Cambio Climático (SEMARNAT-INECC). (2013). Estrategia Nacional de Cambio Climático. Visión 10-20-40 Gobierno de la República, 60 pp.

Servicio Meteorológico Nacional (SMN). (2017). Consulta de la página: http://smn.cna.gob.mx/es/climatologia/monitor-de-sequia/monitor-de-sequia-en-mexico. Consulta el 21 de noviembre de 2017.

WWAP (Programa Mundial de Evaluación de los Recursos Hídricos de las Naciones Unidas). (2017). Informe Mundial de las Naciones Unidas sobre el Desarrollo de los Recursos Hídricos 2017. Aguas residuales: El recurso desaprovechado. París: UNESCO.

Eduardo Ríos Patrón is co-coordinator since 2011 of the Mexican Network of Hydrographic Basins. He has participated in the Organizing and Scientific Committee of the III and IV National Congress of Watershed Management. He is a specialist in watershed planning and management. He has extensive experience in the implementation of environmental policy, in the promotion of coordination processes as Head of the Planning and Environmental Policy Unit, and in the Ministry of Environment and Natural Resources, Delegation in Michoacán. He is co-author of publications on issues of integral management of watersheds. He teaches Public Policy and Environmental Legislation at the National School of Higher Studies of the National Autonomous University of Mexico (UNAM), Morelia campus. He holds a Bachelor's degree and a Master's Degree in Administration from the Universidad Anáhuac del Sur and a Master's Degree in Integrated Watershed Management from the Autonomous University of Querétaro (UAQ). He is a member of the Socioecosystems and Sustainability Network (RedSocioecoS).

Daniel Iura González Terrazas earned a Master of Science with a specialization in Natural Resources Management from the International Institute of Earth Observation and Geo-information (ITC), University of Twente, Netherlands, and a Master in Geography from the Research Center in Environmental Geography. Also a biologist graduated from the Faculty of Sciences of the National Autonomous University of Mexico (UNAM). His research areas include the analysis of supply and demand of environmental services for the processes of adaptation to climate change in the framework of watersheds, incorporation of the basin approach, and climate change in the instruments of territorial planning. He is a specialist in support systems for environmental decisions and geographic information. Since 2012, he is attached to the National Institute of Ecology; currently, he heads the Hydrological Environmental Services Analysis Department of the General Coordination of Adaptation to Climate Change. Among the current projects under his charge include the design of the inter-institutional platform for attention to coastal wetlands, the PAMICs watershed management action plans, and the coordination of the National Atlas of Vulnerability to Climate Change by INECC. He is a member of the Mexican Network of Hydrographic Basins.

Ignacio Daniel González Mora is co-coordinator since 2011 of the Mexican Network of Hydrographic Basins. His academic training was carried out at the UNAM: biologist at the Faculty of Sciences and with subsequent studies of specialization in Applied Statistics at the Research Institute in Applied Mathematics and Systems. He has worked in the National Water Commission (CONAGUA) responsible for the planning and coordination of the National Water Quality Monitoring Network, where he promoted biological and toxicological monitoring and proposed a new system of environmental indicators of water quality. In 2005, he joined the technical team of the World Wide Fund for Nature (WWF). In the development of the WWF-Fundación Gonzalo Río Arronte IAP, he has been responsible for 11 years of the management of the Copalita-Zimatán-Huatulco, Oax. Basins, and coordinator of the interdisciplinary team that determined the ecological flow in 10 representative sites of these basins. Since then he has participated in both the Advisory Council of the Huatulco National Park (CONANP) and the Basin Council of the Coast of Oaxaca and in his auxiliary body the Basin Committee of the Copalita and Tonameca rivers, being a member of the organizations of the civil society. He has participated in projects in other Oaxacan basins, such as Río Verde Atoyac and Tehuantepec. He has collaborated in the Organizing Committees of the National Congresses of Management of Hydrographic Basins III and IV in the capacity building initiatives of CECADESU-SEMARNAT on integrated management of basins. He is a member of the Socio-ecosystems and Sustainability Network (RedsocioecoS).

Chapter 15
Science and Technology for Integrated Water Resources Management in Mexico

Álvaro A. Aldama

Abstract The Global Water Partnership has defined Integrated Water Resources Management (IWRM) as "a process which promotes the coordinated development and management of water, land and related resources in order to maximize economic and social welfare in an equitable manner without compromising the sustainability of vital ecosystems and the environment." To achieve the goals of IWRM, a number of tools must be used, among which science and technology play a crucial role, particularly in river basins where water conflicts are commonplace. Most often, conflict arises between upstream and downstream water users due to the resource scarcity. This has been the case of two of the most strategically important river basins in Mexico: the Lerma-Chapala basin and the Río Bravo/Río Grande basins. In the first of these cases, simulation and optimization models were used to build consensus between water users and to define a set of rules for the integrated operation on the system, according to water availability. In the second case, a number of innovations were introduced to accommodate the volume of water allocated to the USA in the water availability study, the development of a novel approach for monthly water budgets, and a statistical approach that explicitly recognizes trends, cyclical behavior, and randomness in the natural flow regime. These innovations have been adopted by water users in the Rio Bravo basin and have been used to define water policy in the region.

Keywords River basin · Integrated water resources management · Water governance · Science and technology · Innovation · Consensus building · Lerma-Chapala basin · Rio Bravo basin

---

Á. A. Aldama (✉)
Independent Consultant, Jiutepec, Mexico

H. R. Guerrero García Rojas (ed.), *Water Policy in Mexico*, Global Issues in Water Policy 20, https://doi.org/10.1007/978-3-319-76115-2_15

15.1 Introduction

Water resources problems possess a high degree of complexity. They involve engineering, scientific, technological, social, economic, political, public health, and management issues, among others. The complexity of water resources problems has grown as a result of population growth, anthropogenic changes of the environment, global pressure to address development in a sustainable manner, and increased public desire to participate in decision-making. The complexity of water resources problems requires solutions that are beyond the scope of a single discipline (e.g., hydraulic engineering or hydrology). A holistic approach including the concurrent participation of many disciplines is often indispensable.

On the other hand, water is probably the most precious of all natural resources (other than air, but managing air is an almost impossible task!). The Romans were well aware of this, as expressed by Pliny the Elder, the historian of the first century B.C. who wrote the following statement in his celebrated *Natural History* (XXXI,I): "This element commands all the others: waters swallow the earth, suffocate flames; they reach the heights and reclaim the sky for themselves. . .Upon falling, they become the cause of whatever is born from the earth, a marvelous circumstance, given that in order to make cereals grow and trees live, waters travel to the sky and from there they bring plants their vital breath. . . ." It is quite interesting that Pliny not only expressed the preeminence of water among the elements, but he also was acquainted with the concept of the hydrological cycle! We, as "modern" humans, have a lot to learn from ancient wisdom.

In regard to social participation, another pearl of wisdom was expressed by Sima Qian, a second and first century B.C. Chinese historian, who left the following thought in his writing *Shiji* IV: "Silencing the people is equivalent to diking waters. Diked waters overflow and cause many casualties. The same happens to the people. That is why, in the same manner that those who look after the waters should open ditches to let them flow, those who look after the people should establish norms to let them talk." The relevance of the analogy for water management is remarkable. In some way or another, all waters are shared. Most waters are transboundary resources, in the sense that their geographical occurrence in the context of catchments and aquifers does not match municipal, state, provincial, regional, or national boundaries. Water use, particularly in water scarce regions, often generates conflicts. Consensus building is necessary to promote effective governance within catchments and aquifers and to mitigate conflicts between water stakeholders. Consensus building is complicated, on the one hand, by the collective action problem (i.e., as the number of stakeholders increases, it is more difficult to reach an agreement –Olson 1971), and, on the other, by the fact that very often scientific knowledge about the true state of catchments and aquifers and of their probable response to different climatic, water use, and technological scenarios is lacking.

The international water community has responded to the complexity of water resources problems by postulating the concept of *Integrated Water Resources*

Management (IWRM). Several definitions of IWRM have been advanced, but probably the most accepted is the one proposed by the Global Water Partnership (GWP):

> Integrated Water Resources Management is a process which promotes the coordinated development and management of water and related resources, in order to maximize the resultant economic and social welfare in an equitable manner without compromising the sustainability of vital ecosystems. (GWP 2000)

An intimately related concept to that of IWRM is *water governance*. To the effect, it is illustrative to consider the definition of governance offered by the United Nations Development Program (UNDP):

> Governance is the exercise of economic, political and administrative authority to manage a country's affairs at all levels…it comprises the mechanisms, processes and institutions through which citizens and groups articulate their interests, exercise their legal rights, meet their obligations and mediate their differences. (UNDP 1997)

A more specific definition, applicable to water-related issues, has been developed by the GWP and reads:

> Water governance refers to the range of political, social, economic and administrative systems that are in place to develop and manage water resources, and the delivery of water services, at different levels of society. (GWP 2003)

Another concept, which is present in the above definition of IWRM is that of sustainability, which derives from sustainable development. Once again, there are many definitions of sustainable development (and some even consider the concept to be an oxymoron! (e.g., Redcliff 2006)), but the most quoted of all is that of the World Commission on Environment and Development (WCED), which is contained in the so-called Bruntland Report, that is, to say:

> Sustainable development is development that meets the needs of the present without compromising the ability of future generations to meet their own needs. It contains within it two key concepts: the concept of needs, in particular the essential needs of the world's poor, to which overriding priority should be given; and the idea of limitations imposed by the state of technology and social organization on the environment's ability to meet present and future needs. (Brundtland Commission 1987)

One of the problems of dealing with notions such as IWRM, water governance, and sustainability is that they are fuzzy concepts. This explains why there are so many ways to define them. In other words, one could ask the question: When is IWRM, or for that matter water governance or even sustainability, achieved? Without a proper set of indicators or metrics, it is nearly impossible to answer this question. But the problem with the development of indicators to measure the progress toward IWRM is that they must be site-specific. This has a lot to do with the way water is viewed and valued in different countries, regions, states or provinces, and municipalities. It also has to do with the way religions and ethnical groups relate to water. It even has to do with the role that water has played in the history of different peoples. In other words, it has to do with local legislation, culture, and traditions. Culture can be defined as "the transmission from one generation to the next, via teaching and imitation, of knowledge, values and other factors that influence behavior" (Boyd and Richerson 1985). In other words, culture is conformed by the set of values in which a society, community, or group of people believes. Oftentimes one listens to aspirations for the establishment of a "new water culture." Values, and therefore culture, are almost impossible to change. Consequently, in the context of IWRM and water governance, water culture should be considered a "given." This leads to the conclusion that specific strategies and tactics for the achievement of IWRM must be designed and implemented for every site and every case.

In view of the above, one could consider IWRM to be a "syndrome," hopefully a contagious one. In fact, IWRM is in "the eyes of the beholder." Furthermore, IWRM cannot be taught; it must be learned. IWRM should be appropriate, effective, timely, and efficient. Finally, IWRM should incorporate values, principles, strategies, tactics, and lines of action, along with a set of metrics, which are *locally defined and agreed upon.*

15.2 Mexican Water Legal Framework

Article 27 of the Mexican Constitution establishes that, among other natural resources, water is a national asset. In other words, water belongs to the Nation. In addition, it gives the executive branch of government the power of managing this vital resource. The Mexican National Water Law charges the National Water Commission with the management of waters of national property by means of "concessions" for the use of water and "permits" for the discharge of wastewater and the extraction of gravel and sand from natural river channels.

The definition of Nation is *very* elusive. The concept of Nation is related to that of State, but they are not equivalent. The classical definition of State is the one proposed by Weber, which views the State as the "monopoly of the legitimate use of physical force" (Weber, 1946). According to this school of thought, State is the entity that has the power to enforce the rule of law.

Gellner (1983) recognizes that the concept of Nation is more complicated than that of a State. He points out that "having a Nation is not an inherent attribute of humanity, but it has now come to appear as such." He adds: "the State has certainly emerged without the help of the Nation. Some Nations have certainly emerged without the blessings of their own State. It is more debatable whether the normative idea of the Nation, in its modern sense, did not presuppose the prior existence of the State." The same author finally concludes by saying "for two men to be in the same Nation requires two things: a common culture, understandings, meanings etc; and the acknowledgement that the other is a fellow national and the recognition of mutual rights and duties to each other in virtue of shared membership in it." As examples one could mention the Kurds, which constitute a Nation without State, Spain, which is a multinational State, and Mexico, which constitutes a Nation-State (Aldama, Abraham 2017, Personal communication).

It should be clear by now that having water as national property in Mexico is tantamount to saying that everyone and no one owns it. Thus water in Mexico suffers the "tragedy of the commons" (Hardin 1968), i.e., water belongs to all, but no single stakeholder feels responsible for its care.

To make things worse, Mexico has a young and unconsolidated democracy. Although Mexico has held elections for many years, the actual advent of democracy is recent. As has happened in other countries that have become democratic (for instance, Spain), unsatisfied social demands and communal resentments have led to an empowerment of interest groups and a weakening of the State. Thus, governments have often been afraid to enforce legal regulations. Water management has not been the exemption. As will be shown below, this has added to the complexity of implementing IWRM in water basins with acute conflicts. Most interestingly, science and technology have become effective catalysts to build consensus among stakeholders. A couple of IWRM implementation case studies in which the author has been involved are described in the following sections.

15.3 Case Study: The Lerma-Chapala Basin

The Lerma-Chapala basin (Fig. 15.1) is one of the most important regions in Mexico from the economic, political, and social standpoints. The main watercourse in the basin, the Lerma River, is 750 km long. It begins at an altitude of 3000 masl in the central high plateau and ends at an altitude of 1500 masl in the Chapala Lake, the

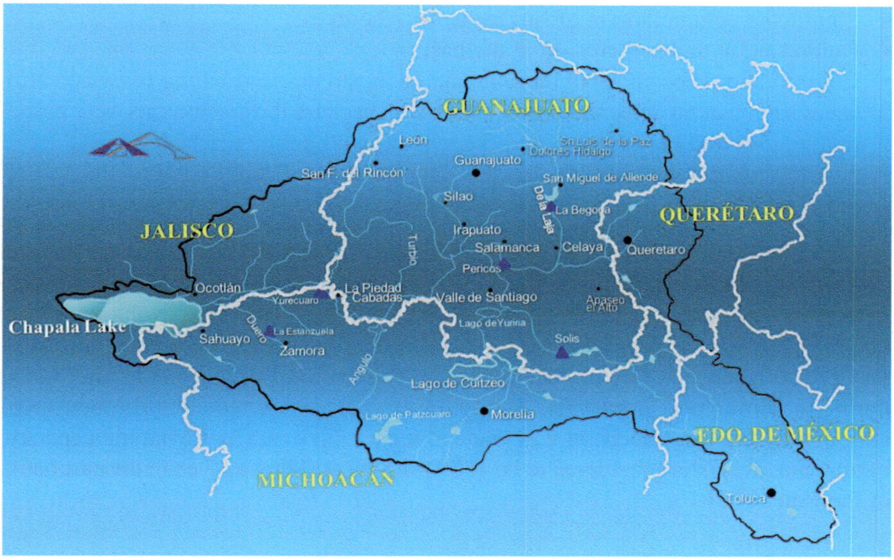

Fig. 15.1 The Lerma-Chapala basin. (Source: Mexican Institute of Water Technology)

largest surface water body in Mexico. The climatology of the basin is such that it can be classified as a subtropical semiarid region. The average temperature is around 21 °C, and its mean annual precipitation is 735 mm. The surface area of the basin is 54,421 km^2, and its mean runoff is 576 km^3. The basin comprises portions of the states of Mexico, Querétaro, Guanajuato, Michoacán, and Jalisco. Even though the basin's surface area represents only 3% of the total surface area in the country, its population represents about 11% of Mexico's population, and its contribution to the nation's GDP is on the order of 9%. Water is withdrawn from aquifers in the Upper Lerma to supply the liquid to more than 2 million inhabitants in Mexico City; the supply to a similar population in Guadalajara, the capital city in the state of Jalisco, comes from the Chapala Lake. Water supply for around 795 thousand ha of irrigated agricultural land, mainly located in the state of Guanajuato, comes from the Lerma River. In addition, close to 10 thousand industries distributed throughout the basin have been increasing their demand for water (Mestre-Rodríguez 2001).

At the beginning of the millennium, a number of challenges for the implementation of IWRM and the improvement of water governance were identified, some of which are listed below:

- Very complex hydrological topology and social fabric
- Increased competition for water and conflicts among users and uses
- Inefficient water use
- Water demand exceeding supply, leading to overexploitation of groundwater resources
- A decrease of 4 m in the level of the Chapala Lake and a loss of 70% of its storage from 1993 to 2003 (Figs. 15.2 and 15.3)
- Intense conflicts during periods of low precipitation
- Significant contamination of surface water resources
- Deforestation and soil erosion in the basin
- Confrontational positions of the Guanajuato (upstream) and Jalisco (downstream) states such that no agreements could be made, because the discussions were based on opinions and not on facts, and consensus could not be reached
- A notorious lack of confidence in the National Water Commission on the part of basin stakeholders

Evidently, the situation was critical. Figure 15.2 shows the decrease in surface area of the Chapala Lake from 1993 to 2003. Actually, the displacement of the Lake's eastern bank was about 10 km. Figure 15.3 shows the time evolution of the volume of water stored in the Lake in 75 years. The second lowest level in recorded history was reached in July of 2002, when the volume went down to 1145 hm^3. Only during the drought of the 1950s slightly lower values were recorded. In fact, soon after that drought, the decision was made to increase the maximum natural capacity of the Lake, which is considered to be 4500 hm^3, by building a levee in the east bank of the water body.

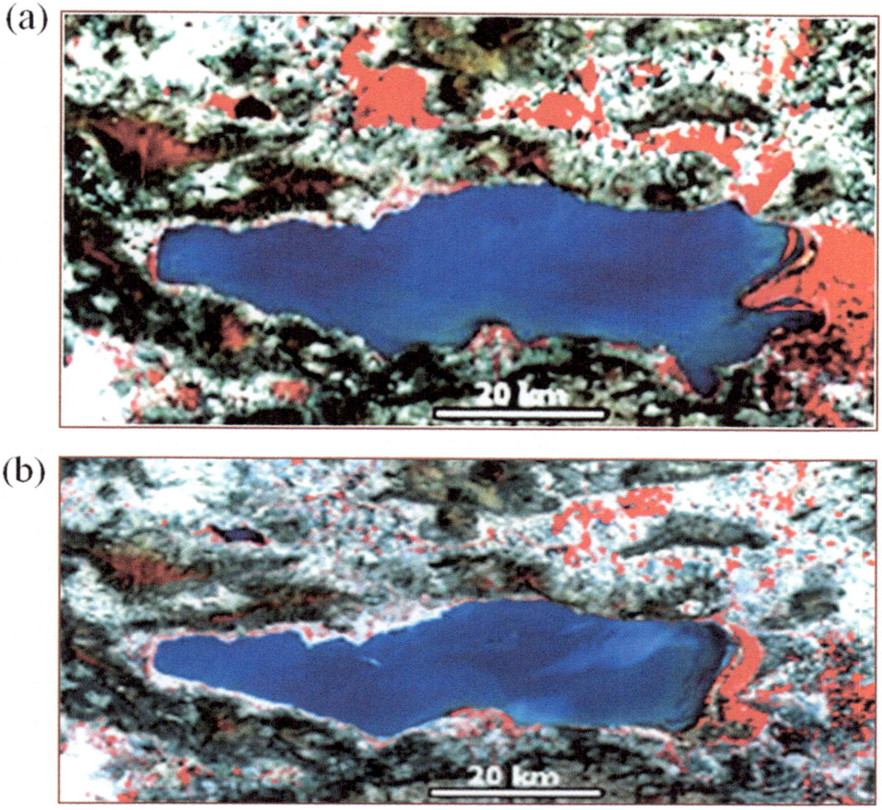

Fig. 15.2 Satellite images of Chapala Lake: (**a**) in 1993 and (**b**) in 2003. (Source: Mexican Institute of Water Technology)

The described situation sounded the alarm in Jalisco and particularly in Guadalajara. It was argued that unless water restrictions were imposed to farmers in Guanajuato, the Lake was indeed in danger of disappearing. Furthermore, water utilities and NGOs in Jalisco demanded that a minimum level for the Lake stored volume should be kept, since there were signs that for very low volumes, water quality was very poor. Farmers in Guanajuato could not care less. They claimed that they had concession titles that gave them the right to exploit certain volumes of water for their crops and even said that it would be more convenient for the Lake to disappear, since most of its volume was lost to evaporation anyway.

Given the described state of matters, the National Water Commission (CNA) retained the services of the Mexican Institute of Water Technology (IMTA, which at that time was headed by the author) to perform a series of scientific studies and to

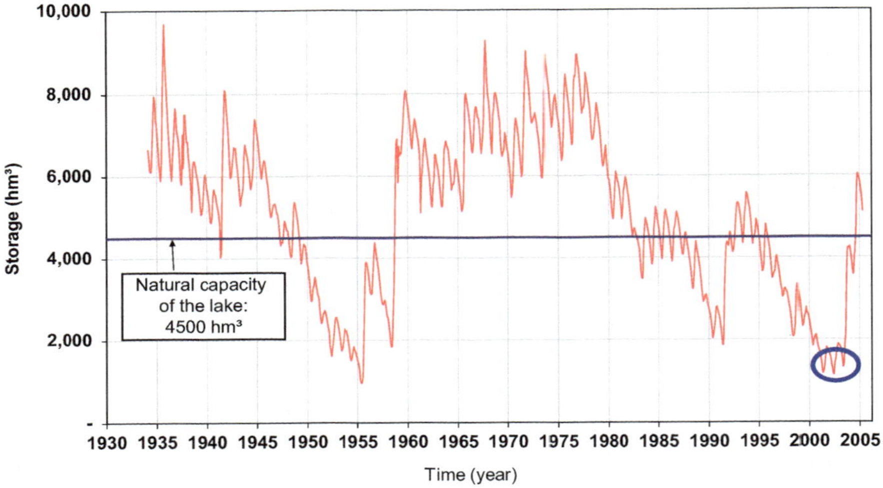

Fig. 15.3 Record of stored volumes in Chapala Lake (the second lowest recorded level is encircled in blue). (Source: Mexican Institute of Water Technology)

develop technological tools in order to contribute to the solution of the basin's complex water-related problems. Thus, the following R and D actions were undertaken:

- Hydrological and hydrometeorological studies of the Lake, with an emphasis on the proper estimation of evaporation and runoff in ungauged catchments, were developed.
- An investigation of the flow and transport processes in the Lerma River and the Chapala Lake was performed.
- An assessment of water pollution in the basin and the Lake was made.
- A regional reforestation program was designed.
- Water consumption, water economics studies, and social conflict studies were performed.
- A system dynamics-based basin-wide simulation model and a genetic algorithm-based optimization model, which includes the river and its tributaries, the existing reservoirs, irrigation units, cities, towns, and industrial complexes, as well as the aquifers and, of course, the Lake, were developed and validated.

The main results of these studies are summarized as follows. It was found that evaporation had been underestimated by 20%. Apparently this conclusion would favor the position adopted by Guanajuato farmers. Nevertheless, through long-term meteorological simulations, by employing the MM5 mesoscale model, it was found that the disappearance of the Lake would mean that the average precipitation in Guanajuato would decrease 30%, which would be catastrophic for the agricultural use of water. It was also found that in order to meet the standards of total dissolved solids and alkalinity for urban water use, the volume stored in the Lake should be no less than 2000 hm^3. The single most significant action was the use of the simulation and optimization models in the consensus building process among stakeholders, as it is described in the following paragraphs.

The simulation model contains hydrological, agro-economical, technological, demographical, social, water quality, and environmental standard modules. It was used to perform scenario assessments with a planning horizon of 50 years, in order to determine the performance of actions on both water allocation (optimal operational rules) and water demand (via an increase of agricultural water use efficiency). The model was also employed to diagnose the current state of water management in the basin. The optimization model was employed to determine an optimal operational policy, whose performance was evaluated through the use of the simulation model.

Each of the five states represented in the basin (Estado de México, Querétaro, Guanajuato, Michoacán, and Jalisco) hired expert consultants to serve as their representatives in a high-level technical group that reported directly to the Basin Council. CNA, as the representative of the Federal Government, coordinated and supervised the overall process. IMTA provided (1) training in the use of the models to the members of the high-level technical group and (2) scientific and technological support and advice.

Once the simulation and optimization models were approved by the high-level technical group as the proper decision-making tools, the same group performed, under IMTA's guidance, the analysis of diverse climatic, water use, and technological scenarios, as well as of solution options. After around 30,000 person-hours, an agreement was reached to define a so-called *Joint Optimal Operational Policy*. The models proved to be essential in the consensus building process, since they created the proper climate for discussions based on facts, rather than on opinions, thus making it possible to overcome the existing impasse that had existed between stakeholders for years.

On the basis of the described scientific and technological actions, a formal and legally binding agreement between the five state governments and the Guanajuato agricultural users was drafted. The agreement included the Joint Optimal Operational Policy in one of its clauses, as well as a reference to the models as decision-making tools. This was the first time that a reference to a mathematical model was made in a legal document in Mexico. The agreement was signed by the five state governors in December 2004, in the presence of the President of Mexico as Honorary Witness. It was also signed by the water users in January 2005. The agreement establishes operational rules for the distribution of water, seeking to satisfy as much of the agricultural demand as is physically possible and to preserve the Chapala Lake. Actually, the 2000 hm^3 minimum acceptable volume recommended by IMTA was included in the agreement. On the basis of the agreement, a basin-wide restoration effort was undertaken by the Ministry of the Environment and Natural Resources and by CNA. Even though water allocation restrictions have been applied by the National Water Commission since the agreement was signed, no complaints have been voiced by the users nor have they taken any legal actions against the application of the said legal instrument.

In conclusion, the scientific and technological studies developed by IMTA were critical in building confidence among the stakeholders, on the basis of which consensus was possible, thus implementing the principles of IWRM and paving the way toward sustainable water governance (Güitrón et al. 2003; Aldama et al. 2006).

15.4 Case Study: The Río Bravo (Grande) Basin

The Río Bravo basin, known as the Rio Grande basin in the USA, has a surface area of 607,975 km^2 (Fig. 15.4). The river has an estimated length of 3040 km. In the USA the river flows through the states of Colorado, New Mexico, and Texas; while

Fig. 15.4 Río Bravo basin. (Source: Musser 2017)

in Mexico, it flows through Chihuahua, Coahuila, Nuevo León, and Tamaulipas. The basin includes portions of the said three states in the USA, as well as portions of the said four states and a small portion of Durango in Mexico.

A portion of the river channel is part of the boundary between Mexico and the USA since 1948. The main tributaries on the American side are the Pecos, Devils, Chama, and Puerco rivers and on the Mexican side, the Conchos, Salado, and San Juan rivers. The Río Bravo is born in the mountains of San Juan in Colorado and ends in the Gulf of Mexico, near Matamoros and Brownsville.

The contributions of the primary (agriculture, stockbreeding, and fishing), secondary (industry), and tertiary (services) sectors of the economy to the regional GDP, respectively, were 9%, 37%, and 47% (with an unspecified 7%) in 2000. The contribution of the portions of the five states in the basin to each state GDP were 66.21% in Coahuila, 97.24% in Chihuahua, 0.40% in Durango, 99.50% in Nuevo León, and 55.05% in Tamaulipas (CNA 2006).

The climate in the region is characterized by low precipitation and high evaporation, resulting in low water availability (CNA 2006). Accordingly, the liquid represents the main and most precious input for the development of the economic activities in the region, and its sustainable management is of the utmost importance.

In 1944 Mexico and the USA established an International Treaty for the allocation of the shared waters of the Colorado and Bravo rivers. The Treaty specifies that the USA must allocate at least 1850 hm^3 of the Colorado River waters to Mexico each year. It also specifies that Mexico must allocate at least an average of 432 hm^3 per year to the USA of the Bravo River waters in periods of 5 years of duration.

At the end of the last century and the beginning of this, a low flow period occurred in the Mexican portion of the basin, making it difficult for Mexico to satisfy the minimum allocation to the USA. An enormous pressure on the Mexican government was exerted by the US government to address the issue. In fact, the acting Texas Governor, Rick Perry, who at the time was running for the first time to be elected as governor, used the topic of "Mexico's water debt to Texas" for political gain, by publicly announcing that he would make "the Mexicans pay." In order to resolve the matter, stern water supply restrictions were made to the agricultural users on the Mexican side. This prompted a great deal of discontent on the part of the farmers, particularly the inhabitants of Tamaulipas (the state located in the lowest part of the basin), who blamed the Federal Government for using waters not contemplated in the Treaty to pay the "water debt" that Mexico owed the USA, and the farmers of Chihuahua (the state located in the highest part of the basin, along with a small part of Durango), for using volumes above those which they had the right to. In fact, the Tamaulipas farmers convinced the state government to present a constitutional controversy against the President of Mexico, the National Water Commission, the Ministry of Foreign Affairs, and the state of Chihuahua, before the Supreme Court of the country. The highest court finally ruled in favor of the defendants, once a number of expert testimonies (among which was that of the author) were considered.

The previous paragraph describes in part the confrontational environment that has surrounded the use of water in the basin for many years, with the added ingredient of water allocation that Mexico has to make to the USA in the terms of the 1944 Treaty.

As was described above, within the Mexican territory, the main conflict was between the upstream users (Chihuahua) and the downstream users (Tamaulipas). Within the realm of the Río Bravo Basin Council (integrated by the Federal Government, through CNA, the five state governments, and user representatives), CNA had unsuccessfully tried to convince users to accept the results of a water availability study (that had been performed in the terms of an official standard) for a very long period of time (8 years!). No consensus had been reached until finally, the members of the Council accepted that an independent expert consultant be hired to redevelop the availability study, by incorporating the water allocation obligation that the country has according to the 1944 Treaty. The members accepted CNA's proposal that a decision be made in accordance with the recommendation of the expert. After considering a number of candidates, the author's services were retained with that purpose.

The author made a rigorous mathematical analysis of the official water availability standard, which is based on the restitution of natural flows and annual water balances. He discovered a few conceptual errors, and, on the basis of his recommendations, the standard was modified. Furthermore, he proposed the division of the basin in sub-basins, which differed from the one originally proposed by CNA. In essence, his proposal is schematically shown in Fig. 15.5. On the basis of this schematization, the author redeveloped the annual water balances and produced a new version of the availability study (Aldama 2007). He made formal presentations to the members of the Río Bravo Basin Council, and he convinced them that his results were scientifically sound, thus prompting an agreement to finally publish the availability study in the *Official Gazette of the Federation*. The whole process lasted

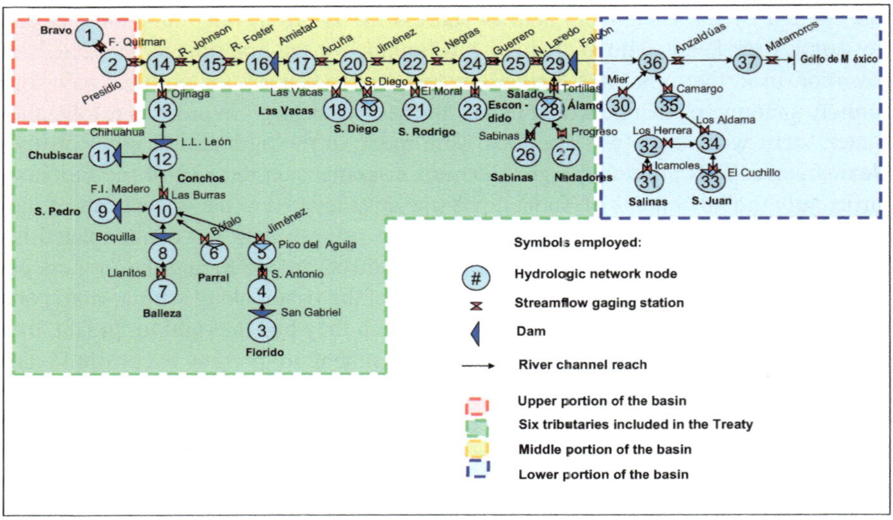

Fig. 15.5 Schematization of the Mexican portion of the Río Bravo basin for the purposes of the water availability study. (Source: Aldama 2007)

only 5 months for the study and the consensus building process and 4 additional months for the publication of the availability study.

The representatives of the state of Chihuahua rightly noted that yearly water balances tend to hide unaccounted for volumes that may be detected when those balances are made on a monthly basis. In consequence, they convinced the Río Bravo Basin Council to hire a consultant to develop a methodology for monthly water balances, as the basis for water availability studies. The services of the author were once again retained to perform such a task. The work was done in collaboration with the Chihuahua state water board. It consisted of developing a parameterization of the water losses and gains in each river channel reach, due to infiltration, unaccounted for withdrawals, and the interaction with subsurface water reservoirs. The basis of this parameterization was a rigorous mathematical analysis of the partial differential equation that governs mass conservation in a cross-section-averaged flow. The end result was a simple, linear expression that involved the flow entering the reach at its upstream end and the flow exiting the reach at its downstream end. The method was calibrated with data collected at the Conchos River basin with excellent results. Furthermore, a methodology was also developed to identify and incorporate the influence of irrigation returns (Aldama 2008). The methodology has been applied to the whole basin by representatives of CNA and the states included in the basin. The next availability study will be published on the basis of the monthly water balances.

During the period 2009–2011, the services of the author were retained by CNA to perform analyses of (a) the flow regime in the six tributaries of the Bravo River mentioned in the 1944 Treaty (Conchos, San Diego, San Rodrigo, Escondido y Salado rivers and Las Vacas Creek), out of which the minimum average yearly volume of 432 hm^3 that Mexico must allocate to the USA, is to be withdrawn, and (b) the flow regime in the Colorado River, out of which the minimum yearly volume of 1850 hm^3 that the USA must allocate to Mexico, is to be withdrawn (Aldama 2009, 2010, 2011). The main reason for the development of such analyses was that the 1944 Treaty establishes that special measures should be applied when an "extraordinary drought" occurs but does not provide a definition for such a term. Accordingly, the Mexican Federal Government wanted to have elements in case there was a need for a negotiation with their US counterpart in regard to the establishment of a definition of extraordinary drought within the realm of the International Boundary and Water Commission. The results obtained for the case of the Río Bravo Basin are briefly described below. The technical details may be found in Aldama (2011).

A record of the monthly evolution of annual natural flows for the sum of volumes supplied by the six tributaries to the Bravo River was assembled. Data were available for the periods 01/1900–12/1943 and 12/1950–12/2008. On the basis of the form of the spectra for those periods, a fractal interpolation scheme was used to complete the record. The resulting record for the period 01/1900–12/2008 is shown in Fig. 15.6. The interpolated data are shown in red.

Since the Treaty establishes that the accounting of water volumes allocated by Mexico to the USA should be done on a 5-year period basis (which implies that the

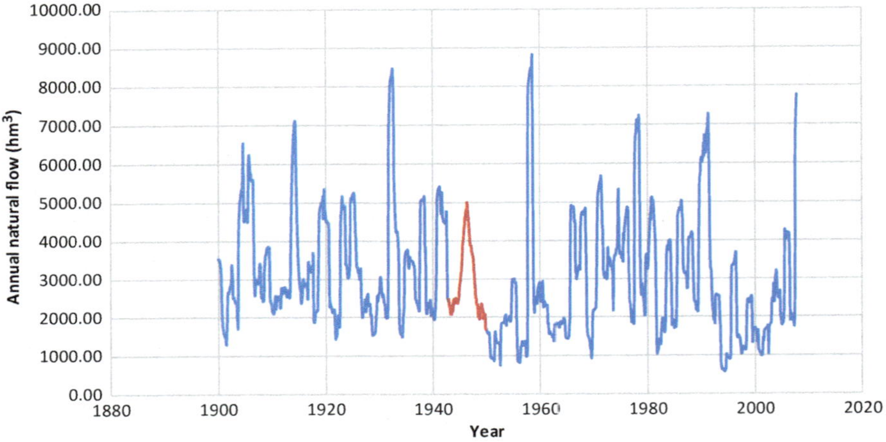

Fig. 15.6 Record of annual natural flows of the six tributaries of the Bravo River mentioned in the 1944 Treaty. (Source: Aldama 2010)

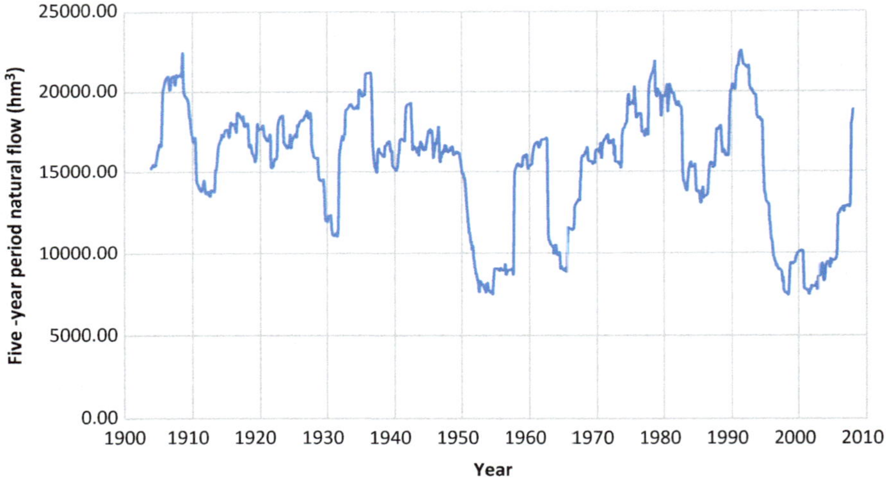

Fig. 15.7 Record of 5-year period natural flows of the six tributaries of the Bravo River mentioned in the 1944 Treaty. (Source: Aldama 2010)

total minimum volume that Mexico must allocate to the USA in 5 years is 2160 hm^3), it was necessary to construct a record of 5-year period natural flow record, by employing the record of annual natural flows. The result is shown in Fig. 15.7.

The traditional way to statistically classify a flow regime would be to try and identify a theoretically probability distribution that fits the observed empirical one. Such an attempt to fit the normal probability distribution to the empirical probability distribution of the record is shown in Fig. 15.8. As may be observed, the fit is very

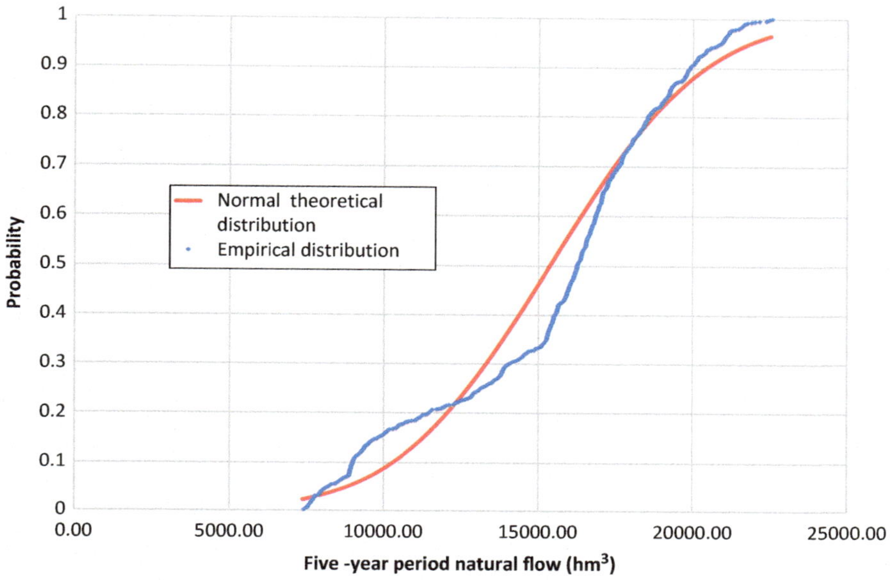

Fig. 15.8 Comparison of the empirical and normal theoretical distribution (least squares fit) of the 5-year period natural flows of the six tributaries. (Source: Aldama 2010)

poor. In fact, by using the Kolmogorov-Smirnov test, it was proven that the empirical distribution was not normal. This was odd, since a 5-year period flow is the sum of 1825 daily flows, and 1825 is a number large enough for the Central Limit Theorem to apply. As a matter of fact, the actual number to be considered was $1825 \times 6 = 10{,}950$, since the natural flows under analysis were the sum of the natural flows for the *six* tributaries. The Central Limit Theorem predicts that the probability distribution of a random variable that is the sum of a large enough set of random variables approaches the normal distribution (for a proof of the theorem, see, for instance, Aldama (2011)). Evidently, a new approach for analyzing the flow record was needed.

First, the linear trend of the 5-year natural flows was identified. The result is shown in Fig. 15.9. Then, a spectral method was developed to jointly identify the cyclical and random components in the de-trended record. The method consists in computing the spectrum of such a record and selecting the dominant Fourier components to construct a filtered spectrum that only contains these components. Once that is done, an estimate of the cyclical component of the de-trended record is computed by applying an inverse Fourier transform. Then, an estimate of the random component is computed by taking the difference between the de-trended record and the estimate of the cyclical component. The process is repeated iteratively until the empirical distribution of the random component follows the normal theoretical distribution, according to what one would expect from the Central Limit Theorem.

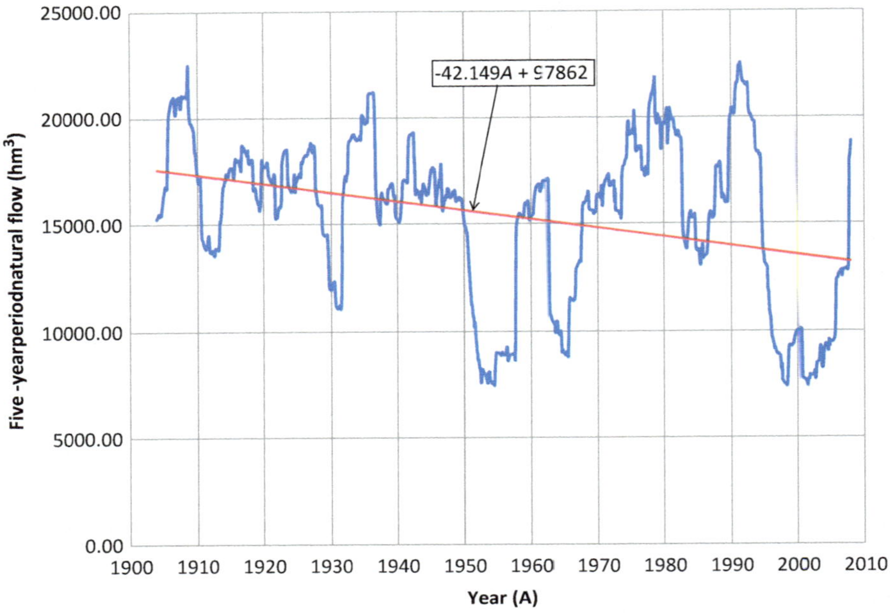

Fig. 15.9 Linear trend in the 5-year period natural flow record for the six tributaries. (Source: Aldama 2010)

The results for the 5-year period natural flows of the six tributaries are shown in Figs. 15.10 and 15.11.

To verify the soundness of the developed approach, the empirical probability distribution of the random component was compared to the normal theoretical distribution. The result is shown in Fig. 15.12. As may be observed, the fit is remarkably good. In fact, by applying the Kolmogorov-Smirnov test, it was shown that the random component values follow a normal distribution, which is in accordance with the Central Limit Theorem. The reason for this is that the linear trend and the cyclical component are not random, nor the values they represent form a set of independent samples, which are prerequisites for the application of the said theorem. Thus, the key was to isolate the true random component in the record.

On the basis of the above findings, a *standardized runoff index* (*SRI*) was developed, as the difference of the random component of the 5-year period natural flow and its mean (which by construction is zero) normalized by the standard deviation of the former. Thus, the classification shown in Table 15.1 was proposed:

A graphical application of the *SRI* classification is shown in Fig. 15.13. One possibility for defining an “extraordinary drought” would be to have a 5-year period flow for which *SRI* < 2. Actually, this has only happened in the following periods: May–August 1932, April–August 1958, July 1974, July–August 1980, October 1983–March 1984, and May–June 1990. Detailed technical details for the described method of analysis may be found in Aldama (2016), where it is applied to a different case study: the Colorado River basin.

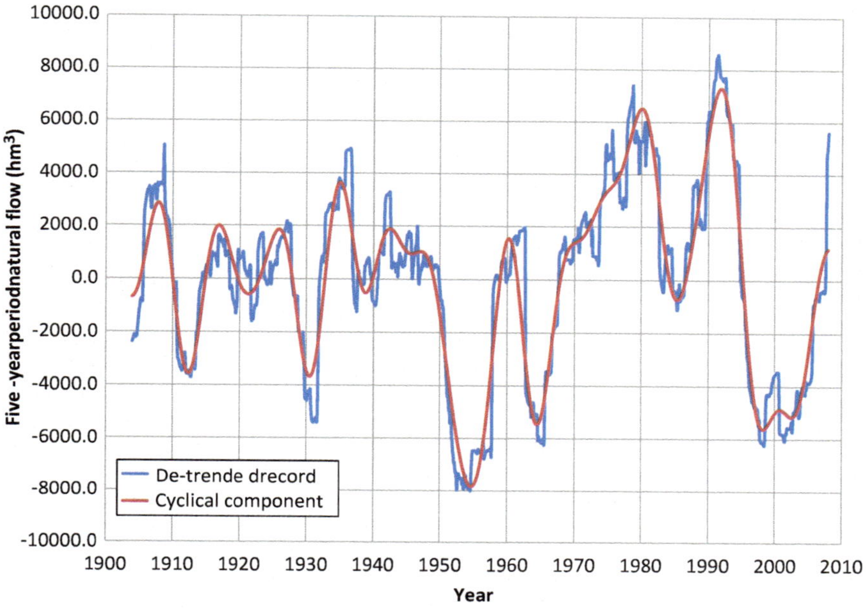

Fig. 15.10 Cyclical component of the de-trended 5-year period for the six tributaries. (Source: Aldama 2010)

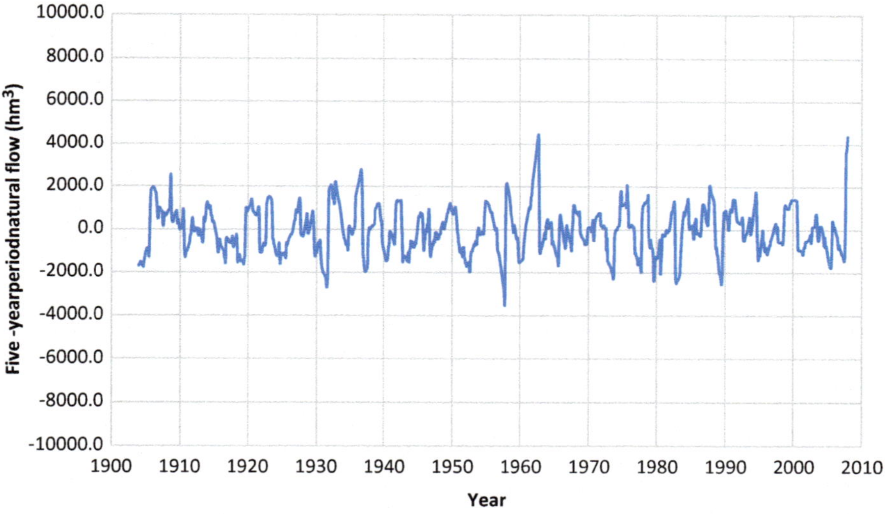

Fig. 15.11 Random component of the de-trended 5-year period for the six tributaries. (Source: Aldama 2010)

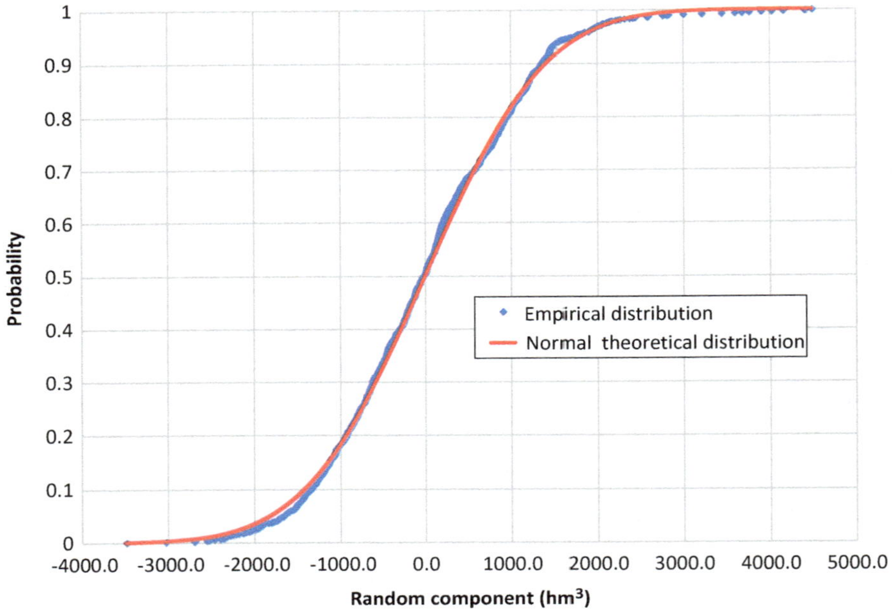

Fig. 15.12 Comparison of the empirical and normal theoretical distribution (least squares fit) of random component of the 5-year natural flow for the six tributaries. (Source: Aldama 2010)

Table 15.1 Classification of standardized runoff index (SRI)

Criterion	Classification	Probability (%)
$SRI > 2$	Very humid	2.28
$1 < SRI \leq 2$	Humid	13.59
$-1 \leq SRI \leq 1$	Normal	68.26
$-2 \leq SRI < -1$	Dry	13.59
$SRI < -2$	Very dry	2.28

Source: Aldama (2010)

Most often, simple rules are applied to classify natural flow regimes, which can be based on statistics of crude streamflow data (mean, standard deviation, or percentiles), but the results shown in this section demonstrate that, strictly speaking, the complexity of natural flow regimes in general and droughts in particular makes it necessary to incorporate trends, cyclical behavior, and randomness into the classification. The author recommended that the *SRI* classification approach be considered by CNA and the International Boundary and Water Commission.

Evidently the application of scientific studies to the natural occurrence of water in the Bravo River basin has been indispensable in building consensus among stakeholders, and it will continue to bring a sound technical basis for negotiations within the realm of the IBWC, in order to continue making a sustainable use of the shared

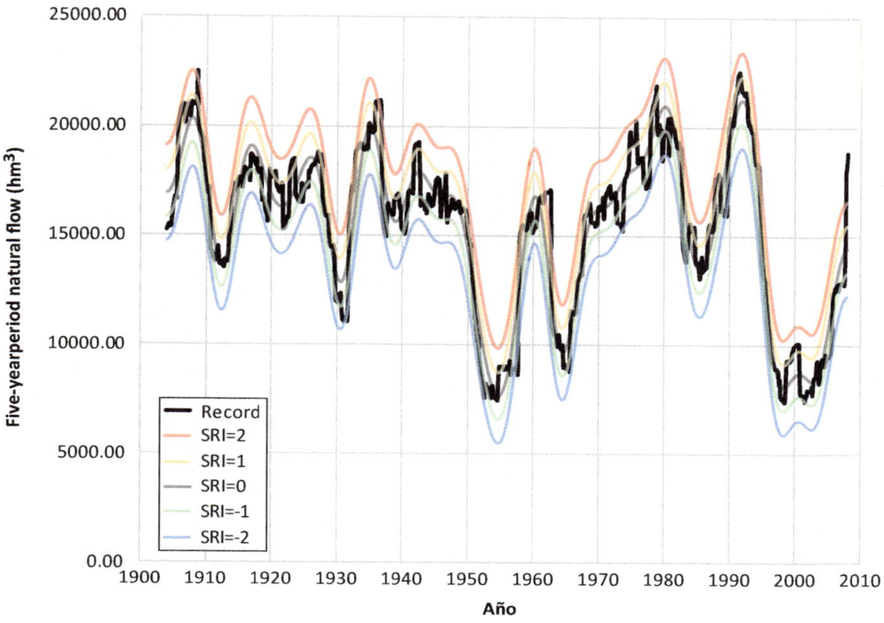

Fig. 15.13 SRI natural flow regime classification for the six tributaries. (Source: Aldama 2010)

waters of the river among the Mexican states, among the American states, and between the two countries.

15.5 Concluding Remarks

It has been shown that science and technology may be true catalysts of IWRM and good water governance. By way of example, the case studies of the Lerma-Chapala basin and the Río Bravo (Grande) Basin have demonstrated that sound scientific results and effective technological tools may be used to end stalemates in negotiations between conflicting water users, particularly when stakeholders base their positions on opinions, rather than facts. The key idea in using science and technology is to shift the discussion to technical issues, which can be settled objectively. This serves as a solid base *to build trust* among stakeholders and in water authorities, thus facilitating the consensus construction process, an essential step in achieving IWRM.

Acknowledgments The author wishes to recognize the National Water Commission, particularly Mario López, Manager of Engineering and Technical Standards, for allowing him to use some of the results of the Río Bravo Basin for the purposes of this publication. The author also wishes to thank his son Abraham Aldama, who is an economist and a political scientist and a Ph. D. Candidate at NYU, for the keen suggestions he made to improve the first two sections of this chapter. This work is dedicated to him.

References

Aldama, Á. (2007). *Disponibilidad de aguas superficiales en la cuenca del río Bravo*. Organización Meteorológica Mundial-Comisión Nacional del Agua.
Aldama, Á. (2008). *Asistencia Técnica Especializada para la Determinación del Balance Hidráulico y Disponibilidad Mensual de las Aguas Superficiales de la Región Hidrológica No. 24*, Bravo–Conchos, Organización Meteorológica Mundial-Comisión Nacional del Agua.
Aldama, Á. (2009). *Asistencia Técnica Especializada para el Análisis del Régimen de Escurrimiento de las Aguas Superficiales de la Región Hidrológica No. 24*, Bravo–Conchos, Organización Meteorológica Mundial-Comisión Nacional del Agua.
Aldama, Á. (2010). *Asistencia Técnica Especializada para el Análisis del Régimen de Escurrimiento de las Aguas Superficiales del Río Colorado*. Organización Meteorológica Mundial-Comisión Nacional del Agua.
Aldama, Á. (2011). *Asistencia Técnica Especializada para la Propuesta de la Postura del Gobierno Mexicano en Cuanto a la Definición del Término "Extraordinaria Sequía" en el Marco del Tratado Internacional de 1944*. Organización Meteorológica Mundial-Comisión Nacional del Agua.
Aldama, A. A. (2016). Three-part decomposition of natural flows: Application to the Colorado River drought characterization. *Tecnol. cienc. agua, 7*(5), 167–195. ISSN 2007-2422.
Aldama, Á., Güitrón, A., Aparicio, J., & Barragán, C. (2006) (marzo). La cuenca Lerma-Chapala y los modelos de simulación. *Ciencia y Desarrollo*. México, México.
Boyd, R., & Richerson, P. (1985). *Culture and the evolutionary process*. Chicago: University of Chicago Press.
Brundtland Commission. (1987). *Our common future*. Oxford: Oxford University Press.
CNA. (2006). *Programa Hídrico Visión 2030 del Organismo de Cuenca Río Bravo*. Comisión Nacional del Agua.
Gellner, E. (1983). *Nations and nationalism*. Ithaca: Cornell University Press.
Güitrón, A., Hidalgo, J., Aparicio, J., & Aldama, Á. (2003). A water crisis management: The Lerma Chapala basin case. *Water Resources Management* 11. Southampton: WIT Press.
GWP. (2000). *Integrated water resources management* (Technical Advisory Committee Background Paper No. 4). GWP, Stockholm, Sweden.
GWP. (2003). *IWRM ToolBox*. Stockholm: Global Water Partnership Secretariat.
Hardin, G. (1968). The tragedy of the commons. *Science, 162*, 1243–1248.
Mestre-Rodríguez, E. (2001). Introduction. In A. Hansen & M. van Afferden (Eds.), *The Lerma-Chapala Watershed: Evaluation and management*. New York: Kluwer Academic-Plenum.
Musser, K. (2017). https://en.wikipedia.org/wiki/File:Riogranderivermap.png
Olson, M. (1971). *The logic of collective action: Public goods and the theory of groups*. Cambridge: Harvard Economic Studies.
Redcliff, M. (2006). Sustainable development (1987–2005) – An oxymoron comes of age. *Horizontes Antropológicos, 12*(25), 65–84. Porto Alegre, Brazil.
UNDP. (1997). *Governance for sustainable development*. A United Nations Development Program Policy Paper, New York.
Weber, M. (1946). *From Max Weber* (H. H. Gerth & C. Wright Mills, Trans & Ed.), New York: Free Press.

Álvaro A. Aldama earned his Civil Engineer and Master of Engineering degrees from the National Autonomous University of Mexico (UNAM). He earned his PhD degree in Fluid Mechanics from the Massachusetts Institute of Technology (MIT). He has had extensive experience in academia, civil service, and private consulting. He was a Professor of the Water Resources Group at the Civil Engineering Department of Princeton University. He has also been a Professor at the Graduate School of Engineering at UNAM. He held various positions in the Mexican Federal Government. Among these, he was the Director General of the Mexican Institute of Water Technology (IMTA). Under his leadership, IMTA became one of the most recognized R&D centers in the subject of water. Currently, he works as an independent consultant for a number of international, foreign, and Mexican institutions and companies, including the World Bank, the World Meteorological Organization, UNESCO, the National Water Commission, and the Federal Electric Commission. His professional interests include computational hydraulics, surface hydrology, and complex water systems. He is a former President of the Mexican Academy of Engineering and of the Mexican Association of Hydraulics. He is a member of the Mexican Academy of Sciences and the Mexican and Panamerican Academies of Engineering. He has produced over 200 publications and has received more than 20 national and international awards for his professional achievements. His works have received almost 900 citations.

Index

H. R. Guerrero García Rojas (ed.), *Water Policy in Mexico*, Global Issues in Water Policy 20, https://doi.org/10.1007/978-3-319-76115-2

H

I

L

R

S

T

Printed by Printforce, the Netherlands